LECTURES ON

NON-PERTURBATIVE CANONICAL GRAVITY

Advanced Series in Astrophysics and Cosmology

ISSN: 1793-1312

Series Editor: Remo Ruffini *(ICRA, Pescara & University of Rome "La Sapienza", Italy)*

---

*Published**

Vol. 2 Galaxies, Quasars and Cosmology
edited by L Z Fang and R Ruffini

Vol. 3 Quantum Cosmology
edited by L Z Fang and R Ruffini

Vol. 4 Gerard and Antoinette de Vaucouleurs: A Life for Astronomy
edited by M Capaccioli et al.

Vol. 5 Accretion: A Collection of Influential Papers
edited by A Treves, L Maraschi and M Abramowicz

Vol. 6 Lectures on Non-Perturbative Canonical Gravity
by A Ashtekar, Notes prepared in collaboration with R S Tate

Vol. 7 Relativistic Gravitational Experiments in Space:
First William Fairbank Meeting
edited by M Demianski and C W F Everitt

Vol. 8 Hawking on the Big Bang and Black Holes
by S Hawking

Vol. 9 Accretion Disks in Compact Stellar Systems
edited by J C Wheeler

Vol. 10 The Chaotic Universe —
Proceedings of the Second ICRA Network Workshop
edited by V G Gurzadyan and R Ruffini

Vol. 13 Exploring the Universe: A Festschrift in Honor of Ricardo Giacconi
edited by H Gursky, R Ruffini and L Stella

Vol. 14 Dark Matter and Cosmic Web Story
by J Einasto

*To view the complete list of the published volumes in the series, please visit
http://www.worldscientific.com/series/asac

Advanced Series in Astrophysics and Cosmology – Vol. 6

Series Editors: **Fang Li Zhi** and **Remo Ruffini**

LECTURES ON NON-PERTURBATIVE CANONICAL GRAVITY

editor

Abhay Ashtekar

Syracuse University

World Scientific

NEW JERSEY • LONDON • SINGAPORE • BEIJING • SHANGHAI • HONG KONG • TAIPEI • CHENNAI • TOKYO

Published by

World Scientific Publishing Co. Pte. Ltd.
5 Toh Tuck Link, Singapore 596224
USA office: 27 Warren Street, Suite 401-402, Hackensack, NJ 07601
UK office: 57 Shelton Street, Covent Garden, London WC2H 9HE

British Library Cataloguing-in-Publication Data
A catalogue record for this book is available from the British Library.

Advanced Series in Astrophysics and Cosmology — Vol. 6
LECTURES ON NON-PERTURBATIVE CANONICAL GRAVITY

ISBN-13 978-981-02-0573-7
ISBN-10 981-02-0573-2

ISBN-13 978-981-02-0574-4 (pbk)
ISBN-10 981-02-0574-0 (pbk)

CONTENTS

PREFACE

The purpose of these notes is to present an up to date status report of a program for non-perturbative, canonical quantization of gravity in 3+1 dimensions. A number of the key ideas in the program were developed just prior to and during a six-month workshop on quantum gravity held at the Institute of Theoretical Physics at Santa Barbara during the first half of 1986. The early developments in this area were reported in a monograph entitled New Perspectives in Canonical Gravity, *published by Bibliopolis in 1988.*

Since then the situation has evolved on several fronts. First, a number of conceptual issues have been clarified. In the classical theory, these include: the Lagrangian formulation leading to the new canonical variables, the polynomiality of all *of Einstein's equations including the reality conditions and the way the reality conditions are implemented in the Hamiltonian framework On the quantum side, there is now a clear quantization program based o n algebraic ideas; the need for sophisticated techniques from geometrical quantum mechanics no longer exists. The program goes beyond the Dirac quantization scheme for constrained systems in one respect: it provides a concrete approach to finding the inner product on the space of physical states. Such an approach is essential in quantum gravity where one cannot rely on the Poincaré (or the diffeomorphism) group to select a vacuum state and provide the inner product. On the technical side, the new Hamiltonian framework was extended to allow for a cosmological constant and also matter sources including Klein-Gordon, Dirac and Yang-Mills fields. On the quantum front, over the last year or so, a number of examples have been analyzed in detail to check the viability of the quantization program and to refine it at various points.*

More significant are the numerous new applications of the framework to classical general relativity, geometry and canonical quantum gravity. In the last part of the monograph mentioned above, several tentative proposals had been made for completion of the quantization program and for other applications. Some of them have now been developed in detail and others have been discarded. A number of interesting, definitive results have emerged. For example, an exhaustive analysis of the moduli space of self dual Einstein metrics is now available. The framework has also shed considerable new light on the structure underlying the space of half flat metrics. In quantum gravity itself, progress has occurred in several directions. The most important among them is the development of the loop representation *for quantum gravity and gauge theories. This representation has provided an infinite dimensional space of solutions to all quantum constraints and they have provided a completely new picture of quantum geometry in the Planck regime. Discrete structures such as knots and link classes appear to play a fundamental role in this regime. Detailed approximation methods are now being developed to determine exactly where and why the perturbation theory breaks down. Ideas from quantum gravity have led to the construction of the loop representation in Maxwell theory, and in lattice and continuum Yang-Mills theory. Thus, a "feed-back" loop has now emerged between general relativity and gauge theories. The quantization program in general and the loop representation in particular were tested in two interesting systems: linearized gravity in 3+1 dimensions and exact general relativity in 2+1 dimensions. The results have provided considerable confidence in the program and physical intuition for the loop variables. On a completely different front, the issue of time was analyzed using the connection representation. It turns out that, in a suitably truncated version of the theory, one can indeed isolate an internal time variable and "derive" the Schrödinger equation for matter plus gravity from the quantum constraints of general relativity. Thus, the rather abstract framework of quantum gravity is indeed capable of reproducing the familiar laboratory physics. Finally, the framework has yielded a number of interesting mathematical results in quantum cosmology.*

In these notes, I will give a comprehensive account of these developments. An early draft of these notes was circulated by the inter-university center for astronomy and astrophyics, Poona, within India. *To prepare the present version, that draft was thoroughly revised, and updated. Three new chapters have been added and the discussion was expanded at a number of points in other chapters. As is evident from the Bibliography which appears at the end, the work I wish to summarize was done by a large number of individuals. I have tried to present a balanced view (although I am sure that not all of these authors would agree with everything I have said!). The treatment is self-contained. While there is considerable new material, I have made a special effort to make it easily accessible to advanced research students. As indicated in the introductions to each of the three parts, some of the material can (and perhaps should be) skipped during the first reading. I also hope that the notes will serve a dual purpose. Specialists should be able to use them as a reference-book; to understand chapter n, it should not be necessary to have gone through all the preceeding n-1 chapters.*

Abhay Ashtekar
Syracuse; December 1990.

Acknowledgements

I am most grateful to Carlo Rovelli and Lee Smolin for long range collaboration, constant intellectual stimulation and for a great deal of fun I have had during the last four years. The impact of their ideas on the non-perturbative quantization program should be evident even to a casual reader of these notes. Evident also should be the influence of the viewpoints, indeed the very way of thinking, that Ted Newman and Roger Penrose have introduced in the field. The results presented here arose through the work of a large number of researchers. In particular, Amitabha Sen and Ted Jacobson played a major role in shaping the initial program. My colleagues at Syracuse, Luca Bombelli, Bernd Brügmann, Josh Goldberg, Gabriela Gonzalez, Viqar Husain, Joohan Lee, Jorge Pullin, Joe Romano, Joseph Samuel, Ranjeet Tate and Charles Torre have made (and continue to make) innumerable contributions to this work. It is a great pleasure and a privilege to be able to work in the stimulating environment they provide everyday. My understanding of the basic problems of quantum gravity was deepened also through many discussions I have had and constructive criticisms I received from a large number of colleagues elsewhere. Among them, I would like to thank especially Ingemar Bengtsson, Miles Blencowe, Steve Carlip, John Friedman, Jim Hartle, Gary Horowitz, Chris Isham, Sucheta Koshti, Karel Kuchăr, Anne Magnon, Steve Martin, Lionel Mason, Vince Moncrief, David Robinson, and Richard Woodard.

These notes could not have been prepared without the help I received from Ranjeet Tate who spent countless hours on this collaboration. He offered numerous constructive criticisms and made suggestions to improve the clarity of presentation in almost every chapter. I also want to thank Joe Romano for his help with proof-reading, Pierre Gravel for his comments on the first draft, and, Peter Hübner and Gabriela Gonzalez for preparing the Bibliography.

The notes grew out of the lectures I gave at a summer school at IUCAA, Poona, India, a workshop at Banff, Canada and the SILARG VII meeting at Mexico City, Mexico. I would like to thank the organizers, Naresh Dadhich and Jayant Narlikar at IUCAA, Robert Mann at Banff and Marcos Rosenbaum and Mike Ryan at SILARG VII, for their hospitality. This work was supported in part by the NSF grants PHY86-12424, PHY90-16733, INT 88-15209 and by research funds provided by Syracuse University.

LIST OF SYMBOLS

The following list includes most of the symbols that are used repeatedly in this book. However, many quantities which are defined and used "locally" in individual chapters may not appear in this list.

$[\ ,\ ]$	commutator
$\{\ ,\ \}$	Poisson bracket
$\langle\ ,\ \rangle$	inner product for $SU(2)$ spinors
$\langle\ \vert\ \rangle$	inner product on Hilbert space
$\dagger$	Hermitian conjugation for spinors
$\ddagger$	Hermitian conjugation for matrices
$\bar{\ }$	complex conjugation
$\approx$	weak equality (modulo constraints)
$:=$	definition
$\equiv$	identity
$\Gamma, \widetilde{\Gamma}$	phase space
$\bar{\Gamma}$	constraint surface in Γ
$\hat{\Gamma}$	reduced phase space $\mathrm{T}/\mathrm{Gauge}$
$\Gamma_{aA}{}^{B}$	spin-connection 1-form of $\sigma^{a}{}_{A}{}^{B}$
Γ_{a}^{i}	spin-connection 1-form of the triad E_{i}^{a}
$\Gamma_{ab}{}^{c}$	Christoffel symbol of the 3-metric q_{ab}
$\epsilon_{AB}, \epsilon_{A'B'}$	alternating tensors in spinor space
$\epsilon_{abc}, \epsilon_{abcd}$	volume elements on Σ and M
η_{ab}	Minkowski metric
$\eta_{\mathbf{ijk}}$	totally antisymmetric symbol
$\widetilde{\eta}^{abc}, \widetilde{\eta}^{abcd}$	Levi-Civita densities on Σ and M
$\underset{\sim}{\eta}_{abc}, \underset{\sim}{\eta}_{abcd}$	inverses of $\widetilde{\eta}^{abc}$ and $\widetilde{\eta}^{abcd}$
$\Lambda_{A}{}^{B}, N_{A}{}^{B}$	Lie algebra-valued smearing function
$\Pi_{ab}, \Pi_{aA}{}^{B}$	difference between $A_{aA}{}^{B}$ and $\Gamma_{aA}{}^{B}$
${}^{4}\sigma$	determinant of $\sigma^{a}{}_{AA'}$
$\sigma^{a}{}_{AA'}$	$SL(2,\mathbb{C})$ soldering form on M
$\sigma^{a}{}_{A}{}^{B}, \widetilde{\sigma}^{a}{}_{A}{}^{B}$	$SU(2)$ soldering form on Σ (densitized)

e^a_I, E^a_i	*tetrad and triad*
Σ	*3-dimensional manifold*
$\tau_{iA}{}^B$	*Pauli matrices*
ω_α	*symplectic potential*
$\Omega_{\alpha\beta}$	*symplectic form*
$\bar{\Omega}_{\alpha\beta}$	*restriction of $\Omega_{\alpha\beta}$ to $\bar{\Gamma}$*
$A_{aA}{}^B$, A^i_a	*$SU(2)$ connection 1-form on Σ*
${}^4A_{aA}{}^B$, ${}^4A^I_a$	*$SL(2,\mathbb{C})$ connection of ${}^4\mathcal{D}$ on M*
$\mathbf{A}, \widetilde{\mathbf{E}}$	*Yang-Mills fields*
C_{ab}, $C_{aA}{}^B$	*perturbation of $A_{aA}{}^B$*
$C_{aA}{}^B$	*difference between two connections on unprimed spinors*
$\mathcal{C}$, $\widetilde{\mathcal{C}}$, $\hat{\mathcal{C}}$	*configuration space*
$\mathbb{C}$	*complex number field*
∂	*reference derivative operator on M or Σ*
D	*derivative operator on Σ defined by q_{ab} or $\sigma^a{}_A{}^B$ or E^a_i*
∇	*derivative operator on M defined g_{ab} or $\sigma^a{}_{AA'}$ or e^a_I*
$\mathcal{D}$	*spatial derivative operator on "internal" indices*
${}^4\mathcal{D}$	*space-time derivative operator on "internal" indices*
$\mathbb{d}$	*infinite dimensional exterior derivative operator*
e	*square-root of the determinant of g_{ab}*
E	*square-root of the determinant of q_{ab}*
e_{ab}	*flat reference metric on Σ*
$F_{abA}{}^B$, F^i_{ab}	*curvature of the spin connection $\mathcal{D}$ or A_a*
${}^4F_{abA}{}^B$ ${}^4F^I_{ab}$	*curvature of the spin connection ${}^4\mathcal{D}$ or ${}^4A_a$*
g_{ab}	*space-time metric*
G_{ab}	*Einstein tensor of g_{ab}*
$G_{A'A}$	*Hermitian inner product for $SU(2)$ spinors*
H	*Hamiltonian*
$\widetilde{h}^a_A{}^B$	*perturbations of $\widetilde{\sigma}^a_A{}^B$*
$\mathcal{H}$	*Hilbert space*
K_{ab}	*extrinsic curvature of Σ (30)*
L	*Lagrangian*
$\mathcal{L}$	*Lagrangian density*
$\mathcal{L}_t$	*Lie derivative with respect to t*
$\mathcal{L}_\Sigma$	*loop space of Σ*
M	*space-time manifold*
$M_{aA}{}^B$	*momentum conjugate to $\sigma^a{}_A{}^B$ in extended phase space*
n_a	*unit normal to Σ_t*

$n_{AA'}$	spinor $n^a\sigma_{aAA'}$, used as $SU(2)$ inner product
$N, (\underset{\sim}{N})$	lapse function (density) on Σ
N^a	shift vector on Σ
$\tilde{p}^{ab}$	momentum conjugate to q_{ab}
q	determinant of q_{ab}
q_{ab}, q^{ab}	spatial metric on Σ
R_{abcd}, R	Riemann tensor and scalar curvature of q_{ab}
${}^4R_{abcd}$, ${}^4R$	Riemann tensor and scalar curvature of g_{ab}
${}^{\mp}R_{abcd}$	(anti-) self-dual part of ${}^4R_{abcd}$
$R_{abA}{}^B$	curvature of D for $SU(2)$ spinors
${}^4R_{abA}{}^B$	curvature of ∇ for unprimed $SL(2,\mathbb{C})$ spinors
$\mathbb{R}$	real number field
t	time function on M
t^a	timelike future-directed vector field on M
T^*	cotangent space (e.g., $T^*\mathcal{C}$, $T^*\Gamma$)
T_{ab}	matter stress-energy tensor
z	complex phase space coordinate, $z = q - ip$
$T^0[\gamma]$, $T^1[\gamma]$	loop observables in 2+1
$T^0[\gamma]$, $T^a[\gamma](s)$, $T^1[\gamma, v]$	loop variables in 3+1 gravity

CONVENTIONS

In the literature on applications of the new canonical variables, a variety of conventions have been used. This makes it somewhat difficult to reconcile equations in one reference with those in another. In these notes, I have tried to standardize conventions in such a way that there is a maximum overlap with those used in the literature on new variables, and more generally, on canonical quantum gravity. As a consequence, the conventions used here do not always agree with those used in New Perspectives in Canonical Gravity.

Index notation

Space-time and spatial tensor indices will be denoted by latin letters $a, b, \ldots$. Spatial tensors will often be regarded as space-time tensors with certain vanishing components. Generally, a stem letter will indicate if we are considering a 4-dimensional, space-time field or a 3-dimensional field defined just on the spatial slice. When ambiguity is likely to arise we will use a prefix 4 to denote the 4-dimensional fields. (Thus ${}^4\omega_a^{IJ}$ is a space-time connection. Its pull-back to a spatial slice will be denoted simply by $\underset{\leftarrow}{\omega}_a^{IJ}$.) Fields with density weight +1 will generally carry a tilde,$\tilde{}$, over the stem letter.
Upper case latin letters, $I, J\ldots$ will denote internal Lorentz indices (and label tetrads), while lower case latin letters, $i, j, \ldots$ will denote three dimensional internal indices (and label triads). Upper case latin letters $A, A', B, B' \ldots$ will denote $SL(2,\mathbb{C})$ or $SU(2)$ spinors.

Geometry

Space-time signature: $(-,+,+,+)$
Definitions of the curvature tensors:

$$R_{abc}{}^{d}k_d := 2D_{[a}D_{b]}k_c, \quad R_{ab} := R_{acb}{}^{c}, \quad R := q^{ab}R_{ab},$$

and similarly for ${}^4R_{abc}{}^{d}$.

$$G\, F_{abA}{}^{B}\lambda_B := 2\mathcal{D}_{[a}\mathcal{D}_{b]}\lambda_A \text{ and}$$
$$G\, \epsilon^{ijk}F^{j}_{ab}\lambda^k := 2\mathcal{D}_{[a}\mathcal{D}_{b]}\lambda^i$$

defines the curvature tensors in terms of the derivative operators, so that,

$$F_{abA}{}^{B} \equiv 2\partial_{[a}A_{b]A}{}^{B} + G\,[A_a, A_b]_A{}^{B} \text{ and}$$
$$F_{ab}{}^{i} \equiv 2\partial_{[a}A_{b]}{}^{i} + G\,\epsilon^{ijk}A_{aj}A_{bk};$$

where the connections are defined by

$$\mathcal{D}_a\lambda_A =: \partial_a\lambda_A + G\, A_{aA}{}^B\lambda_B \text{ and}$$
$$\mathcal{D}_a\lambda^i =: \partial_a\lambda^i + G\epsilon^{ijk}A_{aj}\lambda_k.$$

The $SU(2)$ soldering forms satisfy the normalization conditions

$$\operatorname{tr}\sigma^a\sigma^b = -q^{ab} \qquad \text{and} \qquad [\sigma^a,\sigma^b]_A{}^B = \sqrt{2}\,\epsilon^{abc}\sigma_{cA}{}^B.$$

Canonical Framework

The basic Poisson brackets are:

$$\{q,p\} = 1$$

where p is the momentum canonically conjugate to q. The Hamiltonian H generates time evolution via:

$$\dot{f} = \{f,H\}.$$

The complex coordinate on phase space is:

$$z = q - ip.$$

With the above conventions, the fundamental Poisson brackets between the new variables are:

$$\{A_a^i(x), \widetilde{E}_j^b(y)\} = i\delta_a{}^b\delta_j{}^i\delta^3(x,y),$$
$$\text{and}\quad \{A_a{}^{AB}(x), \widetilde{\sigma}^b{}_{MN}(y)\} = \tfrac{i}{\sqrt{2}}\delta_a{}^b\delta_M{}^{(A}\delta_N{}^{B)}\delta^3(x,y).$$

The spatial metric is obtained from these variables by:

$$\widetilde{\widetilde{q}}{}^{ab} = \widetilde{E}^{ai}\,\widetilde{E}_i^b \quad \text{and} \quad \widetilde{\widetilde{q}}{}^{ab} = \widetilde{\sigma}^a{}_{AB}\,\widetilde{\sigma}^{bAB}.$$

The self dual connection is related to the standard geometrodynamical variables by:

$$A_a^i = \Gamma_a^i - iK_a^i \quad \text{or} \quad A_{aA}{}^B = \Gamma_{aA}{}^B - \tfrac{i}{\sqrt{2}}K_{aA}{}^B,$$

where Γ_a^i ($\Gamma_{aA}{}^B$) is the connection compatible with the triad $\widetilde{E}_i^a$ (or the soldering form $\widetilde{\sigma}^a{}_A{}^B$), and $K_{ab} = K_a^i E_{bi} = \widetilde{\sigma}^a{}_{AB}K_a{}^{AB}$ is the extrinsic curvature of the spatial manifold Σ.

Quantization

We will use the following conventions for the relation between quantum commutators and classical Poisson brackets. For F, G elementary classical variables and $\hat{F}, \hat{G}$ the corresponding operators,

$$[\hat{F}, \hat{G}] = i\hbar \widehat{\{F, G\}}.$$

Thus, for example, in the q representation,

$$\hat{q} \cdot \Psi = q\Psi \quad \text{and} \quad \hat{p} \cdot \Psi = -i\hbar \frac{\partial \Psi}{\partial q}.$$

Time evolution of the wavefunctions is given by the Schrödinger equation:

$$i\hbar \frac{\partial}{\partial t} | \Psi \rangle = H | \Psi \rangle.$$

Fundamental constants

Throughout, the speed of light is set equal to one. Newton's constant (G) is also generally set equal to one. Factors of G will occasionally be restored to illustrate differences from Yang-Mills theories and to take the weak and strong field limits.

I INTRODUCTION

...How then do they manage with these incorrect equations? These equations lead to infinities when one tries to solve them; these infinities ought not to be there. They remove them artificially. ... Just because the results happen to be in agreement with observations does not prove that one's theory is correct. After all, the Bohr theory [of the hydrogen atom] *was correct in simple cases. It gave very good answers, but still the Bohr theory had very wrong concepts.*

- P.A.M. Dirac, *The inadequacies of quantum field theory.*

1 NON-PERTURBATIVE QUANTUM GRAVITY: WHAT AND WHY ?

A fundamental premise of these lecture notes is that gravitational physics is intimately intertwined with the geometry of space and time. In the classical regime, general relativity captures this idea in an elegant way. Indeed, it is the identification of the gravitational field with the curvature of space-time that has led to the most dramatic predictions of general relativity, such as the existence of black holes and the occurrence of the big bang. This leads one to the viewpoint that a primary goal of any quantum theory of gravity should be to probe what one may call, in absence of a better terminology, the quantum structure of space and time.

Let us recall the evolution of the concept of space and time since the days of Newton. Using the modern terminology introduced by Cartan (see, e.g., [1]) Newton's space-time can be described as follows: It is a 4-manifold, topologically $\mathbb{R}^4$, equipped with a preferred foliation by hyperplanes and a contravariant "metric" q^{ab} of signature $(0,+,+,+)$. The existence of the preferred foliation—whose covariant normal constitutes the kernel of q^{ab}—endows space-time with an absolute, observer-independent notion of distant simultaneity. This picture of space-time formed the basis of physics for over two centuries. It is the realization that the speed of light is a universal constant, independent of the choice of an observer, that jolted this comfortable view. The new picture emerged from special relativity. The space-time manifold continued to be topologically $\mathbb{R}^4$ but the preferred foliation and the associated notion of distant simultaneity now disappeared. The space-time metric, η^{ab}, acquired the signature $(-,+,+,+)$. Through each point, we now have a preferred light-cone—whose normal is null with respect to η^{ab}—rather than a spacelike 3-surface. The two notions of space and time are *very* different from each other and, in a sense, it is quite remarkable that, although they differ so dramatically at the conceptual level, one can be approximated by the other in everyday circumstances. Note, however, that the two pictures have a common thread: In both models, the geometry of space-time is fixed once and for all, unaffected by matter. This changed with general relativity and the notion of space-time underwent an even more drastic revision. Space-time ceased to be an inert background on which fields and particles evolve. In this view the space-time metric g^{ab} is *not* fixed once and for all; it is sensitive to matter. Matter curves space-time according to Einstein's field equation. The geometry is a *dynamical* entity, capable of changing in response to changes in matter. What was the stage in the drama of evolution now joins the troupe of actors. Via general relativity, Einstein taught us that the geometry of space and time is very much a *physical* entity and one can ask of it the type of questions we ask of physical systems. It is now generally believed that most properties of physical systems can be understood in terms of the behavior of their constituents. A primary preoccupation of physics of this century has been, therefore, the quest for the ultimate constituents of matter. Hence, it is now natural to look for the building blocks of space-time itself. Thus, the road that Einstein followed in his journey from special to general relativity leads us to ask: What is space-time made of? How do these constituents fit together to give us a continuum picture of space-time on an everyday scale? What concepts can one use to describe the goings-on at a much smaller, fundamental scale? These, we feel, are some of the key questions a quantum theory of gravity should aim to answer.

The two major revisions of our notion of space-time that have occurred in this century were caused by new *physical* inputs and not by just philosophical speculations. One might therefore ask if a similar input now exists motivating the need for yet another revision. The answer is in the affirmative: both general relativity[1] and quantum field theory have internal blemishes, which necessitate a better theory of space-time structure in the small. Let us begin with general relativity. Recall, first, that in Newtonian gravity, a (cold) dust cloud left to itself would shrink due to gravitational attraction and finally collapse to a single point. At this point, physical quantities, such as the mass density, become infinite. The predictions of general relativity are rather similar. The singularity theorems [2] of Penrose and Hawking tell us that a large class of regular, physically reasonable initial data for gravity plus matter, when evolved via Einstein's equation, eventually undergo a gravitational collapse and end up in a singularity, basically because of the universal, attractive nature of gravity. Now, we believe that in the real world, physical quantities do not become infinite. The prediction of such an infinity is just a signal that the theory is being applied beyond its domain of validity. Thus, the singularity theorems are a signal that general relativity—and the associated continuum picture of space-time geometry—cannot be trusted when the space-time curvature grows unboundedly. Under these circumstances, it is crucial to use a better picture of space-time in the small. Similarly, quantum field theory, as we know it, is also incomplete: it predicts that physical quantities such as scattering cross-sections are infinite when *all* radiative corrections are taken into account. The renormalization procedure makes individual terms in perturbative expansions manageable, although, in phys-

1 In all these qualitative arguments, details of Einstein's equations are not relevant. What is important is that we have a theory in which the geometry itself is dynamical, i.e., that there is no non-dynamical background metric.

ically interesting cases, such as QED, the entire series almost certainly diverges. Furthermore, the procedure itself—as was emphasized by its founders—is only a short-cut which enables us to calculate quantities of physical interest to desired accuracy without having to worry about what is "really" going on at extremely small distances (or, high energies). Thus, while one is happy that the procedure is available and that it "works", it is not clear "why" it works. To answer this question, one needs to begin with a better, more realistic model of the micro-structure of space-time geometry and pinpoint those of its features which underlie the success of the renormalization procedure.

Why does one need a non-perturbative framework to analyze these issues? After all, as field theorists are apt to point out, perturbation expansions have been highly successful in quantum theories of other interactions. Why not apply them to the gravitational interaction as well? Unfortunately, in the gravitational case these methods fail even by their own criteria. To emphasize this point, let me present a sketch of the history of these ideas in quantum gravity.

It has been known for some time now that Einstein's theory itself is perturbatively non-renormalizable at two loops for pure gravity and at one loop for gravity interacting with matter. One can change the theory by adding higher derivative terms to the action. Since all the classical tests of general relativity refer to the large-distance behavior of the theory, from a strict experimental viewpoint, we are free to add terms that leave this behavior untouched. The addition of suitable higher derivative terms *does* improve the ultraviolet behavior of the quantum theory. Not only can one make the theory renormalizable but one can even achieve asymptotic freedom; at high energies, the theory behaves as if it were free. However, now, the theory is not unitary. In the perturbative treatment, the Hamiltonian is in fact unbounded from below, signaling a dramatic instability. One might consider supersymmetric extensions of general relativity. The hope here is that the bosonic infinities of gravity would be cancelled by the fermionic infinities of matter, giving us an acceptable theory. Again, the resulting theory, supergravity, *is* better behaved. It has a positive definite Hamiltonian and even in the presence of (supersymmetric) matter, it is two-loop renormalizable. Unfortunately, however, the renormalizability fails at the third loop. A more radical revision is suggested by string theory. Here one abandons local field theories altogether and considers extended objects –strings– as the fundamental entities. It is a very attractive strategy because, unlike in any other attempt, all matter couplings are now fixed automatically; very little is fed in by hand. There is also a tremendous technical improvement. The theory is unitary and, although a conclusive proof is yet to emerge, by now a general consensus has developed among experts that the theory is perturbatively finite. To those who are unfamiliar with the technical jargon, however, this terminology is misleading: finiteness here refers only to each term in the perturbation expansion. To provide finite answers to physical questions, the entire sum must also converge. Unfortunately, it is known that the *sum diverges* and does so rather badly.[2] Now, one might wonder why this is a criticism since, as we saw, even in quantum electrodynamics the renormalized perturbation series, when summed, gives a divergent result. The answer is that theories such as quantum electrodynamics are inherently incomplete. They ignore the micro-structure of space-time and assume that space-time can be represented by Minkowski space at arbitrarily small distance scales. They shift the burden of infinities to the quantum theory of space-time structure. A "theory of everything," on the other hand, cannot escape so easily. It must face the Planck regime squarely. It cannot plead ignorance and shift the burden of infinities to yet another theory.

In the light of the previous discussion, the failure of perturbative quantum gravity is, in a sense, fortunate. For, had a consistent perturbative expansion existed, we could have assumed that the smooth, continuum picture of space-time is viable at arbitrarily small length scales; that, even below the Planck scale, space-time can be approximated by Minkowski space with a few quantum fluctuations which may be treated as small corrections. Then, the success of perturbative treatments in quantum theories of other interactions would have forever remained a mystery. Why is that, even though each term in the series diverges, the infinities that arise are precisely such that they can be systematically absorbed in the few parameters of the theory? Presumably, the reason lies in some peculiarity of the complications associated with the micro-structure of space-time. Presumably, the microstructure is not arbitrary. It is precisely such that one can ignore the complications at the simple cost of infinitely renormalizing a handful of parameters. It is the task of non-perturbative quantum gravity to elucidate this point. From this viewpoint, had a satisfactory perturbative treatment of quantum gravity existed, we would have lost the only avenue we have to understand why renormalization works at all in quantum field theories such as QED. Put differently, the failure of the perturbative approaches is a strong hint that "the buck stops with gravity"; once gravity is brought in, we can no longer ignore the complications associated with the micro-structure of space-time.

Indeed, the general consensus now is that the imaginative field theoretic attempts at quantizing gravity have failed because they assume that even at small distances, where by simple dimensional arguments, gravity should dominate, space-time geometry is smooth. To obtain a viable theory, one must drop this assumption

2 For the case of the bosonic string, this result was first proved by Gross and Periwal [4]. They argued that the perturbation expansion for superstring theory should behave in the same way. By now there are independent calculations supporting this view. (Jan Ambjorn, private communication 1989.)

and let the theory itself determine the micro-structure of space-time. For this, irrespective of whether one prefers general relativity or higher derivative theories, strings or membranes, one must face the problem of quantization *non-perturbatively.* My own view is that once one accepts this premise, the initial rationale for abandoning general relativity as the point of departure for quantization loses much of its force. Furthermore, the question of whether quantum general relativity exists as a consistent theory is of considerable interest in its own right in mathematical physics. Therefore, our program focuses on this issue. Note that I do *not* wish to imply that quantum general relativity would necessarily be the correct physical theory. Rather, the viewpoint is that general relativity is as good a starting point as any to address the key conceptual problems in quantum gravity that arise because of the absence of a background space-time geometry. I believe that one should aim at constructing a robust framework so that the broad qualitative conclusions would continue to hold if and when general relativity is replaced by a more complete theory.

Are there features of classical general relativity that indicate that non-perturbative quantum gravity is likely to be very different from the perturbative one? One can give a number of examples using the fact that non-perturbative general relativity allows non-trivial topologies which are absent in any order of perturbation expansion. However, here, we shall restrict ourselves to a simpler example, which is specifically geared to illustrate how spurious infinities can arise if one insists on making expansions in powers of Newton's constant. Consider the problem of self-energy of a point charge in classical field theory. We can regard the point charge as the limit of a sequence of spherical shells of radius e, with uniform charge and mass density, as ϵ tends to zero. If we ignore gravity altogether, the self-energy of the shell is given by:

$$m(\epsilon) = m_0 + \frac{e^2}{\mathrm{e}}, \tag{1a}$$

where m_0 and e are, respectively, the (bare) mass and charge. Clearly, $m(\mathrm{e})$ diverges as e tends to zero. Let us now bring in Newtonian gravity and include the appropriate gravitational self-energy. Then, we have:

$$m(\mathrm{e}) = m_0 + \frac{e^2}{e} - \frac{G m_0^2}{e}, \tag{1b}$$

which again diverges in the limit unless the parameters m_0 and e are fine-tuned (in which case, $m(e)$ equals m_0 for all e). Now, let us bring in general relativity. The key idea of Einstein's theory is that *all* energy contributes to the source of the gravitational field. We are therefore led to replace Eq.1b by

$$m(\epsilon) = m_0 + \frac{e^2}{\epsilon} - \frac{G\,m^2(\epsilon)}{\epsilon}. \tag{1c}$$

This equation can be solved for $m(\epsilon)$ in terms of other parameters. The positive root is given by

$$m(\epsilon) = \frac{-\epsilon}{2G} + \frac{\epsilon}{2G}\sqrt{1 + \frac{4G}{\epsilon}\left(m_0 + \frac{e^2}{\epsilon}\right)}, \tag{1d}$$

which, in the limit as ϵ tends to zero, yields

$$m(\epsilon{=}0) = \frac{e}{\sqrt{G}}. \tag{1e}$$

This answer has three remarkable features. First, it is *finite*; the universality of gravity automatically regularizes the limit. Second, the limit does not depend on the value of the bare mass m_0; no fine-tuning is involved. Finally, the result is *essentially* non-perturbative. For, if we first expand $m(\epsilon)$ in powers of Newton's constant G around $G = 0$ and then set ϵ to zero, each term in the expansion is divergent, although the sum is finite! We do *not* wish to suggest that Eq.1e gives the correct value of the mass of physical particles with charge e. The calculation is too naive from a number of standpoints—it ignores all quantum effects as well as weak and strong interactions—to enable a direct physical prediction (see, however, [5]). Rather, the calculation is meant to point out that there are certain *mathematical* mechanisms in general relativity which automatically regularize physical quantities, such as the self-energy in the example, which, however, are lost when one carries out a perturbation expansion in powers of Newton's constant.

To understand this mechanism more clearly, let us go through the self-energy calculation carefully. In the framework of general relativity, we need to first calculate the gravitational and electromagnetic fields of a spherical, uniformly charged shell of radius ϵ, compute the *total* energy of this configuration, and then take the limit of this expression as ϵ tends to zero. In fact, this precise calculation was carried out by Arnowitt, Deser and Misner (ADM) in the early sixties to explore the difference between perturbative and non-perturbative gravity [6]. Their analysis can be summarized as follows. Since the system under consideration is time-independent, it suffices to construct the required field configuration at one instant of time. For this, we need to solve the initial value equations of the combined Einstein-Maxwell system. These equations constrain the values of the canonically conjugate variables, the Maxwell potential $\mathbf{A}_a$, its conjugate momentum, $\widetilde{\mathbf{E}}^a$, the 3-metric, q_{ab}, on the slice Σ of space-time representing the instant of time and its conjugate momentum, $\tilde{p}^{ab}$. The electromagnetic constraints are the familiar ones:

$$D_a \tilde{\mathbf{E}}^a - \tilde{\rho}_{(c)} = 0 \qquad \text{and} \qquad D_a \tilde{\mathbf{B}}^a = 0, \tag{2}$$

where, $\tilde{\rho}_{(c)}$ is the charge density. The analogous gravitational constraints are:

$$C_a(q, p, \mathbf{A}, \mathbf{E}) := q_{am}(D_n \tilde{p}^{mn} - \tilde{J}^m) = 0 \tag{3a}$$
$$C(q, p, \mathbf{A}, \mathbf{E}) := q^{1/2} G^{-1} R - q^{-1/2} G\,(\tilde{p}^{ab} \tilde{p}_{ab} - \tfrac{1}{2} \tilde{p}^2) - \tilde{\rho} = 0, \tag{3b}$$

where $\tilde{J}^a$ and $\tilde{\rho}$ are, respectively, the matter momentum and matter energy density *including* the contributions from the electromagnetic field; and q, D and R are, respectively, the determinant, the derivative operator and the scalar curvature of the metric q_{ab}. ADM found the unique spherically symmetric solution to these equations, satisfying the appropriate boundary conditions, which correspond to static fields in space-time. To calculate the total energy of this field configuration, one needs the expression of the Hamiltonian. Now, it is a remarkable fact that, *modulo* the constraints (Eqs.3) the Hamiltonian of the entire system is always given by just a surface integral, evaluated at infinity:

$$H(q, p, \mathbf{A}, \mathbf{E}) = \int d^3x\, C(q, p, \mathbf{A}, \mathbf{E}) + \oint_{r=\infty} d^2S\, f(q), \tag{4}$$

where $f(q)$ is a simple, local function of the 3-metric q_{ab}. *This is a general result which stems simply from the fact that there is no non-dynamical background metric; it does not depend on details of Einstein's equation.* At an intuitive level, this is not surprising: total energy in general relativity is analogous to the total charge in Maxwell's theory, which is also expressible as a surface integral. However, it is precisely this simple fact that lies at the root of the finiteness of self-energy. For, in the limit as ϵ tends to zero, the values of the fields at infinity continue to be regular, whence the surface integral giving the total energy continues to be well-defined. In fact, as ADM showed, the limit is precisely the one we obtained in (1.e); the simple-minded calculation does give the right answer. How can we reconcile the finiteness with the fact that the energy in the electromagnetic field, $\int d^3V\,(\vec{\mathbf{E}}^2 + \vec{\mathbf{B}}^2)$, diverges because $\vec{\mathbf{E}}^2$ diverges like ϵ^{-4}, as ϵ goes to zero? Roughly, the divergent contribution to the Hamiltonian from the gravitational part just cancels the divergence from the electromagnetic part, because we have imposed the constraint equation Eq.3 of general relativity. The remainder, being the surface term at *infinity*, continues to remain finite just because of the boundary conditions. Thus, the "mechanism" responsible for finiteness is the fact that the volume integral contribution to the

Hamiltonian is just a constraint function. This is a ***non-perturbative*** feature of general relativity which *cannot* be recovered to any finite order in a perturbation expansion in powers of G. It is such features, we would like to argue, that make non-perturbative quantum gravity promising.

Indeed, the form of the full Hamiltonian suggests that if one could solve the quantum constraints exactly, the dynamics of the combined, Einstein-plus-matter quantum theory will be straightforward to handle. This is in striking contrast to Minkowskian quantum field theories where constraints, if any, are relatively straightforward to solve, and the crux of the problem lies in making quantum dynamics well-defined. To bring out the contrast, let us compare Yang-Mills theory in Minkowski space with Einstein-Yang-Mills theory. In the pure Yang-Mills case, the phase space variables are the connection 1-form $\mathbf{A}_{a\mathbf{A}}{}^{\mathbf{B}}$ and the conjugate momentum $\widetilde{\mathbf{E}}^{a}{}_{\mathbf{A}}{}^{\mathbf{B}}$, both of which take values in the Lie algebra of the gauge group under consideration. The constraints are given by the Gauss law:

$$\mathbf{D}_a\widetilde{\mathbf{E}}^a = 0, \tag{5}$$

where $\mathbf{D}$ is the gauge covariant derivative operator on the 3-plane in Minkowski space representing an instant of time. The Hamiltonian is given by:

$$H(\mathbf{A},\mathbf{E}) := \int d^3x \; \mathrm{tr}\,(\vec{\mathbf{E}}^2 + \vec{\mathbf{B}}^2). \tag{6}$$

Now, in quantum theory, one can begin by considering wave functions $\Psi(\mathbf{A})$ of the configuration variable $\mathbf{A}_{a\mathbf{A}}{}^{\mathbf{B}}$ and require that physical states be annihilated by the operator version of the Gauss law constraint. At least in principle, this step is relatively easy to carry out: the solutions to the Gauss law constraint are those $\Psi(\mathbf{A})$ which are gauge-invariant. The key difficulty arises in the next step. One has to introduce a Hermitian inner product on the space of physical states, define the quantum Hamiltonian on the resulting Hilbert space, and show that it is a self-adjoint operator. One has not yet been able to carry out this task. In particular, if one chooses Fock space as the Hilbert space, one finds that the Hamiltonian cannot be made into a well-defined self-adjoint operator. Let us now consider Einstein-Yang-Mills theory. Here the constraints are given by Eqs.5 and 3, where $\widetilde{J}^a$ and $\widetilde{\rho}$ are now the momentum and energy density of the Yang-Mills field. The Hamiltonian, however, is *still* given by Eq.4. If we were to pass to quantum theory, we would now begin by considering wave functions $\Psi(\mathbf{A}, q)$ of both Yang-Mills and Einstein configuration variables. The first task is the imposition of quantum constraints. *If* we were to find wave functions which are annihilated by these, we would have physical states. On these, the quantum Hamiltonian would reduce to just the surface term and its action would be straightforward. Indeed, the task of introducing *some* inner product which makes this Hamiltonian self-adjoint would be rather simple, although, there would still remain the problem of the physical interpretation of the framework. Nonetheless, if we focus just on the *mathematical* problem of constructing the quantum description, we see that the situation is quite different from that in Yang-Mills theory in Minkowski space; in the Einstein-Yang-Mills case, the difficulty of the task lies in solving the quantum constraints. Thus, the issues that arise and the strategies that are called for in non-perturbative quantum gravity are *very* different from those that arise in conventional field theories. We are not suggesting that quantum Einstein-Yang-Mills theory would be necessarily simpler than quantum Yang-Mills theory in Minkowski space. We are pointing out that the two theories are *structurally* very different. Therefore, one's experience with Minkowskian field theories is of rather limited use in the analysis of quantum theories such as Einstein-Yang-Mills. At present, we simply do not know if such theories are manageable non-perturbatively. It seems well worth exploring if they are.

The purpose of this work is to suggest an avenue for this task.

References

[1] E. Cartan, *On Manifolds with Affine Connection and Theory of General Relativity* (Bibliopolis, Naples 1985).

[2] S.W. Hawking and G.F.R. Ellis, *The large scale structure of space-time* (Cambridge University Press 1973).

[3] *Quantum Theory of Gravity: Essays in Honor of Bryce De Witt*, edited by S.M. Christensen (Adam Hilger, Bristol 1984).

[4] D. Gross and V. Periwal, Phys. Rev. Lett. **60**, 2105-8 (1988).

[5] R. Woodard, Ph. D. thesis (Harvard University 1984).

[6] R. Arnowitt, S. Deser and C.W. Misner, Phys. Rev. **120**, 313-20 (1960); Ann. Phys. (N.Y.) **33**, 88-107 (1965).

2 OVERVIEW

1 Introduction

The purpose of this chapter is to provide a brief status report of the program that will be discussed in detail in these notes. I will therefore omit details and present only the main ideas. The hope is that this discussion will serve a dual purpose: For readers who are unfamiliar with the program, this chapter should serve as an introduction and to experts it should serve as an up to date summary of the main developments.

The basic ideas underlying this program arose as follows. At the (Padova) GRG conference in '83, a new avenue to non-perturbative, canonical quantum gravity was suggested [1]. It was based on techniques introduced by by Amitabha Sen [2] in his work on the zero modes of the Dirac operator and Edward Witten [3] in his proof of the positivity of energy. While considerable progress was made in the following two years, (see, e.g., the proceedings of the '84 Oxford Symposium [4]), it was during a six month workshop at Santa Barbara in '86, that these ideas finally blossomed into a broad program aimed at analyzing the structure of classical and quantum general relativity from a somewhat unusual standpoint [5,6]. In particular, using Wilson loops, Theodore Jacobson and Lee Smolin [7] obtained a large class of solutions to the difficult quantum scalar constraint of general relativity. This work then served as the starting point for the construction of the "loop representation" by Carlo Rovelli and Lee Smolin [8,9]. By now, over two dozen individuals have contributed to the various aspects of this program. Although I will try to present an objective summary in this chapter, it is inevitable that not everyone who is working in this field will agree with all the views expressed here. Also, the discussion is far from being exhaustive. I will concentrate on issues which I feel are directly relevant to the main stream of development and only mention other interesting results.

The key line of thought underlying this program is to shift the emphasis from geometrodynamics to connection dynamics. In the classical theory, the new viewpoint merely complements the traditional one in which the metric, rather than a connection, is taken as the fundamental variable. We do obtain a fresh perspective that simplifies certain issues and suggests new ways of tackling unresolved problems. However, as far as the basic features of the theory are concerned, nothing is really altered conceptually. It is in the quantum regime that the shift of emphasis plays a major role. More precisely, there are indications that connection dynamics is indeed a better tool to analyze the micro-structure of space-time in a non-perturbative way. The emphasis on connections opens up new windows; a number of unanticipated concepts begin to play a major role in the formulation of the basic questions and new technical machinery becomes available to analyze these issues. Finally, the resulting mathematical framework is closely related to the one which has been successful in the study of other basic interactions in physics. Thus, if these ideas succeed, quantum general relativity would no longer remain isolated; a unifying mathematical framework would bind it to other theories of Nature. It is interesting to note that, in a certain sense, this approach strikes a chord with views that have been expressed both by relativists and quantum field theorists. Relativists have long felt that a *non-perturbative* approach is needed; any attempt that begins by "steam-rollering general relativity into flatness and linearity" cannot ultimately succeed [10]. Field theorists, on the other hand, have felt that the emphasis on the space-time metric and its geometrical properties have "driven a wedge between general relativity and the theory of elementary particles" (in which quantum mechanics of connections plays a deep role) [11]. A non-perturbative analysis of connection dynamics appears to bridge this gap.

The material in this chapter is divided as follows. Section 2 summarizes the new Hamiltonian formulation of classical general relativity on which the program is based. I also point out the close relation between this formulation and the standard canonical framework for Yang-Mills fields. Section 3 outlines the quantization program. I will first indicate how the familiar Schrödinger equation arises from the quantum scalar constraint in a suitable approximation, once "time" (or, an internal clock) is singled out from among the components of the connection. While it is still unclear whether this procedure can be extended to the full theory without any approximation, the result nonetheless gives one considerable confidence that the everyday quantum physics we are used to *does* arise from the basic equations of quantum gravity. I will then turn to full quantum gravity and discuss the "loop representation" introduced by Rovelli and Smolin [8,9] in which quantum states arise as suitable functions of closed loops on a 3-manifold and present the infinite dimensional space of their solutions to all quantum constraints. Although these developments represent substantial progress, a number of important problems remain unresolved. To gain insight into the remaining problems, the program has been carried to conclusion in some simplified models. I will conclude with a brief discussion of these results. In section 4, I will return to the shift of emphasis from geometrodynamics to connection dynamics and suggest that the space of connections can serve as the kinematical arena in full quantum gravity where one must do physics without a background space-time.

For conciseness, in this summary as well as in the detailed notes that follow, I will have to leave out a number of interesting results. These include: A new characterization of self dual metrics [12,13] and of Einstein metrics with self dual Weyl tensor [14-17]; BRST-formulation of general relativity [18]; a discussion of the relation between the Einstein and the Yang-Mills instantons [14]; classical and quantum theory of gravitational fields with two Killing fields [19]; lattice approach to quantum gravity [20,21]; and some ideas on quantum cosmology [22,23].

2 Hamiltonian framework

For simplicity, I shall restrict myself to source-free general relativity. The framework is, however, quite robust: all its basic features remain unaltered by the inclusion of a cosmological constant and coupling of gravity to Klein-Gordon fields, (classical or Grassmann-valued) Dirac fields and Yang-Mills fields with any internal gauge group [24,25]. For brevity of presentation, I will begin with complex general relativity –i.e. by considering complex Ricci-flat metrics g_{ab} on a real 4-manifold M– and take the "real section" of the resulting phase-space at the end.

The idea is to use a first-order formalism.[1] The basic space-time fields will consist of a pair, $(e^a{}_I\ ,{}^4A_a{}^{IJ}\ )$, of a (complex) tetrad $e^a{}_I$ and a connection 1-form ${}^4A_a{}^{IJ}$ which is *self dual* in the "internal" (or Lorentz) indices 'IJ'; we have: ${}^4A_a^{IJ} = \frac{i}{2}\epsilon^{IJ}{}_{KL}{}^4A_a^{KL}$. (If one is interested in Euclidean general relativity, one should drop the i in the right hand side while defining self duality and consider real tetrads and connections throughout.) The action is given by ([26-28]) :

$$S(e, {}^4A) := \int d^4x\ (e)e^a{}_Ie^b{}_J{}^4F_{ab}{}^{IJ}, \tag{1}$$

where ${}^4F_{ab}{}^{IJ}$ is the curvature of ${}^4A_a{}^{IJ}$, and e is the determinant of the co-tetrad. This action is rather like the one introduced by Palatini except that, since the

1 In this chapter (and except where noted, in the rest of the book) lower case latin letters from the beginning of the alphabet, $a, b, \ldots$ will denote space-time as well as spatial indices. Upper case latin letters from the middle of the alphabet $I, J, \ldots$ will denote internal Lorentz (Lie algebra) indices; the corresponding lower case latin letters, $i, j, \ldots$ will denote internal $SO(3)$ indices. If there is a possibility of confusion between 4-dimensional space-time fields and 3-dimensional fields defined intrinsically on a spatial slice, we will use a prefix 4 to denote space-time fields. Thus, ${}^4A_a{}^{IJ}$ is a space-time connection while $A_a{}^i$ is its pull-back to a spatial slice.

connection ${}^4A_a{}^{IJ}$ is now required to be self dual in the internal indices, so is the curvature tensor in Eq.1. By setting the variation of the action with respect to ${}^4A_a{}^{IJ}$ to zero we obtain the result that ${}^4A_a{}^{IJ}$ is the self dual part of the (torsion-free) connection ${}^4\Gamma_a{}^{IJ}$ compatible with the tetrad $e^a{}_I$. Thus, ${}^4A_a{}^{IJ}$ is completely determined by $e^a{}_I$. Setting the variation with respect to $e^a{}_I$ to zero and substituting for the connection from the first equation of motion, we obtain the result that the space-time metric $g^{ab} = e^a{}_I e^b{}_J \eta^{IJ}$ (where η_{IJ} is the fixed, Minkowski metric on the internal space) satisfies the vacuum Einstein's equation. Thus, even though we are using a self dual connection, Eq.1 is completely equivalent to the standard Palatini action as far as the classical equations of motion are concerned. The reason behind this can be traced back to the Bianchi identities.

One can carry out the Legendre transform of this action by carrying out a 3+1-decomposition in a fairly straightforward way. The resulting canonical variables are then complex fields on a ("spatial") 3-manifold Σ. The configuration variable turns out to be a $SO(3)$-connection 1-form $A_a{}^i$ and its canonical momentum turns out to be a triad $\widetilde{E}^a{}_i$ with density weight one, where 'a' is the (co)vector index and 'i' is the triad or the $SO(3)$ internal index. The (non-vanishing) fundamental Poisson brackets are:

$$\{\widetilde{E}^a{}_i(x),\ A_b{}^j(y)\} = -i\delta^a{}_b\delta_i^j\delta^3(x,y). \tag{2}$$

The geometrical interpretation of these canonical variables is as follows. In any solution to the field equations, $A_a{}^i$ turns out to be a potential for the self dual part of the Weyl curvature and $\widetilde{E}^a{}_i$ the "square-root" of the 3-metric (times its determinant) on Σ. The relation of these variables to the familiar geometrodynamical variables, the 3-metric q_{ab} and the extrinsic curvature K_{ab} on Σ, is as follows:

$$GA_a{}^i = \Gamma_a{}^i - iK_a{}^i \quad \textit{and} \quad \widetilde{E}^a{}_i\widetilde{E}^{bi} = (q)q^{ab} \tag{3}$$

where G is Newton's constant, $\Gamma_a{}^i$ is the spin connection determined by the triad, $K_a{}^i$ is obtained by transforming the space index 'b' of K_{ab} into an internal index by the triad and q is the determinant of q_{ab}. Note, however, that, as far as the mathematical structure is concerned, we can also think of $A_a{}^i$ as a (complex) $SO(3)$-Yang-Mills connection and $\widetilde{E}^a{}_i$ as its conjugate electric field. Thus, the phase space has a dual interpretation. It is this fact that enables one to import into general relativity and quantum gravity ideas from Yang-Mills theory and quantum chromodynamics and may, ultimately, lead to a unified mathematical framework underlying the quantum description of all fundamental interactions. In what follows, we shall alternate between the interpretation of $\widetilde{E}^a{}_i$ as a triad and as the electric field canonically conjugate to the connection $A_a{}^i$.

Since the configuration variable $A_a{}^i$ has nine components per space point and since the gravitational field has only two degrees of freedom, we expect seven first class constraints. This expectation is indeed correct. The constraints are given by:

$$\mathcal{G}_i := \mathcal{D}_a \widetilde{E}^a{}_i = 0, \tag{4a}$$
$$\mathcal{V}_a := F_{ab}{}^i \widetilde{E}^b{}_i = 0, \tag{4b}$$
$$\mathcal{S} := \epsilon^{ijk} F_{abk} \widetilde{E}^a{}_i \widetilde{E}^b{}_j = 0, \tag{4c}$$

where $F_{ab}{}^i := 2\partial_{[a}A_{b]}{}^i + G\epsilon^{ijk}A_{aj}A_{bk}$ is the field strength constructed from $A_a{}^i$. Note that all these equations are simple polynomials in the basic variables; the worst term occurs in the last constraint and is only quadratic in each of $\widetilde{E}^a{}_i$ and $A_a{}^i$. The three equations are called, respectively, the Gauss constraint, the vector constraint and the scalar constraint. The last two are familiar; they arise also in the older canonical formulation of general relativity based on 3-metrics q_{ab} and their conjugate momenta $\widetilde{p}^{ab}$. The first, Gauss law, arises because we are now dealing with triads rather than metrics. It simply tells us that the internal $SO(3)$ triad rotations are "pure gauge."

From geometrical considerations we know that the "kinematical gauge group" of the theory is the semi-direct product of the group of local triad rotations with that of spatial diffeomorphisms on Σ. This group has a natural action on the canonical variables $A_a{}^i$ and $\widetilde{E}^a{}_i$ and thus admits a natural lift to the phase-space. It turns out that this is precisely the group formed by the canonical transformations generated by the Gauss and the vector constraints. Thus, six of the seven constraints admit a simple geometrical interpretation. What about the scalar constraint? Note that, being quadratic in momenta, it is of the form $G^{\alpha\beta}P_\alpha P_\beta = 0$ where, the connection supermetric $\epsilon^{ijk}F_{abk}$ plays the role of $G^{\alpha\beta}$. Consequently, the motions generated by the scalar constraint in the phase space correspond precisely to the *null geodesics of the "connection supermetric"*. It is well-known that the space-time interpretation of these canonical transformations is that they correspond to "multi-fingered" time-evolution. Thus, we now have an attractive representation of the Einstein evolution as a null geodesic motion in the (connection) configuration space.

In the asymptotically flat situation, asymptotic (space and time) translations are generated on the phase-space by Hamiltonians. As in geometrodynamics, these are obtained by adding suitable surface terms to constraints. Given a lapse $\underset{\sim}{N}$ and a shift N^a, the Hamiltonian is given by:

$$H(A,\widetilde{E}) = -i\int_\Sigma d^3x\,(N^aF_{ab}{}^i\widetilde{E}^b{}_i - \tfrac{i}{2}\underset{\sim}{N}\epsilon^{ijk}F_{abk}\widetilde{E}^a{}_i\widetilde{E}^b{}_j) + \oint_{\partial\Sigma} d^2S_a\,(\underset{\sim}{N}\epsilon^{ijk}A_{bk}\widetilde{E}^a{}_i\widetilde{E}^b{}_j + 2iN^{[a}\widetilde{E}^{b]}{}_iA_b{}^i). \tag{5}$$

(In this framework, the lapse naturally arises as a scalar density $\underset{\sim}{N}$ of weight -1. It is $\underset{\sim}{N}$ that is the basic, metric independent field. The "geometric" lapse function N is metric dependent and given by $N = \sqrt{q}\underset{\sim}{N}$.)

So far, we have discussed *complex* general relativity. To recover the Lorentzian theory, we must now impose reality conditions, i.e., restrict ourselves to the real, Lorentzian section of the phase-space. Let me explain this point by means of an example. Consider a simple harmonic oscillator. One may, if one so wishes, begin by considering a complex phase-space spanned by two complex co-ordinates q and p and introduce a new complex co-ordinate $z = q - ip$. (q and p are analogous to the triad $\widetilde{E}^a{}_i$ and the extrinsic curvature $K_a{}^i$, while z is analogous to $A_a{}^i$.) One can use q and z as the canonically conjugate pair, express the Hamiltonian in terms of them and discuss dynamics. Finally, the real phase-space of the simple harmonic oscillator may be recovered by restricting attention to those points at which q is real and $ip = q - z$ is pure imaginary (or, alternatively, $\dot{q}$ is also real.) In the present phase-space formulation of general relativity, the situation is analogous. In terms of the familiar geometrodynamic variables, the reality conditions are simply that the 3-metric be real and the extrinsic curvature –the time derivative of the 3-metric– be real. If these conditions are satisfied initially, they continue to hold under time-evolution. In terms of the present canonical variables, these become [25]: *i*) the (densitized) 3-metric $\widetilde{E}^a{}_i\widetilde{E}^{bi}$ be real, and, *ii*) its Poisson bracket with the Hamiltonian H be real, i.e.,

$$(\widetilde{E}^a{}_i\widetilde{E}^{bi})^\star = \widetilde{E}^a{}_i\widetilde{E}^{bi} \tag{6a}$$
$$[\epsilon_{ijk}\widetilde{E}^a{}_i\mathcal{D}_c(\widetilde{E}^c{}_j\widetilde{E}^b{}_k)]^\star = -\epsilon_{ijk}\widetilde{E}^a{}_i\mathcal{D}_c(\widetilde{E}^c{}_j\widetilde{E}^b{}_k)). \tag{6b}$$

(In Euclidean relativity, these conditions can be further simplified; they only require that we restrict ourselves to real triads and real connections.) As far as the classical theory is concerned, we could have restricted to the "real slice" of the phase-space right from the beginning. In quantum theory, on the other hand, it is simpler to first consider the complex theory, solve the constraint equations and then impose the reality conditions as suitable Hermitian-adjointness relations. Thus, the quantum reality conditions would be restrictions on the choice of the inner-product on physical states.

Could we have arrived at the phase-space description of *real* general relativity in terms of ($A_a{}^i$, $\widetilde{E}^a{}_i$) without having to first complexify the theory? The answer is in the affirmative. This is in fact how the new canonical variables were first introduced [5,6]. The idea is to begin with the standard Palatini action for real tetrads and real Lorentz-connections, perform the Legendre transform and obtain the phase-space of real relativity *a la* Arnowitt, Deser and Misner. The basic canonical variables in this description can be taken to be the density weighted triads $\widetilde{E}^a{}_i$ and their canonical conjugate momenta Π^i_a. The interpretation of Π^i_a is as follows: In any solution to the field equations, i.e., "on shell", $K_{ab} := \Pi^i_{(a}E_{b)i}$ turns out to be the extrinsic curvature. Up to this point, all fields in question are real. On this real phase space, one can make a (complex) canonical transformation to pass to the new variables: $(\widetilde{E}^a_i, \Pi^i_a) \to (\widetilde{E}^a_i, A^i_a := \Gamma^i_a - i\Pi^i_a \equiv (\delta F/\delta \widetilde{E}^a_i) - i\Pi^i_a)$, where the generating function $F(\widetilde{E})$ is given by: $F(\widetilde{E}) = \int_\Sigma d^3x \widetilde{E}^a_i \Gamma^i_a$, and where Γ^i_a are the spin coefficients determined by the triad $\widetilde{E}^a{}_i$. Thus, $A_a{}^i$ is now just a complex coordinate on the traditional, real phase space. This procedure is completely analogous to the one which lets us pass from the canonical coordinates (q,p) on the phase space of the harmonic oscillator to the coordinates $(q, z = dF/dq - ip)$, with $F(q) = \frac{1}{2}q^2$, and makes the analogy mentioned above transparent. Finally, the second of the reality conditions, Eq.6b, can now be re-expressed as the requirement that $A^i_a - \Gamma^i_a$ be purely imaginary, which follows immediately from the expression of $A_a{}^i$ in terms of the real canonical variables $(\widetilde{E}^a_i, K^i_a)$.

A number of remarks are in order.

i) Note that all equations of the theory –the constraints, the Hamiltonian and hence the evolution equations and the reality conditions– are simple polynomials in the basic variables $\widetilde{E}^a{}_i$ and $A_a{}^i$. The framework may therefore be of considerable interest to numerical relativists [29]. In this connection, it is especially worth noting that a simple approach is in fact available to obtain the "free data", i.e. solutions to the constraints (Eqs.4) [30]. The strategy is the following. Choose any connection $A_a{}^i$ such that its magnetic field $\widetilde{B}^a{}_i := \widetilde{\eta}^{abc}F_{bci}$, regarded as a matrix, is non-degenerate. (We shall denote by $\widetilde{\eta}^{abc}$ the metric independent Levi-Civita density.) A "generic" connection $A_a{}^i$ will satisfy this condition; it is not too restrictive an assumption. Now, we can expand out $\widetilde{E}^a{}_i$ as $\widetilde{E}^a{}_i = M_i{}^j \widetilde{B}^a{}_j$ for some matrix $M_i{}^j$. The pair $(A_a{}^i\ , \widetilde{E}^a{}_i\ )$ then satisfies the vector and the scalar constraints *if and only if* $M_i{}^j$ is of the form $M_i{}^j = [\phi^2 - \frac{1}{2}\operatorname{tr}\phi^2]_i{}^j$, where $\phi_i{}^j$ is an arbitrary trace-free, symmetric field on Σ. Thus, as far as these four constraints are concerned, the free data consists of $A_a{}^i$ and $\phi_i{}^j$. It only remains to solve the Gauss constraint which simply reduces to: $\widetilde{B}^a{}_i \mathcal{D}_a M_i{}^j = 0$. Although it is straightforward to invent procedures to solve this remaining equation, a simple, elegant method has not yet emerged; input from numerical relativists is needed to determine the most useful strategy.

ii) The phase-space of general relativity is now identical to that of complex-valued Yang-Mills fields (with internal group $SO(3)$). Furthermore, one of the constraint equations is precisely the Gauss law that one encounters on the

Yang-Mills phase-space. Thus, we have a natural imbedding of the constraint surface of Einstein's theory into that of Yang-Mills theory: Every initial datum $(A_a{}^i, \widetilde{E}^a{}_i)$ for Einstein's theory is also an initial datum for Yang-Mills theory which happens to satisfy, in addition to the Gauss law, a scalar and a vector constraint. From the standpoint of Yang-Mills theory, the additional constraints are essentially the simplest diffeomorphism and gauge invariant expressions one can write down in absence of a background structure such as a metric. Note that the degrees of freedom match: the Yang-Mills field has 2 (helicity) $\times 3$ (internal) $= 6$ degrees and the imposition of four additional first-class constraints leaves us with $6-4=2$ degrees of freedom of Einstein's theory. I want to emphasize, however, that in spite of this close relation of the two initial value problems, the Hamiltonians (and thus the dynamics) of the two theories are *very* different. Nonetheless, the similarity that does exist can be exploited to obtain interesting results relating the two theories [14-17,31].

iii) Since all equations are polynomial in $A_a{}^i$ and $\widetilde{E}^a{}_i$ they continue to be meaningful even when the triad (i.e. the "electric field") $\widetilde{E}^a{}_i$ becomes degenerate or even vanishes. Results obtained in 2+1 gravity by Witten [32] indicate that this fact would play a significant role in the "unbroken," diffeomorphism invariant phase of quantum gravity where the vacuum state would correspond to just the zero, rather than Minkowskian, metric. In fact, recently, Capovilla, Dell and Jacobson [30] have introduced a Lagrangian framework which reproduces the Hamiltonian description discussed above but which *never even introduces a space-time metric*! This formulation of "general relativity without the metric" lends strong support to the viewpoint that the traditional emphasis on metric-dynamics, however convenient in classical physics, is not indispensable.

3 Quantum theory

I will now outline the quantization program and summarize its status in full quantum gravity in $3+1$ dimensions. We will see that although significant progress has been made on a number of issues, several important and difficult problems remain to be investigated; even as a mathematical framework, the theory is incomplete. To gain insight into these problems, we have carried out the entire program to its conclusion in certain simplified contexts. In this overview, I will simply state these results. Details will be found in Part III of these notes.

3.A The program

We shall use a slightly modified version of the Dirac procedure for quantization of constrained systems. The key steps of the program may be stated as follows:

1. Select a subspace $\mathcal{S}$ of functions on the classical phase-space, closed under the operation of taking Poisson-brackets. Each element of $\mathcal{S}$ is to be promoted to a quantum operator unambiguously; it represents *an elementary classical variable*. $\mathcal{S}$ has to be "small enough" so that this Dirac-quantization procedure can be carried out unambiguously (i.e. without factor ordering problems) and yet "large enough" so that they can serve as (complex) coordinates on the phase-space. For the harmonic oscillator discussed in section 2, for example, $\mathcal{S}$ can be taken to be the 2-dimensional space spanned by (complex-valued) q and p. In general relativity, $\mathcal{S}$ could be the space of linear functionals of $A_a{}^i$ and $\widetilde{E}^a{}_i$.
2. Associate with each element F of $\mathcal{S}$ an abstract operator $\hat{F}$. These are the *elementary* quantum operators. Construct the (free) algebra generated by these elements and impose the canonical commutation relations: $[\hat{F}, \hat{G}] - i\hbar\widehat{\{F, G\}} = 0$. (If there are algebraic relations between the elementary classical variables, these have to be incorporated in the quantum theory by suitable anti-commutation relations. For details, see chapter 10.) Denote the resulting algebra by $\mathcal{A}$.
3. For later use, let us also *define* a $\star$-relation (i.e. an involution) on this algebra by requiring that, if two elementary classical variables F and G are complex conjugates of one another, i.e., if $\bar{F} = G$, then $\hat{G} = (\hat{F})^\star$, and that the $\star$ operation satisfies the properties of an involution. (Recall that these are: $(\hat{F} + \lambda\hat{G})^\star = \hat{F}^\star + (\bar{\lambda})\hat{G}^\star$, $(\hat{F}\hat{G})^\star = \hat{G}^\star\hat{F}^\star$, and, $(\hat{F}^\star)^\star = F$.) Denote the resulting $\star$-algebra by $\mathcal{A}^{(\star)}$.
4. Find a representation of the algebra $\mathcal{A}$ by operators on a complex vector space V. Note that V is *not* equipped with any inner-product and that the $\star$-relations are ignored at this stage. This representation may be obtained by any convenient means. Possible candidates are: geometric quantization techniques and group theoretical methods [33,34].
5. Obtain the quantum analogs of the classical constraints. This requires the choice of a factor-ordering *and* regularization. Find the linear subspace V_{phy} of V which is annihilated by all quantum constraints. This is the space of physical quantum states.
6. Introduce an inner-product on V_{phy} such that the $\star$-relations –ignored so far– become Hermitian adjoint relations on the resulting Hilbert space. Note

that the full $\star$-algebra $\mathcal{A}^{(\star)}$ itself does *not* have a well-defined action on the physical subspace V_{phy}; a general element of $\mathcal{A}$ would not weakly commute with the constraints and would therefore throw physical states out of V_{phy}. One must therefore find a "sufficiently large" set of operators which weakly commute with the constraints whose $\star$-adjoints also weakly commute with the constraints. It is the $\star$-relations between *these* "physical" operators that are to be taken over to Hermitian adjointness relations by the inner product. This is a rather involved procedure. In simple quantum mechanical systems where constraints are either absent or play a minimal role, the $\star$-relations –quantum versions of the classical reality conditions– select the inner-product uniquely. In particular, this is the case in $2+1$-gravity and the weak field limit of the $3+1$ theory. It is not known if the situation in the full $3+1$ theory is analogous.

7. Interpret a sufficiently large class of self-adjoint operators; devise methods to compute their spectra and eigenvectors; analyse if there is a precise sense in which the 1-parameter family of transformations generated by the Hamiltonian can be interpreted as "time evolution"; ...

If this program can be completed, one would have available a coherent mathematical framework. Important conceptual issues from measurement theory will still have to be faced.

3.B The connection representation.

The obvious candidate for the subspace $\mathcal{S}$ of functions on the phase-space is the space of functionals which are linear in $A_a{}^i$ and $\widetilde{E}^a{}_i$. With this choice, it is straightforward to complete the first four steps given above using, as the representation space, functionals of either $A_a{}^i$ or $\widetilde{E}^a{}_i$. The problem arises in step 5. The quantum Gauss constraint poses no problem. However, the vector and the scalar constraints do. As in the traditional canonical framework, while it is straightforward to write down a large class of solutions to the vector constraint in the E-representation, not a single solution is known to the scalar constraint. In the A-representation, the situation is somewhat different. An infinite dimensional space of solutions to the scalar constraints is available (see Jacobson & Smolin [7]). These are essentially the "Wilson-loops" , i.e., traces of holonomies, $h[\gamma, A]$, (regarded as functions of $A_a{}^i$, parametrized by γ.)[2] However, none of these is diffeomorphism

2 Thus, $h[\gamma, A] := \mathrm{tr}\, \mathcal{P} \exp \oint_\gamma A_a dS^a$ is the trace of the path ordered exponential of the integral of the connection $A_a{}^i$ around the closed loop γ .

invariant since, under the action of diffeomorphisms, each γ is mapped to another loop. One can *formally* integrate these functionals over the diffeomorphism group of the 3-manifold and the resulting "functional" of $A_a{}^i$ *will* satisfy all constraints. But this construction is only formal since one does not know how to carry out the required integration with any degree of rigor. Alternatively, one can solve the Gauss and the vector constraints as follows. Consider connections $A_a{}^i$ for which the magnetic field, $\widetilde{B}^a{}_i$, regarded as a 3×3 matrix, is non-degenerate. One can then use it as a "triad" and construct wave functionals (which have support only on those $A_a{}^i$ with non-degenerate magentic fields and) which are the analogs of the solutions to the Gauss and the vector constraints in the triad representation. None of these, however, seem to satisfy the scalar constraint.[3]

Significant progress has, however, been made in a "truncated theory" in which one expands the constraints around a classical background (both $\widetilde{E}^a{}_i$ and $A_a{}^i$ flat) and keeps terms up to *second* order [35]. In this theory, not only can one solve all constraints but it is also possible to identify the component of the connection, A_T, that is to play the role of time; the true, dynamical degrees of freedom of the theory, A_D, evolve with respect to this "internal clock." Let me focus on the scalar constraint. Recall that, in the classical theory, this constraint plays a dual role. On the one hand, it restricts the physically allowable states and, on the other, it generates time evolution. In the truncated theory, the situation is completely analogous in the quantum picture as well. Physical states are annihilated by the scalar constraint. In this sense, the constraint is just a restriction on the physical states; nothing "happens." This is analogous to the relativists' view that a solution of Einstein's equation in the classical theory is simply an entire space-time; intrinsically, there is no dynamics. Things begin to "happen" when we identify a time variable and slice space-time into space and time. Again, the situation in quantum theory is similar. If we identify one of the connection components, A_T, as "internal time" –and the choice is suggested by the form of the scalar constraint itself– then, the quantum constraint equation can be regarded as an "evolution equation" in the infinite-dimensional space of connections; the constraint simply tells us how to "evolve" the wave function $\Psi(A_D, A_T)$, given its value, $\Psi(A_D, A_T = 0)$, at the "initial time" $A_T = 0$. It turns out that the generator of this "evolution" is *precisely* the standard Hamiltonian of linearized gravity plus matter. Therefore, the "evolution," which takes place in the space of connections, can be re-interpreted

3 In quantum cosmology, one often handles the Gauss and vector constraints by gauge fixing. Several solutions to the scalar constraints are then available both in the triad and the connection representations [22]. Very recently, by performing an appropriate transform on one of these solutions, Moncrief and Ryan have obtained the first exact solution to the Wheeler-DeWitt equation of quantum geometrodynamics in Bianchi IX models. (M.Ryan, private communication, 1990.)

as taking place in Minkowski space. Thus, in this truncated theory, the quantum scalar constraint actually reproduces the familiar Schrödinger equation for evolution! It is not yet clear whether this "derivation" of the Schrödinger equation from quantum gravity will extend to the full, non-truncated theory. Nonetheless, it is gratifying that at least in the weak field limit, the general framework outlined here reproduces the familiar, flat space quantum mechanics. While the flatness of the chosen background point $(A_a{}^i,\ \widetilde{E}^a{}_i)$ in the phase space simplifies calculations, I believe it is not essential. The main ideas underlying this truncated theory should provide new insight into the nature of quantum fields on curved backgrounds, including the origin of Hawking radiation. In this picture, quantum field theory in curved space-times would "descend from above" as an interesting approximation to full quantum gravity; traditionally, it has "emerged from below" as a first order correction to the classical description.

3.C The loop representation

In the full 3+1 theory, the most promising approach to date is to use the "loop representation" in which quantum states are represented by functions of loops on the 3-manifold Σ [8,9, 36]. This construction rests heavily on the shift of emphasis in the classical canonical theory from geometrodynamics to connection dynamics. The motivation comes from the fact that the Wilson loops solve the scalar constraint. The intuitive idea is the following. If one formally writes the integral:

$$\mathcal{A}(\gamma) := \int d\mu[A]\, h[\gamma, A]\Psi[A], \tag{7}$$

(where "$d\mu[A]$" is a diffeomorphism invariant "measure" on the space of connections), one finds that, because the kernel $h[\gamma, A]$ satisfies the scalar constraint in the A-representation, the left hand side, $\mathcal{A}(\gamma)$, satisfies the scalar constraint for *any* choice of $\Psi(A)$ in the integral on the right. This is analogous to the fact that, in the Klein-Gordon theory in Minkowski space, the function $\mathcal{A}(x)$ on space-time given by

$$\mathcal{A}(x) := \int d^4k\, [\delta^4(k\cdot k - \mu^2)e^{ik\cdot x}]\Psi(k)\,, \tag{7'}$$

satisfies the Klein-Gordon equation automatically for *any* $\Psi(k)$ because the kernel in the square brackets does so. In general relativity, then, it only remains to impose the vector constraint on the loop-functions $\mathcal{A}(\gamma)$ of Eq.7. Since under the action of the diffeomorphism group of Σ, a loop is mapped to another loop in the same knot class, it follows that loop functions will satisfy *all* constraints if they are obtained via

Eq.7 *and* they depend only on the knot-classes of loops (rather than on individual loops.)

With these formal constructions as motivation, Rovelli and Smolin [8,9] carry out the first four steps in the quantization program as follows. First they introduce a brand new space of functions on the phase-space which is to serve as the space $\mathcal{S}$ of step 1. Associated with a loop γ, there is an element of $\mathcal{S}$ called $T^0[\gamma](A)$, and associated with a loop γ, and a point x on γ there is another element, $T^1[\gamma, x](A, \widetilde{E})$ (which takes values in the space of vectors of density weight one at the point x). These are defined as follows:

$$T^0[\gamma](A) = h[\gamma, A] \equiv \mathrm{tr}\, \mathcal{P}\ \exp\left(\oint_\gamma A_a ds^a\right) \tag{8a}$$

$$\text{and} \quad T^1[\gamma, x](A, \widetilde{E}) = \mathrm{tr}\, \widetilde{E}^a(x)\ \mathcal{P} \exp\left(\oint_\gamma A_a ds^a\right), \tag{8b}$$

where $A_a = A_a{}^i \sigma_i$ and $\widetilde{E}^a = \widetilde{E}^a{}_i \sigma^i$ with σ^i the 2×2 Pauli matrices. The T^0 are the configuration variables and the T^1, being linear in $\widetilde{E}^a_i$, the momentum variables. (One can define in a similar way functions $T^{(n)}[\gamma, x_1, ... x_n]$ which are of order n in the momenta $\widetilde{E}^a{}_i$. These are useful but not essential to the construction.) One can check that the vector space $\mathcal{S}$ spanned by T^0 and T^1 is closed under the Poisson bracket and is large enough to qualify as the space $\mathcal{S}$ in the first step of the quantization program. This space is referred to as the (classical) T-algebra. This Poisson algebra carries the full information of the symplectic structure of general relativity. What is remarkable is that the basic Poisson brackets can be fully specified in terms of operations of breaking and joining of the loops γ labelling the T^0 and the T^1 variables.

Step 2 of the quantization program is now easily carried out by promoting T^0 and T^1 to quantum operators $\hat{T}^0$ and $\hat{T}^1$. To carry out step 3, one uses for V the space of functions $\mathcal{A}(\alpha)$ on the loop space of Σ satisfying certain properties. $\hat{T}^0$ and $\hat{T}^1$ are then represented as linear operators on V. Although their expressions are rather simple, I would have to introduce new notation to write them out explicitly. Therefore, I will just explain their general structure. The action of $\hat{T}^0[\gamma]$ on a state $\mathcal{A}(\alpha)$ gives the state $\mathcal{A}(\gamma \cup \alpha)$ where the loop $\gamma \cup \alpha$ is obtained from the individual loops α and γ by a specific prescription. $\hat{T}^1[\gamma, x]$ operating on $\mathcal{A}(\alpha)$ is zero unless α intersects γ at the point x. If it does, then the result is a linear combination of the value of $\mathcal{A}$ on the loops obtained by composing (at x) the two loops γ and α in two different ways. Using their definitions, one can explicitly check that the commutator of the two operators mirrors the Poisson algebra of their classical counterparts, T^0 and T^1.

The next task is to find regularized operators corresponding to the constraints. Since the entire T-algebra is gauge-invariant, the Gauss constraint has been taken care of already. The vector constraint now requires that the states $\mathcal{A}(\alpha)$ should depend only on the knot-class of α. Thus, in the loop representation, we know the *general* solution to the vector constraint. In the traditional metric representation, by contrast, while we know that the imposition of the quantum vector constraint is equivalent to demanding diffeomorphism invariance of the wave functional, the problem of explicitly finding the *general* solution is difficult and unresolved. The fact that one is led to represent quantum states by functions of a *discrete* space –the space of knots– may cause some unease since in field theories states normally arise as functionals of fields. After all, in some sense, quantum states should carry the information that the gravitational field has two degrees of freedom per space point. Is that not intrinsically impossible if states are functions on the discrete set of knots? I believe that this unease is not warranted. Consider the quantum theory of an hydrogen atom. In the position or momentum representation, states are functions of a continuous variable, $\vec{x}$ or $\vec{k}$, reflecting the fact that the classical system has 3 degrees of freedom. The fact that these variables take values in a continuum is comforting because the classical configuration space is continuous. However, in quantum theory, we can also go to the basis in which the Hamiltonian, the total angular momentum and the z-component of the angular momentum are all diagonal. Then, a generic state is given by a complex-valued function $\psi(n,l,m)$ of three integers! A similar representation dependence arises also in the quantum theory of a free field confined to a box. Therefore, in quantum gravity, the discreteness of the space on which physical states are defined may just be a reflection of the choice of basis that happens to be well-adapted to the diffeomorphism constraint.

It remains to impose the scalar constraint. In the loop representation, one can first construct, at the classical level, a function $T^{(2)}(\gamma; x_1, x_2)$ whose appropriately taken limit, as the loop γ shrinks to a point x, gives us the scalar constraint evaluated at x. One can promote $T^{(2)}$ to a quantum operator $\hat{T}^{(2)}$ on V in a precise way and require that the physical quantum states be those elements $\mathcal{A}(\alpha)$ of V which satisfy : $Lim(\hat{T}^{(2)} \cdot \mathcal{A}(\alpha)) = 0$. The availability of $\hat{T}^{(2)}$ provides a way of both regularizing and imposing the scalar constraint on quantum states.[4] It is known that if the state $\mathcal{A}(\alpha)$ has support on only those (differentiable) loops which contain no self-intersections, it is annihilated by the quantum scalar constraint given above. Thus, a function

4 This regularization *is* unconventional in that it does not use an inner-product on the space V on which $\hat{T}^{(2)}$ operates. As a result, there is some controversy as to whether this procedure is fully satisfactory. However, in every instance where this issue has come up so far, a background geometry is used in an essential way to construct the inner-product on the space of unconstrained states. This would be unacceptable in non-perturbative quantum gravity. Is there a more satisfactory procedure?

$\mathcal{A}(\gamma)$ which has support only on non-intersecting loops γ *and* which depends only on the knot class of these loops is in the physical subspace V_{phy}. While this space is infinite-dimensional, there are several reasons to believe that this is *not* the complete set of physical states. (In fact, there is some question as to whether this sub-space of V_{phy} is physically interesting at all.) In the case when the spatial topology is non-trivial, additional solutions *are* known [36]. However, one would expect entire sectors of new solutions even when the topology is trivial. For example, there must exist loop space analogs of the solutions discovered by Jacobson and Smolin in the connection representation which are parametrized by self-intersecting loops. Are there more general solutions? Recently, a systematic search for these solutions was made using a computer [37] and a number of additional solutions have been found at a formal level, i.e., without facing the issue of regularization.

The loop representation is quite unconventional. In particular, the space of states is not constructed by introducing a polarization on the phase-space; states are functions on the loop-space of the 3-manifold rather than functionals of classical fields. Consequently, interesting quantum operators have to be constructed by indirect means. For example, the quantum scalar constraint was not obtained by simply replacing each classical field in the expression of the scalar constraint by an operator-valued distribution and then regularizing the resulting object. In this sense, the procedure adapted in the loop representation takes the suggestions that Dirac made in his last paper[5] ([38], published posthumously) very seriously: in the passage to quantum theory, one does not copy the classical expressions but allows more general dynamical variables. It is for this reason that the final physical variables –operators on the space of physical quantum states– do not have an obvious physical interpretation in terms that we are familiar with from the classical theory. As advocated by Dirac, the task is to first construct a consistent mathematical framework and then let it guide us to the appropriate interpretation of the dynamical variables.

This is an unconventional viewpoint and one might therefore wonder if such a description is viable at all. Are there some intrinsic difficulties associated with this unfamiliar procedure which are masked just by the technicalities of 3+1-dimensional general relativity? To ensure that there are no inherent problems, the loop representation was constructed in two tractable, complementary cases, each of which

5 "The theory of Heisenberg is more powerful than classical mechanics because its dynamical variables can be of a more general nature. ... work should be concerned with finding the correct Hamiltonian, making use of the vast possibilities of noncommuting quantities which need not be suggested by classical mechanics. That would mean some kind of degrees of freedom occurring in a fundamental way in the equations of quantum theory. The trend followed by most physicists of keeping ideas suggested by classical mechanics and then supplementing them by certain groups is a very restricted one."

captures an aspect of general relativity. We shall conclude this section with a brief discussion of these models.

The first model is general relativity in 2+1-dimensions [39]. In 3 dimensions, the vanishing of the Ricci tensor implies the vanishing of the Riemann tensor, whence every solution to the vacuum equations is flat. Thus, there are no gravitational waves or "gravitons." Yet the theory is not empty: It has a finite number of topological degrees of freedom. Furthermore, structurally it is identical to the 3+1 theory. It therefore serves as a useful toy model in which one can test various ideas. Since the notion of self duality of the curvature tensor does not go over to 2+1-dimensions, one might wonder if the ideas introduced in section 2 have any place at all in this model. It turns out that a completely analogous Hamiltonian framework can in fact be constructed [32,39,40]. The basic configuration variable is again a connection $A_a{}^i$ on a Cauchy surface (which, however, is real and takes values in the $SO(2,1)$ Lie-algebra). It is therefore possible to introduce both the connection and the loop representations explicitly and exhibit the Rovelli-Smolin transform relating them. (In fact, in asymptotically flat space-times, the 2+1-analogs of T^0 and T^1 have a direct physical interpretation: they are related, in a simple way, to the total mass and angular momentum of the space-time.) In the connection representation, physical states turn out to be functions on the (finite-dimensional) moduli-space of flat connections. In the loop representation, they are suitable functions of *homotopy classes* of closed loops on the spatial 2-manifold. Since the homotopy classes in 2-dimensions are the analogs of knot classes in 3-dimension, this picture is perhaps not too surprising. In both representations, the reality conditions suffice to pick out the inner-product on physical states. In the loop representation, three distinct, concrete procedures are available to make this selection. One of these, based on the Gelfand-Naimark-Segal construction [41], appears to be well suited for extension to the 3+1-theory. Work is now in progress to see if this is indeed the case.

While the 2+1 theory captures many features of 3+1 general relativity, it does have only a finite number of degrees of freedom. To gain insights into the infinite dimensional problems, the loop representation was constructed also for source-free Maxwell theory in 3+1-dimensional Minkowski space [42] and for the linearized gravitational field [43]. Again, it was possible to show rigorously that the Rovelli-Smolin transform exists and that the reality conditions suffice to select the inner-product. Results on linearized gravity have also opened up the avenue of developing approximation methods. Certain physical states in the full theory can be re-interpreted as the physical states of the linearized theory. This interpretation in turn suggests relations between the operators in the linear theory and those in the full theory. Using these relations, one can introduce, e.g., the notion of gravitons in the full theory. The hope is to use such notions to probe in detail the domain of validity and the limitations of perturbation theory.

4 Outlook

In this section, I would like to return to the key feature of the present canonical framework: the shift of emphasis from geometrodynamics to connection dynamics.

We saw in section 2 that the scalar constraint $\mathcal{S}$ is of the form

$$\mathcal{S} = G_{ab}{}^{ij}\widetilde{E}^a{}_i\widetilde{E}^b{}_j = 0, \tag{9}$$

where the "connection supermetric", $G_{ab}{}^{ij}$, is given by $\epsilon^{ijk}F_{abk}$. Since $\mathcal{S}$ is purely quadratic in momenta, it follows that the Hamiltonian flow it generates on the phase-space corresponds precisely to null geodesics of $G_{ab}{}^{ij}$ on the space of connections. By contrast, in terms of geometrodynamical variables, the 3-metric q_{ab} and its conjugate momentum $\widetilde{p}^{ab}$, the scalar constraint has the form:

$$\mathcal{S}' := G'_{abcd}\widetilde{p}^{ab}\widetilde{p}^{cd} + \frac{\sqrt{q}}{G}\ {}^3R = 0, \tag{9'}$$

where $G'_{abcd}(q) = \frac{G}{2\sqrt{q}}(q_{ac}q_{bd} + q_{bc}q_{ad} - q_{ab}q_{cd})$ is the Wheeler-DeWitt supermetric and G, Newton's constant. Not only is the functional dependence of $\mathcal{S}'$ on q_{ab} and $\widetilde{p}^{ab}$ more complicated than that of $\mathcal{S}$ on $\widetilde{E}^a{}_A{}^B$ and $A_a{}^i$, but $\mathcal{S}'$ also contains a "potential term" ${}^3R$. Consequently, the trajectories in the space of metrics defined by the Einstein evolution do not have a natural geometrical interpretation. (One may divide $\mathcal{S}'$ by ${}^3R$ and regard $({}^3R)^{-1}G'_{abcd}$ as the new supermetric to make $\mathcal{S}'$ resemble $\mathcal{S}$. However, the new supermetric is ill-defined at points where ${}^3R$ vanishes and this leads to problems even in simple models such as Bianchi cosmologies.) In particular, Misner [44] has pointed out that, since dynamical orbits in the space of 3-metrics can start out in, say, a time-like direction of G'_{abcd} and then become space-like in the course of evolution, the supermetric G'_{abcd} appears to play only a secondary role in the geometrodynamic description of the Einstein evolution. In the space of connections, on the other hand, it is the supermetric $G_{ab}{}^{ij}$ that completely determines the evolution. It is therefore possible that many of the interesting properties of space-times are captured in the intrinsic, geometric structure of the connection supermetric.

An example is provided by Bianchi cosmologies. In the type I model, the connection supermetric turns out to be flat except at certain points where it is degenerate. Points at which it vanishes correspond to the flat Kasner solutions. If we excise degenerate points, the remainder of the connection space is isometric to the three dimensional Minkowski space. It is trivial to integrate null geodesics in this portion. These correspond to general Kasner space-times. The end-points of these null geodesics (at which the gauge-fixed connection diverges) correspond to space-time singularities. Thus, singularities in space-time are represented by points at (future) null-infinity of the connection supermetric! The connection supermetric has a rather simple form also in the type IX model [22,23]. In the diagonal gauge, the space of connections is again 3-dimensional and the diagonal connection components, A_I, provide a natural chart. The A_I turn out to be the null coordinates of the supermetric. Consequently, the $A_I = const.$ curves trivially provide a family of null geodesics, i.e., solutions to Einstein's equations. A number of other features could be similarly explored. For example, it is of considerable interest to find out if the type IX supermetric is asymptotically flat at null infinity. If it is, one could analyze the asymptotic behavior of solutions near space-time singularities using local, differential geometric techniques at null infinity of the connection space. In this case, Lyapounov exponents associated with these geodesics will almost certainly go to zero as one approaches null infinity. Therefore, it is likely that the chaotic behavior of the exact type IX solutions –if it exists– will be coded in the lack of asymptotic flatness of the supermetric at null infinity. Thus, the emphasis would be shifted: *One would investigate invariant properties of the connection supermetric rather than the space-time behavior of specific solutions to type IX equations.* Similarly, in the full theory, the asymptotic behavior of the null geodesics of $G_{ab}{}^{ij}$ –i.e. of space-time geometries near singularities– is fully coded in the conformal structure of $G_{ab}{}^{ij}$. Therefore, a detailed analysis of this structure will shed new light on a number of issues in classical relativity.

The use of the space of connections as the arena for dynamics is especially attractive in quantum gravity. Already in classical general relativity, since we do not have access to a background geometry, the issue of the kinematical arena becomes more subtle than it is in other field theories. In other theories, the presence of a fixed, kinematical space-time metric makes it relatively straightforward to introduce notions such as time evolution, causal propagation, and scattering matrices. In general relativity, on the other hand, one can introduce such notions only *after* one has a complete solution to Einstein's equation. The notion of a black-hole, for example, is not well-defined before one has constructed the entire space-time. In quantum gravity, the situation gets even more complicated. To see this, recall first that due to the uncertainty principle, in non-relativistic quantum mechanics particles do not have well-defined trajectories; time-evolution produces only a wave function $\Psi(\vec{x}, t)$ rather than a specific trajectory $\vec{x}(t)$. Similarly, in quantum gravity, even after a complete evolution, one would *not* recover a classical space-time. In absence of a space-time, how would one introduce familiar physical notions such as causality, dynamics, black holes and scattering matrices? Although these notions arise from classical considerations, presumably they do have *some* counterparts in the quantum theory as well. I believe that the space of connections would be the natural arena to formulate and analyze these counterparts. Is there any reason to prefer this space over the space of 3-metrics of geometrodynamics? After all, inspite of the advantages of the scalar constraint of Eq.9 over that of Eq.9′, one should note that Eq.9 has its own complications: While the Wheeler-DeWitt supermetric G'_{abcd} is an ultra-local metric on a real configuration space, $G_{ab}{}^{ij}$ is defined on the space of complex connections and, due to the presence of derivatives of the connection in its expression, fails to be ultra-local. These differences *are* important and consequently a number of issues in classical relativity are easier to analyze in the geometrodynamical framework. However, because of the simple, geometric picture of Einstein evolution that connection dynamics yields, the connection superspace appears to be a better candidate for the new arena for physics. Indeed, we saw in section 3 that one can single out time and re-interpret the quantum scalar constraint as the Schrödinger equation in an approximate way on the space of connections. In geometrodynamics, on the other hand, where the wave functions are functions of 3-metrics, it has not been possible to carry out this procedure. In fact, to extract time in the traditional variables in the weak field limit, one is forced to step outside the space of 3-metrics and construct a "mixed" representation in which states are functionals of conformal metrics and traces of the extrinsic curvatures [45]. Furthermore, it has not been possible to extend this mixed representation to the full theory since it is cumbersome to express the full scalar constraint in terms of these variables. The connection representation, on the other hand, was first introduced for the full theory and the truncation was carried out subsequently. This difference arises precisely because the connection contains information about both the 3-metric and the extrinsic curvature, in just the right combination. The recent work of Capovilla, Dell and Jacobson [30] on "general relativity without metric" lends additional support to the idea of using the self dual connection as the basic variable and the metric as a secondary quantity.

Now, it may well be that the construction outlined in section 3 does not extend to the full theory. If that should happen, one would simply say that in a generic situation one has to work entirely in the space of connections. Presumably, the solutions to constraints will exist. These will be the physical states. Physical observables will act on this space. Their algebraic properties, eigenvalues and eigenvectors would suggest their interpretation. In general, it may not be possible to speak of time and evolution at least in terms that we are familiar with. It may be that only for certain solutions of constraints can we introduce the notion of time following, e.g., the ideas of section 3. In these cases, it would be superfluous to work in the infinite dimensional space of connections; it would be possible to construct a four dimensional space-time and, to a good approximation, describe everything that "really" takes place in the infinite dimensional space of connections in terms of structures and constructs in four dimensions. In more general situations, the four dimensional approximate structure would be unavailable and we would be forced to deal with the infinite dimensional space directly. The challenge, then, is to develop intuition for structures in this space, learn to pose physically interesting questions in terms of these structures and answer them.

References

[1] E. T. Newman, In *Proceedings of the 10th International Conference on General Relativity and Gravitation*, (ed. B. Bertotti, F. de Felice and A. Pascolini), D. Reidel, Amsterdam (1984).

[2] A. Sen, J. Math. Phys. **22**, 1718 (1981).

[3] E. Witten, Comm. Math. Phys. **80**, 381 (1981).

[4] A. Ashtekar, in *Quantum Concepts in Space and Time*, (ed. C. J. Isham and R. Penrose), Oxford University Press (1986).

[5] A. Ashtekar, Phys. Rev. Lett. **57**, 2244 (1986).

[6] A. Ashtekar, Phys. Rev. **D36**, 1587 (1987).

[7] T. Jacobson, and L. Smolin, Nucl. Phys. **B299**, 295 (1988).

[8] C. Rovelli, and L. Smolin, Phys. Rev. Lett. **61**, 155 (1989).

[9] C. Rovelli, and L. Smolin, Nucl. Phys. **B331**, 80 (1990).

[10] R. Penrose, Gen. Rel. & Grav. **7**, 31 (1976).

[11] S. Weinberg, *Gravitation and Cosmology*, John Wiley, New York (1972).

[12] A. Ashtekar, T. Jacobson and L.Smolin, Comm. Math. Phys. **115**, 631 (1988).

[13] L. Mason, and E. T. Newman, Comm. Math. Phys. **121**, 659 (1989).

[14] J. Samuel, Class. & Quant. Grav. **5**, L123 (1988).

[15] A. Ashtekar, In *Mathematics and General relativity* (ed. J.Isenberg), American Mathematical Society Providence, RI (1988).

[16] S. Koshti, and N. Dadhich, Class. & Quant. Grav. **7**, L223 (1990).

[17] C. G. Torre, Phys. Rev. **D41**, 3620 (1990).

[18] A. Ashtekar, P. Mazur and C. G. Torre, Phys. Rev. **D36**, 2955 (1987).

[19] V. Husain and L. Smolin, Nucl. Phys. **B327**, 205 (1989).

[20] P. Renteln and L. Smolin, Class. & Quant. Grav. **6**, 275 (1989).

[21] P. Renteln, Ph. D. Thesis, Harvard University (1989).

[22] H. Kodama, Prog. Theo. Phys. **80**, 1024 (1988); H. Kodama, Phys. Rev. **D42**, 2548 (1990).

[23] A. Ashtekar and J. Pullin, Ann. Israel Phy. Soc. **9**, 66 (1990).

[24] T. Jacobson, Class. & Quant. Grav. **5**, L143 (1988).

[25] A. Ashtekar, J. D. Romano and R. S. Tate, Phys. Rev. **D40**, 2572 (1989).

[26] J. Samuel, Pramana J. Phys. **28**, L429 (1987).

[27] T. Jacobson, and L. Smolin, Phys. Lett. **196B**, 39 (1987).

[28] T. Jacobson, and L. Smolin, Class. & Quant. Grav. **5**, 583 (1988).

[29] A. Ashtekar, and J. D. Romano, Syracuse University Report (1989).

[30] R. Capovilla, T. Jacobson, and J. Dell, Phys. Rev. Lett. **63**, 2325 (1989).

[31] R. Capovilla, T. Jacobson, and J. Dell, Class. & Quant. Grav. **7**, L1 (1990).

[32] E. Witten, Nucl. Phys. **B311**, 46 (1988).

[33] N. J. M. Woodhouse, Geometric Quantization, Oxford University Press (1981).

[34] C. J. Isham, In *Relativity Groups and Topology II*, (ed. B. S. De Witt and R. Stora), North-Holland, Amsterdam (1984).

[35] A. Ashtekar, In *Conceptual Problems of Quantum Gravity* (ed. A. Ashtekar and J. Stachel), Birkhäuser, Boston, MA (1989).

[36] M. P. Blencowe, Nucl. Phys. **B 341**, 213 (1990).

[37] B. Brügmann and J. Pullin, Syracuse University pre-print (1990).

[38] P. A. M. Dirac, In: *Paul Adrien Maurice Dirac*, (ed. B. N. Kursunoglu and E. P. Wigner), Cambridge University Press, Cambridge (1987).

[39] A. Ashtekar, V. Husain, C. Rovelli, J. Samuel and L. Smolin, Class. & Quant. Grav. **6**, L185 (1989); L. Smolin, In *Proceedings of the '89 Johns Hopkins Workshop* (World Scientific, Singapore, 1990); See also chapter 17 of these notes.

[40] I. Bengtsson, Phys. Lett. **220B**, 51 (1989).

[41] I. M. Gelfand and M. A. Naimark, Mat. Sobernik **12**, 197 (1943); I. E. Segal, Bull. Am. Math. Soc. **53**, 73 (1947).

[42] A. Ashtekar and C. Rovelli, Syracuse University pre-print (1990).

[43] A. Ashtekar, C. Rovelli, L. Smolin (in preparation).

[44] C. W. Misner, private communication (1989).

[45] K. Kuchař, J. Math. Phys. **11**, 3322 (1970).

II CLASSICAL THEORY

A point of view for studying $\mathcal{H}$*-spaces* [i.e. self dual solutions of Einstein's equation] *could thus be – push this extrapolation to its extreme, find what structures arise naturally and then study their relationships. ... It would be a cruel and unaesthetic God who would lay such a scheme before us and not have it mean something.*

- E.T. Newman *et al*, *The theory of* $\mathcal{H}$*-space.*

In this Part, we shall present in detail the introduction of the new canonical variables for classical general relativity, possibly coupled with matter. The new configuration variable is a complex-valued $SO(3)$ connection A_a^i. In any solution to the field equations, A_a^i is a potential for the self dual part of the Weyl curvature. The canonically conjugate momentum, $\widetilde{E}_i^a$, which would be the electric field in Yang-Mills theory, now has the interpretation of a (density weighted) triad. The advantages of using these variables over the more traditional, geometrodynamical ones are three-fold:[1]

i) All field equations are simple polynomial in the new variables. As we noted in chapter 2, this simplicity has been exploited to obtain a generic solution to the vector and scalar constraints of the classical theory and a large class of solutions to all constraints in the quantum theory. Also, since none of the equations requires us to raise the vector index in A_a^i or to lower the one in $\widetilde{E}_i^a$, the entire framework continues to be meaningful even when the triad -and hence the 3-metric- becomes degenerate. One thus obtains a slight generalization of the traditional ADM framework which may be useful both in the analysis of certain space-time singularities in the classical theory and in the issue of topology change in quantum theory.

ii) The use of new variables brings out an unexpected relation between general relativity and Yang-Mills theories. This relation has turned out to be extremely useful. First, there are a number of mathematical applications. The relation between the two theories has led to the discovery of relations between their instanton solutions. For example, the one-instanton solution of the $SU(2)$ Yang-Mills theory gives, just by a re-interpretation of the mathematical variables, a gravitational instanton with a cosmological constant [1]. Furthermore, the powerful mathematical machinery available in the Yang-Mills theory to analyze the structure of the moduli space of instantons can now be taken over to the analysis of the half-flat as well as the self-dual Einstein metrics [2]. In quantum theory, the relation between the two theories

1 In the following discussion, I will provide explicit references only to those papers whose results are not discussed in detail in these notes.

has had a bigger impact. A number of techniques from QCD could be taken over to quantum gravity and these have deeply influenced the approach to both the mathematical *and* the conceptual problems. The most outstanding example of this influence is the use of the Wilson loops in quantum gravity which has led to the introduction of the loop representation. Recently, it was shown that the loop representation is a useful tool also in the Yang-Mills theory [3,4]. Thus, an exciting cross-fertilization of ideas between QCD and quantum gravity has resulted from the use of the new variables.

iii) All constraints now have a simple geometric interpretation. This is a significant point because the polynomial character of constraints by itself is not unique to the new variables (at least in the source-free theory).[2] For example, in the geometrodynamical variables, if one multiples the scalar constraint by the square of the determinant of the 3-metric, the resulting equations *are* polynomial in the canonical variables. However, the polynomial involved is of a high order: it is a density of weight six which is quadratic in $\tilde{P}^{ab}$ and of order eight in q_{ab}. Consequently, the equations are still difficult to solve both classically and quantum mechanically. Perhaps a more significant point is that, irrespective of such strategies, due to the presence of the "potential term" the scalar constraint fails to have a simple geometric interpretation. In terms of the new variables, $(A_a^i, \tilde{E}_i^a)$, on the other hand, as was pointed out in chapter 2, *all* constraints have a simple geometrical meaning. The Gauss and the vector constraints generate the kinematical symmetry group, the semi-direct product of the group of local triad rotations with the diffeomorphism group of the 3-space, on the entire phase space. Finally, since the scalar constraint is now free of the "potential" term, it generates null geodesics (of the connection supermetric) in the connection superspace, thereby providing an attractive geometrical interpretation to the Einstein dynamics.

Most of the discussion contained in this Part has appeared in the literature before. However, there has been much confusion in various papers about the reality conditions. Technical confusion arose because the earlier papers incorrectly reported that the reality conditions are non-polynomial in the basic variables. (There was no computational error. Rather, these papers failed to realize that the apparent non-polynomial form resulted because the reality conditions were expressed in a form that is tailored to the geometrodynamical variables.) Conceptual confusion arose because the idea of separating these conditions from the constraints is unconventional from the standpoint of the Dirac theory of constrained systems. I have therefore tried to clarify these issues in detail at several points in this Part.

---

2 I thank Steve Martin for correspondence on this point.

Chapters 3 and 4 introduce the new canonical framework starting from a Lagrangian. (The second of these chapters is somewhat technical and can be skipped without loss of continuity.) The emphasis is on explaining the difference between the well known Palatini framework and the new framework, based on self dual connections. The new framework lies somewhat outside Dirac's theory basically because the self dual action is complex (even for real general relativity). To recover the Hamiltonian framework from the action, for example, one begins with *complex* general relativity for which the Legendre transform is quite simple to perform, obtains the Hamiltonian framework and passes to the real theory by simply restricting oneself to the appropriately defined real section of the complex phase space. This procedure is quite unusual. However, as shown later in chapters 8 and 10, it does serve as an adequate basis for canonical quantization of the theory.

Chapter 5 provides a quick transition from triads to spinors. Here, the emphasis is on translating calculations from one framework to another; a more complete introduction to spinors can be found in appendix A. Chapters 6 and 7 discuss in detail the Hamiltonian framework. Here we arrive at the new variables by performing a canonical transformation on the geometrodynamical phase space and discuss in detail the structure of the constraint algebra, Hamiltonians and dynamics. Throughout this procedure, one remains in the confines of real general relativity; the self dual connection arises simply as a complex variable on the real phase space of the real theory. This final picture is the same as that obtained from the self dual action; the two procedures are equivalent. However, the procedure adopted in chapters 6 and 7 is simpler to use for quantization. Therefore, I have made a special attempt to make these chapters self-contained. ***Readers can skip chapters 3 and 4 entirely if they so desire.*** Chapter 8 discusses the role of reality conditions both in the classical and the quantum theory. I have devoted a separate chapter to this discussion because of the confusion I mentioned earlier. Chapter 9 extends the entire framework to allow for a cosmological constant, and for scalar, spin-$\frac{1}{2}$ and Yang-Mills sources. While the status of this issue is quite satisfactory in the classical theory, almost no work has been done along these lines in the quantum domain. Indeed, One of the outstanding open problems in the program is to use the results of this chapter to extend the connection and the loop representations to allow for matter couplings in quantum theory as well.

References

[1] J. Samuel, Class. & Quant. Grav. **5**, L123 (1988), A. Ashtekar, T. Jacobson and L. Smolin, Comm. Math. Phys. **115**, 631 (1989); A. Ashtekar, In:

Mathematics and general relativity (ed. J. Isenberg) (American Mathematical Society, Providence, 1988); L. Mason and E.T. Newman, Comm. Math. Phys. **121**, 659 (1989); R. Capovilla, T. Jacobson and J. Dell, Class. & Quant. Grav. **7**, L1 (1990); S. Koshti and N. Dadhich, Class. & Quant. Grav. **7** L5 (1990).

[2] C.G. Torre, Phys. Rev. **D41**, 3620 (1990).

[3] C. Rovelli and L. Smolin, 'Loop representation for lattice gauge theory', Syracuse University pre-print (1990).

[4] B. Brügmann, 'The method of loops applied to lattice gauge theory', Phys. Rev. **D**, in press (1990).

3 LAGRANGIAN FRAMEWORK

1 Introduction

The purpose of this chapter is to introduce the reader to the new canonical framework starting from a first order action. This framework was first obtained by performing a canonical transformation directly on the phase space of (real) general relativity [1]. Soon afterwards, Samuel [2] and Jacobson and Smolin [3] provided a manifestly covariant basis for it by first introducing a new action principle and then deriving the key equations of [1] via a Legendre transform. Since then, over half a dozen papers have appeared which give different versions of this action and perform the Legendre transform in different ways (see, e.g., [4]).[1] Some of these authors suggest that their version is better because it clarifies or simplifies some aspect of the Legendre transform. My own view is that the specific choice depends primarily on one's taste and on what one regards as being "simple". All these formulations have one feature in common: strictly speaking, the Legendre transform of the new action leads to the canonical formulation of *complex* general relativity. To recover the real theory, one must impose certain reality conditions by hand.

In this chapter, I will present yet another procedure which to my knowledge has not appeared in the literature, but which I found, from a pedagogical viewpoint, to be best suited to the audience I was addressing. The procedure is based on tetrads, rather than on spinors, which were used in the early constructions [2,3].[2] There

1 Remarkably, one of the most elegant of these Lagrangian formulations, due to Plebanski [5], has been available in the literature since 1977! However, no one seems to have bothered to perform the Legendre transform on this action at the time. The reason, I believe is that the Lagrangian appeared in one of Plebanski's papers on self dual solutions and up until the mid-eighties, the sets of researchers in canonical gravity and of experts in self dual solutions had almost no overlap. The relation between Plebanski's framework and the new Hamiltonian formulation of [1] was discovered only this year [6].

2 The relation between the two is discussed briefly in chapter 5 and in detail in appendix A. The tetrad framework has also been worked out by a number of other authors although there are minor differences in their treatments; see, e.g. [4]. The Lagrangian formulation in terms of spinors is discussed in chapter 9.

are two reasons for choosing to work with tetrads. First, this will emphasize the fact that the "internal gauge group" in the resulting phase-space description of the source-free theory is really $SO(3)$ rather than $SU(2)$. Indeed, even when spinors are used, the effective gauge group is $SO(3)$ because any $SU(2)$ transformation and its negative have the same effect on the canonical variables of the source-free theory. (It is only in the presence of fields with half-integer spin that the "internal gauge group" gets enlarged to $SU(2)$.) This fact is somewhat obscured in the spinorial treatments of pure gravity. The second reason is that the use of tetrads brings out explicitly the fact that the framework exploits, in a crucial way, the isomorphism between the *self dual* sub-Lie algebra of $SL(2,\mathbb{C})$ and the complexification of the Lie algebra of $SO(3)$. 2-component spinors are so well-tailored to the existence of this isomorphism that it is absorbed in the notation itself; the isomorphism is never displayed. As a consequence, one can easily miss the point that the availability of this isomorphism lies at the heart of the construction.

The action we will use is closely related to the one given by Palatini [7]. Since most of the literature on general relativity uses the Einstein-Hilbert action, we begin in section 2, by recalling the Palatini framework for real general relativity. When one makes the Legendre transform, however, the Palatini action leads to the *same* phase-space framework as the Einstein-Hilbert, in which constraints have a rather complicated dependence on the basic canonical variables. The situation is the same for complex general relativity. If, on the other hand, we use the Samuel-Jacobson-Smolin type of action based on connections which are *self dual* in internal indices, the situation improves significantly. In section 3 we introduce this action and discuss the Legendre transform. The result is a Hamiltonian formulation for complex general relativity in which all equations are low order polynomials in the basic canonical variables. Finally, we specify the reality conditions which must be imposed by hand to recover real general relativity. A number of subtleties arise in this way of constructing the Hamiltonian formulation of the real theory which were overlooked in the early work on the subject.

The aim of this chapter is to present only the main ideas of the Lagrangian formulation. A detailed discussion of the difference between the Legendre transforms of the Palatini and the self dual actions will be given in chapter 4. Chapters 6-8 will present an alternate derivation of the resulting Hamiltonian framework in which one works entirely in the confines of the real theory following the strategy adopted in [1].

2 Palatini action

The Einstein-Hilbert action that one normally uses for general relativity is second order; the basic dynamical variable is assumed to be just the 4-metric and the Lagrangian density contains its *second* derivatives. Palatini [7] provided a new action in which the basic dynamical variables are a 4-metric and a 4-connection. While this action appears to depend on too many dynamical variables at first, the field equations are such that the connection is completely determined by the metric and the metric in turn satisfies just the field equations which emerge from the variation of the Einstein-Hilbert action. Thus, at least as far as the classical regime is concerned, the two actions lead to the same theories. In this section, we shall derive the field equations starting from the Palatini action. This discussion will serve as an introduction to our use of the self dual action in the next section.

Fix a 4-manifold M, topologically $\Sigma \times \mathbb{R}$ for some 3-manifold Σ (which is either compact or has the property that the complement of a compact set in it is diffeomorphic to the complement of a closed ball in $\mathbb{R}^3$). The basic dynamical variables in the Palatini framework are tetrads e^a_I and Lorentz connections ${}^4\omega_a^{IJ}$ on M.[3] Let me make a small detour to recall, from chapter 2, the notation we will use. The lower case latin letters $a, b, ...$ refer to the tangent space of M while the upper case latin letters $I, J, ...$ denote the internal Lorentz group. The internal space is equipped with a Minkowskian metric η_{IJ} (of signature $- + + +$) which is fixed once and for all. Consequently, one can freely raise and lower the internal indices; their position does *not* depend on the choice of dynamical variables. To raise or lower the "world indices" $a, b, ...$, on the other hand, one needs a space-time metric g_{ab} which *is* a dynamical variable, constructed from the cotetrads e_a^I via: $g_{ab} = \eta_{IJ}\, e_a^I e_b^J$. Since at any point p of M the cotetrad e_a^I provides a (vector space) isomorphism between the tangent space at p and the fixed internal space, and since the internal, kinematic metric η_{IJ} has signature $- + + +$, so does the metric g_{ab}. The connection ${}^4\omega_a^{IJ}$ acts only on internal indices; it defines a derivative operator ${}^4D$ via: ${}^4D_a\, K_I := \partial_a\, K_I + {}^4\omega_{aI}{}^J K_J$. Since ${}^4D$ is a Lorentz connection, i.e. since it annihilates the fiducial Minkowskian metric η_{IJ} on the internal space, the connection 1-forms ${}^4\omega_a^{IJ}$ are antisymmetric in I and J; they take values in the Lorentz Lie algebra. Note that, *a priori*, ${}^4D$ does not know how to act on fields with

3 Generally, a stem letter will indicate if we are considering a 4-dimensional, space-time field or a 3-dimensional field defined just on the spatial slice. However, when ambiguity is likely to arise in future calculations, we will use a prefix 4 to denote the 4-dimensional fields. Thus ${}^4\omega_a^{IJ}$ is a space-time connection. Its pull-back to a spatial slice will be denoted simply by ω_a^{IJ}.
Since a number of new ideas are being introduced in this chapter, I will ignore the issue of boundary conditions and omit surface integrals here. These will be discussed in detail in chapter 6.

space-time indices $a, b, ...$; strictly speaking, objects like ${}^4D_a\, e^I_b$ are not defined and will not be needed.[4] Finally, the curvature ${}^4\Omega_{ab}{}^{IJ}$ of the connection ${}^4\omega_a^{IJ}$ is given by: ${}^4\Omega_{ab}{}^{IJ} = 2\partial_{[a}{}^4\omega_{b]}^{IJ} + [{}^4\omega_a, {}^4\omega_b]^{IJ}$, where [,] stands for the commutator in the Lorentz Lie algebra.

The Palatini action is given by:

$$S_P(e, {}^4\omega) := \frac{1}{2}\int d^4x(e)\, e^a_I\, e^b_J\, {}^4\Omega_{ab}{}^{IJ}\,, \tag{1}$$

where e is the square root of the determinant of the 4-metric g_{ab}.[5] The field equations are obtained by varying this action with respect to e^a_I and ${}^4\omega_a^{IJ}$. To carry out the variation with respect to the connection, it is convenient to introduce the unique (torsion-free) connection ∇ on both space-time and internal indices determined by the the tetrad e^a_I via $\nabla_a\, e^b_I = 0$. The difference between the actions of ∇ and ${}^4D$ on internal indices is characterized by a field $C_{aI}{}^J$: $({}^4D_a - \nabla_a)V_I = C_{aI}{}^J\, V_J$. The difference between their curvatures is given by:

$${}^4\Omega_{ab}{}^{IJ} - R_{ab}{}^{IJ} = 2\nabla_{[a}C_{b]}^{IJ} + 2C_{[a}^{IM}C_{b]M}{}^J\,, \tag{2}$$

where $R_{ab}{}^{IJ}$ is the internal curvature of ∇. The variation of the action S_P with respect to ${}^4\omega_{aI}{}^J$ is simplest to compute if one first re-expresses S_P in terms of ∇ and $C_{aI}{}^J$ and then notices that the variation with respect to ${}^4\omega_{aI}{}^J$ (keeping the tetrad fixed) is the same as the variation of the resulting action with respect to $C_{aI}{}^J$. The first step yields:

$$S_P(e, {}^4\omega) = \frac{1}{2}\int d^4x\, (e)e^a_Ie^b_J\, (R_{ab}{}^{IJ} + 2\nabla_{[a}C_{b]}{}^{IJ} + 2C_{[a}{}^{IM}C_{b]M}{}^J)\,. \tag{3}$$

Let us now extremize this action with respect to variations in $C_{aI}{}^J$. The first term involves only $R_{ab}{}^{IJ}$ and is independent of $C_{aI}{}^J$. Since ∇ annihilates the tetrad, the second term is a pure divergence and therefore does not contribute to the variation.

4 The action of ${}^4D$ can, however, be extended to the world indices by choosing a space-time connection, $\Gamma_{ab}{}^c$. One can then set ${}^4D_ae^b_I := \partial_ae^b_I + {}^4\omega_{aI}{}^Je^b_J - \Gamma_{ac}{}^be^c_I$. One may occasionally need, in the intermediate stages of calculations, the action of ${}^4D$ on world indices. Unless specified otherwise, the results will, however, be *independent* of the choice of the (torsion-free) extension. (Recall that $\Gamma_{ac}{}^b$ is torsion-free iff $\Gamma_{[ac]}{}^b = 0$.)

5 It would be more elegant to regard the covariant triad e^I_a as the basic variable and write the integrand as a 4-form, $3!\epsilon_{IJKL}e^I_{[a}e^J_b\, {}^4\Omega_{cd]}{}^{KL}$, which is polynomial in both the cotetrad and the connection. (See chapter III.1 in [8].) However, I have not followed that route here since the form (Eq. 1) for the action is the one that features more directly in the Legendre transform.

The last term yields: $(e^{[a}_M e^{b]}_N \delta^M_{[I} \delta^K_{J]}) C_{bK}{}^N = 0$. It is easy to check that $(e^{[a}_M e^{b]}_N \delta^M_{[I} \delta^K_{J]})$ is non-degenerate, whence the only solution to the above equation is:

$$C_{aI}{}^J = 0. \tag{4}$$

Thus, the equation of motion for the derivative operator ${}^4D$ is simply that it equals ∇ while acting on objects with only internal indices. Thus the connection ${}^4D$ is completely determined by the tetrad. The second equation of motion is straightforward to obtain. By carrying out the variation of S_P with respect to the tetrad, one obtains:

$$e^c_I \, {}^4\Omega_{cb}{}^{IJ} - \tfrac{1}{2} \, {}^4\Omega_{cd}{}^{MN} e^c_M e^d_N \, e^J_b = 0\,. \tag{5}$$

Substitution of the first equation of motion Eq.4 in Eq.2 implies that ${}^4\Omega_{ab}{}^{IJ} = {}^4R_{ab}{}^{IJ}$. Using the fact that the internal curvature of ∇ is related to its space-time curvature by ${}^4R_{abI}{}^J = {}^4R_{abc}{}^d e^c_I e^J_d$ and multiplying Eq.5 by e_{aJ} tells us that the Einstein tensor $G_{ab} := R_{ab} - \frac{1}{2} R g_{ab}$ of the metric $g^{ab} := e^a_I e^b_J$, vanishes.

Thus, we have recovered the vacuum field equations starting from the first order action. The connection is the restriction to internal indices of the connection compatible with the tetrad and the metric constructed from the tetrad is Ricci-flat.

3 Self dual action

If one carries out the Legendre transform of the Palatini action discussed above, one recovers the triad version of the Hamiltonian description of general relativity first found by Arnowitt, Deser and Misner [9]. (For details, see chapter 4.) Each of the resulting constraints, as well as the evolution equations, have a geometric interpretation in terms of the intrinsic metric and the extrinsic curvature of the 3-manifold. However, their form is rather complicated. As a result they are somewhat inconvenient to use in quantum theory. The situation is exactly the same for real and complex general relativity. In the complex case, however, one can start from a *different* action and obtain an equivalent but different system of equations. As noted in chapter 2, this form of the equations has several attractive features which have turned out to be useful both in the classical and the quantum theory. The real theory can be recovered at the end by restricting oneself to the appropriately defined real section of the resulting complex phase space. In this section, we will see how this procedure can be implemented.

Thus, to begin with, we are interested in *complex* solutions g_{ab} to the Einstein equation on a real 4-manifold $M = \Sigma \times \mathbb{R}$. The idea is to use, in place of the connections ${}^4\omega_a^{IJ}$ of the Palatini theory, (complex) Lorentz connections ${}^4A_a^{IJ}$ which are *self dual* in the internal indices I and J, i.e., which satisfy $\frac{1}{2}\epsilon_{IJ}{}^{MN}{}^4A_a^{IJ} = i{}^4A_a^{MN}$. Thus, while the Palatini connection 1-forms ${}^4\omega_a^{IJ}$ of complex general relativity would take values in the complexified Lorentz Lie algebra, the connection forms ${}^4A_a^{IJ}$ now used will take values in the *self dual sub-algebra* of the complexified Lorentz Lie algebra. Since the metrics of interest are complex, now the tetrads e_I^a will also be complex. However, the internal metric, η_{IJ} will continue to be real; g_{ab} is complex because the tetrads are complex. The action is given by:

$$S(e, {}^4A) := \int d^4x\, (e) e_I^a\, e_J^b\, {}^4F_{ab}{}^{IJ}\ , \tag{6}$$

where ${}^4F_{ab}{}^{IJ}$ is the curvature of ${}^4A_{aI}{}^J$. (To compare the self-dual action with the Palatini, may think of ${}^4F_{ab}{}^{IJ}$ as being the self-dual part of ${}^4\Omega_{ab}^{IJ}$: ${}^4F_{ab}{}^{IJ} = \frac{1}{2}{}^4\Omega_{ab}{}^{IJ} - \frac{i}{4}\epsilon_{KL}{}^{IJ}\ {}^4\Omega_{ab}{}^{KL}$. However, in this section, the self-dual action itself will be fundamental; we will not need to refer to the Palatini framework at all.) Using the fact that the commutator of self dual matrices is self dual it is straightforward to verify that the curvature is also self dual in the internal indices. Note that the action is now complex rather than real. This is quite unusual. Indeed, in Minkowskian field theories, even while dealing with complex fields one normally uses real actions which involve both the fields and their complex conjugates. By contrast, now our action itself is complex valued and the complex conjugates of the connection ${}^4A_{aI}{}^J$ and the tetrads e_I^a will never enter the formalism. These features have given rise to considerable confusion in the literature. I shall therefore try to discuss in detail the points at which this framework differs from the standard treatment of constrained systems *a la* Dirac and point out the pitfalls that one may encounter in this framework.

The variational calculations are quite similar to those carried out in the last section. By setting the variation of the action with respect to ${}^4A_a^{IJ}$ equal to zero, one obtains the result that ${}^4A_a^{IJ}$ is the *self dual part* of the internal, Lorentz connection determined by the tetrad. (That is, if one repeats the calculation of the last section, one now finds that only the self-dual part of $C_a{}^{IJ}$ vanishes.) Consequently, when this equation is satisfied, ${}^4F_{ab}{}^{IJ}$ equals the self dual part of the curvature $R_{ab}{}^{IJ}$ of the connection ∇ compatible with the tetrad. The variation with respect to the tetrad provides the analog of Eq.5 where the curvature ${}^4\Omega_{ab}{}^{IJ}$ is replaced by ${}^4F_{ab}{}^{IJ}$. When we substitute for ${}^4F_{ab}{}^{IJ}$ in terms of $R_{ab}{}^{IJ}$, we obtain: $(R_{ab}{}^{IJ} - \frac{i}{2}\epsilon^{IJ}{}_{MN}R_{ab}{}^{MN})e_I^a = 0$. Note, however, that the second term is identically zero due to the algebraic Bianchi identity. Hence, the second field equation just says that the metric g_{ab} constructed from the triads is Ricci-flat. This completes the discussion of the covariant field equations obtained from the self-dual action. It is quite remarkable that the action of the complex theory can be written in a first order form just using the self dual connection, without any reference at all to the anti-self dual part.

Let us now perform a Legendre transform of this action to pass to a Hamiltonian framework. Introduce a foliation of the space-time (M, g_{ab}) by a family of space-like hypersurfaces labelled by $t = const$. Let t^a be a real vector field whose integral curves intersect each leaf of the foliation precisely once and which is normalized such that $t^a \nabla_a t = 1$. This t^a is the "dynamical vector field"; Lie derivatives along t^a will be identified with "time-derivatives". Let us decompose it into normal and tangential parts with respect to the foliation: $t^a = N\, n^a + N^a$, with $N^a\, n_a = 0$, where n^a is the unit normal to the leaves of the foliation. The function N is called the *lapse* and the vector field N^a is called the *shift*. Since the metrics g_{ab} under consideration are complex, the fields n^a, N and N^a will in general also be complex. The idea now is to decompose the various fields in the expression of the action in a similar way using n^a and the projection operator $q_a^b := \delta_a^b + n_a\, n^b$. Setting $E_I^a := q_b^a\, e_I^b$, we obtain[6]

$$\begin{aligned} S(e, {}^4\!A) &= \int d^4x\, (e)\, (E_I^a E_J^b\, {}^4\!F_{ab}{}^{IJ} - 2(e)\, E_J^b n^a n_I\, {}^4\!F_{ab}{}^{IJ}) \\ &= \int d^4x (\underset{\sim}{N}\, \widetilde{E}_I^a \widetilde{E}_J^b\, {}^4\!F_{ab}{}^{IJ} + i\, N \widetilde{E}_J^b n^a n_I\, \epsilon^{IJ}{}_{MN}\, {}^4\!F_{ab}{}^{MN}) \\ &= \int d^4x\, (\underset{\sim}{N} \widetilde{E}_I^a \widetilde{E}_J^b\, {}^4\!F_{ab}{}^{IJ} - i\, \widetilde{E}_J^b (t^a - N^a)\, \epsilon^J{}_{MN}\, {}^4\!F_{ab}{}^{MN}) \\ &= \int d^4x\, ([-i\, \widetilde{E}_J^b\, \epsilon^J{}_{MN}\,][\mathcal{L}_t {}^4\!A_b{}^{MN} - N^a\, {}^4\!F_{ab}{}^{MN}] \\ &\quad - i({}^4\!A_a{}^{MN}\, t^a) {}^4\!\mathcal{D}_b [\widetilde{E}_J^b\, \epsilon^J{}_{MN}] + \underset{\sim}{N} \widetilde{E}_I^a \widetilde{E}_J^b\, {}^4\!F_{ab}{}^{IJ})\, . \end{aligned} \tag{7}$$

Here, we have set $n_I := e_I^a\, n_a$, $\epsilon^{IJK} := \epsilon^{IJKL} n_L$, $\underset{\sim}{N} = q^{-\frac{1}{2}} N$ and $\widetilde{E}_I^a := q^{\frac{1}{2}} E_I^a$. The calculation has been carried out in the following steps: In the first step, we have decomposed the two tetrad vectors e_I^a and e_J^b in Eq.6 into parts which are tangential and normal to Σ; in the second step we have used the fact that $e = N q^{\frac{1}{2}}$ and that the curvature ${}^4\!F_{ab}{}^{IJ}$ is self dual in the internal indices; in the third step, we have used the expression $t^a = N\, n^a + N^a$ of the dynamical vector field t^a and the definition of

6 Note that we are regarding E_I^a as a 4-dimensional field which happens to be orthogonal to n_a. This is a general strategy: Given a sub-manifold of M, we will regard tensor fields on it as space-time tensor fields whose contractions (on any index) with any of the normals to the sub-manifold vanish. That is, as is customary in the abstract index notation, we will identify tensor fields defined intrinsically on the sub-manifolds with tensor fields in space-time which have certain vanishing components.

ϵ^{IJK}; and, in the last step, the identity $t^a\,({}^4F_{ab}{}^{IJ}) = \mathcal{L}_t({}^4A_b{}^{IJ}) - {}^4\mathcal{D}_b({}^4A_a\,t^a)$, where ${}^4\mathcal{D}_a$ is the derivative operator defined by ${}^4A_{aI}{}^J$.

We now note that in the last step in Eq.7, ${}^4F_{ab}{}^{IJ}$ always appears with its space-time indices projected into the 3-surface Σ; we can replace it everywhere by its pull-back $F_{ab}{}^{IJ}$. Now, the pull-back is in fact the curvature of the pull-back $A_a{}^{IJ}$ of the connection ${}^4A_a{}^{IJ}$ to the 3-surface. Similarly, the derivative operator ${}^4\mathcal{D}$ in the term ${}^4\mathcal{D}_b\,\widetilde{E}^b_J\,\epsilon^J{}_{MN}$ can be replaced by its pull-back, $\mathcal{D}$, defined by $A_a{}^{IJ}$. Finally, using $\mathcal{L}_t q^b_a = 0$, a simple calculation shows that, since the index a of $\mathcal{L}_t{}^4A_a{}^{MN}$ is projected into Σ, it can also be replaced by $\mathcal{L}_tA_a{}^{MN}$. Consequently, the Lagrangian can now be written as:

$$L(\widetilde{E},\underset{\sim}{N},N^a,n^A;A,{}^4A.t) = \int d^3x\,\Big([-i\,\widetilde{E}^a_I\,\epsilon^I{}_{MN}\,][\mathcal{L}_tA_a{}^{MN}] + i\,[\widetilde{E}^b_J\,\epsilon^J{}_{MN}]N^aF_{ab}{}^{MN} - i({}^4A_a{}^{MN}\,t^a)\mathcal{D}_b[\widetilde{E}^b_J\,\epsilon^J{}_{MN}] + \underset{\sim}{N}\widetilde{E}^a_I\widetilde{E}^b_J\,{}^4F_{ab}{}^{IJ}\Big). \tag{8a}$$

This is the form we were seeking. The Lagrangian is now expressed in the form $p\dot{q} - H$ and we can read-off the configuration and momentum variables and the Hamiltonian. The configuration variable is just a connection 3-form $A_a{}^{MN}$ on the 3-manifold Σ which takes values in the (3-complex dimensional) *self dual* Lorentz Lie algebra. The conjugate momentum is given by $\widetilde{\Pi}^a_{MN}$:= *self dual part of* $-i\widetilde{E}^a_I\,\epsilon^I{}_{MN} \equiv \widetilde{E}^a_{[M}n_{N]} - \frac{i}{2}\widetilde{E}^a_I\,\epsilon^I{}_{MN}$. Thus the basic non-vanishing Poisson bracket is simply:

$$\{A_b{}^{IJ}(y)\,,\widetilde{\Pi}^a_{MN}(x)\} = \frac{1}{2}\delta^b_a\,\delta^3(x,y)[\delta^I_{[M}\delta^J_{N]} - \tfrac{i}{2}\epsilon_{MN}{}^{IJ}]. \tag{9}$$

Let us now express the Lagrangian in terms of these "canonical variables". We obtain:

$$\begin{aligned} L(\widetilde{\Pi},\underset{\sim}{N},N^a,A,\,{}^4A.t) &= \int d^3x\,\Big(\widetilde{\Pi}^a_{MN}\mathcal{L}_tA_a{}^{MN} - [\widetilde{\Pi}^b_{MN}]N^a\,F_{ab}{}^{MN} \\ &\qquad + ({}^4A_b{}^{MN}\,t^b)\mathcal{D}_a[\widetilde{\Pi}^a_{MN}] - \underset{\sim}{N}[\widetilde{\Pi}^a_M{}^N][\widetilde{\Pi}^b_N{}^P]\,F_{abP}{}^M\Big) \\ &= \int d^3x\ \mathrm{tr}\,\Big(-\widetilde{\Pi}^a\,\mathcal{L}_tA_a + N^a\widetilde{\Pi}^b\,F_{ab} - ({}^4A\cdot t)\mathcal{D}_a\,\widetilde{\Pi}^a - \underset{\sim}{N}\widetilde{\Pi}^a\widetilde{\Pi}^b\,F_{ab}\Big), \end{aligned} \tag{8b}$$

where, we have replaced ${}^4F_{ab}{}^{IJ}$ and ${}^4A^{IJ}_a$ by their pull-backs $F_{ab}{}^{IJ}$ and A^{IJ}_a to Σ whenever their co-vector indices are projected into Σ. (Note incidently that the internal vector n^I has disappeared in this form of the action.) We can now read-off the equations of motion. First, since the momenta conjugate to the lapse, the shift and the time-component of the 4-connection are absent, extremization of the action with respect to $\underset{\sim}{N}, N^a$ and ${}^4A \cdot t$ gives us constraint equations. These are:

$$\mathrm{tr}\ \widetilde{\Pi}^a\widetilde{\Pi}^b\ F_{ab} = 0, \quad \mathrm{tr}\ \widetilde{\Pi}^a\ F_{ab} = 0, \quad \text{and} \quad \mathcal{D}_a\widetilde{\Pi}^a = 0. \tag{10}$$

Thus, we find that the Hamiltonian is just a sum of constraints. This is a general feature of theories in which there is no non-dynamical metric; it is a reflection of the underlying diffeomorphism invariance. The evolution equations can be obtained by evaluating the Poisson brackets of A_a^{MN} and $\widetilde{\Pi}^a_{MN}$ with the Hamiltonian. Since the constraints are polynomial in the canonical variables, so is the Hamiltonian and hence also the evolution equations. (We will not need the explicit form of these equations in this chapter.) Furthermore, as the notation suggests, all equations have a rather simple interpretation in terms of notions that naturally arise in Yang-Mills theories. These are the useful features of the new canonical formalism we referred to in chapter 2 and the introduction to Part I.

In contrast with the Legendre transform of the Palatini framework that leads to the ADM canonical description (see, e.g., chapter 4), the procedure adopted above is remarkably simple. In the final picture, the configuration and the momentum variables each have 3 (space) $\times$ 3 (internal) $=$ 9 *complex* degrees of freedom per space-point. Eq.10 gives us $1 + 3 + 3 = 7$ *complex* constraints per space point. A direct calculation (which we will see in some detail in chapter 7) shows that these constraints are first class. Consequently, the system has two true complex degrees of freedom. These are the degrees of freedom of complex general relativity. Thus, the Legendre transform we have performed *does* give us a complete and simple phase space description for complex general relativity.

We will conclude this discussion of complex general relativity by recasting the basic equations of the canonical framework in the notation used in chapter 2. (For a more detailed discussion, see chapter 4.) For this, we first note that the self dual Lorentz Lie algebra is isomorphic to the Lie algebra of complexified $SO(3)$. Therefore, in the above Hamiltonian description, we can let the canonical variables take values in the Lie algebra of $\mathbb{C}SO(3)$. The transformation between the two pictures can be carried out as follows. Fix any internal vector n^I with $n^I n_I = -1$. Now, every self dual internal 2-form f_{IJ} can be characterized completely by its "electric part" $2f_{IJ}n^I$, which is an internal spatial vector orthogonal to n^I. Hence, out basic canonically conjugate pair admits an expansion of the type:

$$\widetilde{\Pi}^a{}_{IJ} = \widetilde{E}^a_{[I}n_{J]} - \frac{i}{2}\widetilde{E}^a_M\epsilon_M{}^{IJ}, \quad \text{and} \quad A_a^{IJ} = i(A_a^{[I}n^{J]} - \frac{i}{2}A_a^M\epsilon_M{}^{IJ},$$

where the factors of i have been adjusted to simplify the reality conditions later on. Substituting in the expression of the symplectic structure,

$\Omega = \int d^3x\, \mathrm{d\!I}\widetilde{\Pi}^a_{IJ} \wedge \mathrm{d\!I}A^{IJ}_a$, we obtain $\Omega = -i\int d^3x \mathrm{d\!I}\widetilde{E}^a_I \wedge \mathrm{d\!I}A^I_a \equiv -i\int d^3x \mathrm{d\!I}\widetilde{E}^a_i \wedge \mathrm{d\!I}A^i_a$, where $\mathrm{d\!I}$ denotes the infinite dimensional exterior derivative and where in the last step we have used lower case letters for internal indices to emphasize the fact that, being orthogonal to n^I, they are 3-dimensional, $SO(3)$-indices. Thus, now the canonical variables of complex general relativity are simply a complex-valued, $SO(3)$ connection on Σ and a (density weighted) complex triad $\widetilde{E}^a_i$ defined intrinsically on Σ. The constraints of Eq.6 translate to Eq.4 of chapter 2 and the Hamiltonian is given by Eq.5 of that chapter. In spinorial treatments (e.g., [2,3]) the fact that additional structure has to be used to pass from $(A^{MN}_a, \widetilde{\Pi}^a_{MN})$ to $(A^i_a, \widetilde{E}^a_i)$ is concealed by the notation. In the Legendre transform, until one arrives at the equivalent of Eq.7, one thinks of the spinorial indices $A, B, \ldots$ as the unprimed $SL(2,\mathbb{C})$ indices. Then, in the analog of Eq.8*b* one simply re-interprets them as the $SU(2)$ indices associated with the 3-surface Σ. Of course, for calculations the fact that one can thus switch the interpretation automatically is a strength of the notation. It is only when when one is first trying to understand all the conceptual steps involved that it is inconvenient to use the spinorial notation. The situation is similar to the use of Dirac's bra-ket notation in quantum mechanics. The fact that every Hilbert space is its own dual is built into it. Therefore, for calculations, it is extremely convenient. However, if one works with this notation from the beginning, it is difficult to appreciate the fact that this isomorphism is a non-trivial property of Hilbert spaces.

Finally, let us consider the problem of extracting a Hamiltonian framework for real general relativity. The idea is that the phase space of the real theory can be obtained by restricting oneself to an appropriate real section of the complex phase space constructed above. The first reality condition is that the metric $q_{ab} = g_{ab} + n_a n_b$, constructed from the tetrad e^a_I be real. In terms of the canonical momenta, we have: $(q)q^{ab} := \widetilde{E}^a_I \widetilde{E}^{bI} \equiv \mathrm{tr}\, \widetilde{\Pi}^a \widetilde{\Pi}^b$. Hence, on the phase space, this reality condition can be expressed simply as:

$$\mathrm{tr}\ \widetilde{\Pi}^a\ \widetilde{\Pi}^b \text{ is real.} \tag{11a}$$

To ensure that the reality condition is preserved under time-evolution, we have to impose the condition that the time-derivative of this metric is also real. It is straightforward to compute the Poisson bracket of the 3-metric with the Hamiltonian. This gives us another condition, which is again polynomial in the canonical variables:

$$\mathrm{tr}\, \widetilde{\Pi}^{(a}\ \mathcal{D}_c \left[\widetilde{\Pi}^{|c|}, \widetilde{\Pi}^{b)}\right] \text{ is real.} \tag{11b}$$

One can now check explicitly that if these conditions are satisfied initially, they continue to be satisfied along the entire dynamical trajectory. That is, the Hamiltonian vector field which gives us the dynamical flow is in fact tangential to the "real section" of the phase-space defined by Eqs.11. This is not surprising: We do know from the initial value formulation of Einstein's equation that if the initial 3-metric and its time derivative (i.e. the extrinsic curvature) are real initially, their evolution by Einstein's equation yields a real 4-metric which necessarily induces real initial data on *any* Cauchy surface. The phase space of the real theory can thus be obtained by simply restricting oneself to the part of the complex phase space on which Eqs.11 hold. In this sense, then, real general relativity can be recovered from the canonical description given above.

It is however important to emphasize that the these considerations are meant to be only plausibility arguments. A substantial amount of work is needed to rigorously show that we can indeed extract a satisfactory canonical description of real general relativity from that of the complex theory. First, one has to show by explicit calculations that the constraints do in fact form a closed, first class system. Second, one has to verify that the real section, given by Eqs.11, is indeed left invariant under dynamics. Finally, one has to show that the pull-back of the symplectic structure to the real section is in fact real. To do this, it seems unavoidable that we exhibit manifestly real coordinates on the real section. Thus, from the analysis as it stands, it is not even clear that there is an underlying real phase-space for the real theory.[7] As far as I can see, the additional work needed to show that this is in fact the case is roughly the same as that required if one works directly in the real, Hamiltonian framework and then performs the canonical transform to the new variables (as in [1]). We shall carry out this task in chapter 6.

We will end this discussion with a few remarks:

i) Note that, in the Hamiltonian framework, the 3-metric is a *derived* variable; the fundamental variables are the connection A_a^i and the conjugate momentum $\widetilde{E}_i^a$. None of the equations of the theory –the constraints, the evolution equations or the reality conditions– require us to raise the vector index of the connection or lower that of its conjugate momentum. Consequently, the framework continues to be meaningful even if the 3-metric is allowed to become degenerate. Thus, what we have here is a slight generalization of Einstein's theory which reduces to Einstein's if one restricts oneself to non-degenerate triads. (The form that we used for the action $S(e, {}^4A)$ seems to require a non-degenerate tetrad. Had we used the covariant tetrad as the basic variable

7 This point was overlooked in the early work, e.g. [2,3], where it was felt that the desired goal can be met simply by restricting oneself from the beginning to real tetrads. That strategy produces a "hybrid" phase space –which, strictly speaking is not a phase space at all– in which the configuration variables are real but momenta are complex. Consequently, in these treatments, the structure of the underlying real phase space remains obscure.

and expressed the action as an integral of a 4-form, the assumption could have been dropped. For details, see, e.g., [8].)

ii) For complex general relativity, we have a satisfactory canonical formulation. The situation is the same in the case of Euclidean (i.e., +,+,+,+) general relativity, where all fields can be assumed to be real (and, say, analytic) from the beginning. (Recall that in the positive definite case, real self dual 2-forms exist and every real 2-form can be decomposed in to its real self dual and anti-self dual parts.) To obtain the canonical framework for the real, Lorentzian general relativity, on the other hand, one must further impose reality conditions. Now, most field theorists and experts in constrained dynamics follow the path laid down by Dirac to analyze theories with gauge freedom such as general relativity. This procedure would require that we impose the reality conditions as constraints and deal with them using Dirac's theory. We are *not* following this route. This difference has given rise to some confusion that I alluded to in the beginning of this chapter.

iii) Indeed, even if we wanted to, it is difficult to follow the Dirac route in the present (and in most other) discussions of the Legendre transform. The reason is that the constraints corresponding to the reality conditions would involve the canonical variables and their *complex conjugates* and nowhere in the framework does one have the Poisson brackets involving the complex conjugate fields. One of the central points of the whole framework is precisely that another route is available both for classical general relativity and for its canonical quantization. In the classical theory, we can regard real general relativity as being imbedded in the larger, complex theory. Given a physical or mathematical problem, we can first formulate it in the larger theory and *then* impose the reality conditions *by hand.* Such a procedure does shed new light on a number of issues (e.g., the dynamics of Bianchi models in general relativity.) In the quantum theory, we can first ignore the reality conditions, solve the quantum constraints, and then incorporate the reality conditions as constraints on the admissible inner product by requiring that the real classical observables should become *self-adjoint* operators. Thus, the idea of using a complex action and considering the real theory as a sub-theory of the complex theory exploits the freedom available in choosing an inner product, freedom which is ignored in the standard Dirac procedure.

References

[1] A. Ashtekar, Phys. Rev. Lett. **57**, 2244 (1986); Phys. Rev. **D36**, 1587 (1987).

[2] J. Samuel, Pramana J. Phys. **28**, L429 (1987).

[3] T. Jacobson and L. Smolin, Phys. Lett. **B196**, 39 (1987); Class. & Quant. Grav. **5**, 583 (1988).

[4] A. Ashtekar, A.P. Balachandran and S. Jo, Int. J. Theo. Phys. **A4**, 1493 (1989), Appendix; M. Henneaux, J.E. Nelson and C. Schomblond, Phys. Rev. **D39**, 437 (1989); J.L. Friedman and I. Jack, Phys. Rev. **D37**, 3495 (1988).; H. Kodama and M. Seriu, Prog. Theo. Phys. **83**, 7 (1990).

[5] J. Plebanski, J. Math. Phys. **18**, 2511 (1977).

[6] R. Capovilla, J. Dell, T. Jacobson and L. Mason (University of Maryland preprint, 1990).

[7] A. Palatini, Rend. Circ. Mat. Palermo **43**, 203 (1919).

[8] J. Romano, *Connection dynamics Vs Geometrodynamics*, Ph.D. thesis, Syracuse University (1991).

[9] S. Deser and C.J. Isham, Phys. Rev. **D14**, 2505 (1976); M. Henneaux, Phys. Rev. **D27**, 986 (1983); A. Ashtekar, In: *The Proceedings of the 1990 Banff School*, (ed. R. Mann), (World Scientific, Singapore, 1991).

4 LEGENDRE TRANSFORM: DETAILS

1 Introduction

In chapter 3, we introduced both the Palatini and the self dual actions and outlined how one can perform the Legendre transform on the self dual action to obtain a Hamiltonian formulation of complex general relativity. We saw that this procedure as well as the resulting canonical description is rather simple. We also remarked that the Legendre transform of the Palatini action, by contrast, is rather complicated to perform. The purpose of this chapter is to carry out this transform in detail and to clarify the reasons underlying the difference between the two frameworks.

In section 2 we begin with the Palatini action for real general relativity and perform the Legendre transform. We will find that, to begin with, *all* constraints and evolution equations are polynomial in the basic variables. Thus, the simplicity of these equations is not unique to the use of the self dual action or the new canonical variables. However, two of the constraints are now *second-class* in Dirac's terminology. Therefore, we have to solve them explicitly and introduce reduced canonical variables using the solution. It is at this stage that non-polynomial dependences comes in: the remaining, first class constraints are no longer simple functions of the new canonical variables. This overall situation persists if one uses the Palatini action for complex general relativity. In section 3, we consider the self dual action for the complex theory. It turns out that there are now *no* second-class constraints. The first class constraints are again simple polynomials in the canonical variables, and, since there are no second class constraints to solve, remain so in the final description. Thus, as far as complex general relativity is concerned, the self dual action is simpler to use because of the absence of second class constraints.[1] As remarked in chapter 3, however, to recover the real theory from this simple canonical

1 This point was overlooked in the early literature (see, e.g. [1,2]) and was first emphasized in the appendix of [3]. However, the treatment of the self dual action in this appendix is also not fully satisfactory.

description, one has to step outside the standard Dirac theory of constrained systems. While this may seem to be a handicap at first, we will see in chapters 8 and 10 that our modified procedure is in fact well suited for canonical quantization.

2 Palatini action

Recall that the Palatini action, S_P, for real general relativity is a real-valued functional of a tetrad, e_I^a , and a connection 1-form, ${}^4\omega_a^{IJ}$, which takes values in the Lorentz Lie-algebra:

$$S_P = \frac{1}{2}\int d^4x\,(e)e_I^a e_J^b\,{}^4\Omega_{ab}^{IJ} \tag{1}$$

where e denotes the square-root of the determinant of the space-time metric g_{ab} and where ${}^4\Omega_{ab}^{IJ} = \partial_a{}^4\omega_b^{IJ} - \partial_b{}^4\omega_a^{IJ} + [{}^4\omega_a, {}^4\omega_b]^{IJ}$ is the curvature of the connection ${}^4\omega_a^{IJ}$. Since we require that the connection must annihilate the fixed Lorentzian metric $\eta_{IJ} \equiv diag(-1,1,1,1)$ on internal indices, the connection 1-form and its curvature are both anti-symmetric in the internal indices.

To perform the Legendre transform, we introduce, as before, a foliation in space-time and a time-like vector field t^a whose integral curves intersect each leaf of the foliation precisely once. Let n^a denote the unit normal to the foliation. We can then decompose the "time-evolution vector field" t^a normal and tangential to the foliation:

$$t^a = Nn^a + N^a, \quad n_a N^a = 0. \tag{2}$$

The function N is called the *lapse* and the vector field N^a is called the *shift*. Using n^a and $(g^{ab} + n^a n^b)$ to project fields normal and tangential to the foliation, we can now decompose the action as follows:

$$S_P = \frac{1}{2}\int dt \int d^3x\,(E)\Big(NE_I^a E_J^b \Omega_{ab}^{IJ} + 2n_{[I}E_{J]}^a D_a({}^4\omega^{IJ}\cdot t) - 2n_{[I}E_{J]}^a \dot{\omega}_a^{IJ} + 2N^a n_{[I}E_{J]}^b \Omega_{ab}^{IJ}\Big), \tag{3}$$

where, $E_I^a = e_I^b(g_b^a + n^a n_b)$ is the spatial projection of e_I^a; $E \equiv q^{\frac{1}{2}}$ is the square-root of the determinant of the spatial metric q_{ab}; ${}^4\omega^{IJ}\cdot t$ is the time component, $t^a\,({}^4\omega_a^{IJ})$, of ${}^4\omega_a^{IJ}$; $\dot{\omega}_a^{IJ}$ is the Lie derivative[2] of ω_a^{IJ} with respect to t^a, and $n_I = n_a e_I^a$. To

2 The internal indices are treated as scalars in this Lie-derivative.

further simplify the action, let us define:

$$\underset{\sim}{N} := (E)^{-1} N, \qquad \widetilde{E}^a_I := (E)\, E^a_I \tag{4a}$$

$$\widetilde{\alpha}^a_{IJ} := \widetilde{E}^a_{[I}\, n_{J]}. \tag{4b}$$

In terms of these fields, the action S_P becomes

$$S_P \equiv \int dt \int d^3x \, \mathrm{tr} \Big(-2\underset{\sim}{N}\, \widetilde{\alpha}^a \widetilde{\alpha}^b \Omega_{ab} - ({}^4\omega \cdot t) D_a \widetilde{\alpha}^a + N^a \widetilde{\alpha}^b \Omega_{ab} - \widetilde{\alpha}^a \dot{\omega}_a \Big), \tag{5}$$

where, 'tr' denotes trace over the (suppressed) internal indices. To arrive at Eq.5, we have used the fact that $\eta^{IJ} E^a_I n_J$ vanishes.

From the form of the action, it is clear that there are no dynamical equations for $\underset{\sim}{N}, N^a, {}^4\omega_t^{IJ}$. They therefore serve as Lagrange multipliers. Variation of the action with respect to these three fields yields the constraints:

$$\mathrm{tr}\ \widetilde{\alpha}^a \widetilde{\alpha}^b \Omega_{ab} \equiv S \approx 0 \tag{6a}$$

$$\mathrm{tr}\ \widetilde{\alpha}^b \Omega_{ab} \equiv V_a \approx 0 \tag{6b}$$

$$D_a \widetilde{\alpha}^a_{IJ} \equiv G_{IJ} \approx 0 \tag{6c}$$

Secondly, it follows from Eq.5 that $-\widetilde{\alpha}^{aIJ}$ is canonically conjugate to ${\omega_{aJ}}^I$, or, $\widetilde{\alpha}^a_{IJ}$ is conjugate to ω_a^{IJ}:

$$\{\omega_b^{MN}(y)\, , \widetilde{\alpha}^a_{IJ}(x), \} = \delta^a_b \delta^M_{[I} \delta^N_{J]} \delta^3(x,y) \tag{7}$$

Finally, since the action is now reduced to the form $\int dt[P_j \dot{q}^j - H]$, we can simply read-off the Hamiltonian. Upto surface terms (which we ignore also in this chapter), the Hamiltonian H is given by:

$$H = \int d^3x \, \Big(2\underset{\sim}{N} S - N^a V_a + \mathrm{tr}\, ({}^4\omega \cdot t)\, G \Big), \tag{8}$$

where S, V_a and G are defined in Eq.6.

Note that the constraints (Eq.6) and hence also the Hamiltonian (Eq.8) are polynomial in these canonical variables and are quite similar to the ones obtained using new variables. (Compare with Eqs.4 and 5 in chapter 2 and Eq.10 of chapter 3). One may therefore be tempted to say that we have a polynomial formulation of the theory in the Palatini framework too. This conclusion, however, is incorrect. To see this, let us count the degrees of freedom of this theory. We have $3 \times 6 = 18$ configuration variables $\widetilde{\alpha}^a_{IJ}$ and there are $1 + 3 + 6 = 10$ constraints. If they were closed and first class, the Hamiltonian description would have $18 - 10 = 8$ degrees of freedom per space point. (If some of the constraints were second class, the number of degrees of freedom would be greater.) General relativity, on the other hand, has only 2 degrees of freedom. Therefore, even if we assume that the canonical description given by Eqs.6 and 7 is consistent and complete, we know that the resulting theory cannot be general relativity.

What is missing from general relativity? As one might expect, Eqs.6 do not constitute the full set of constraints and therefore fail to be closed under Poisson brackets. For, we have ignored a primary constraint that $\widetilde{\alpha}^a_{IJ}$ is not arbitrary but has the specific form given by Eq.4*b*; it is a "pure boost" with respect to *some* internal, unit time-like vector n^I. We will show later in this section that this constraint is equivalent to the requirement that $\widetilde{\alpha}^a_{IJ}$ should satisfy:

$$\epsilon^{IJKL}\widetilde{\alpha}^a_{IJ}\widetilde{\alpha}^b_{KL} \equiv \phi^{ab} \approx 0 \tag{9}$$

and that tr $\widetilde{\alpha}^a\widetilde{\alpha}^b$ be positive definite (since we want $\widetilde{\widetilde{q}}^{ab} = \widetilde{E}^{aI}\widetilde{E}^b_I$ to be so):

$$\text{tr }\widetilde{\alpha}^a\widetilde{\alpha}^b > 0. \tag{10}$$

Since Eq.10 is an inequality, it is a non-holonomic constraint; it does not reduce the dimension of the allowed portion of the phase space. Eq.9, on the other hand, is a regular constraint and must be added to the system Eq.6. Our task now is to determine if the entire system is closed or if there are further constraints.

Using Eq.7, the Poisson brackets between the constraints are straightforward to evaluate. One obtains:

$$\begin{aligned}
\{S(x), S(y)\} =& - \tfrac{1}{2}(\text{tr }[\widetilde{\alpha}^b,\ \widetilde{\alpha}^c][\Omega_{bc},\ \widetilde{\alpha}^a](x)\ \partial^x_a\ \delta^3(x,y) \\
&- (x \leftrightarrow y))\ , && (11a)\\
\{S(x), V_a(y)\} =& S(y)\ \partial^x_a\ \delta^3(x,y)\ , && (11b)\\
\{S(x), G_{IJ}(y)\} =& 0\ , && (11c)\\
\{S(x), \phi^{ab}(y)\} =& 4\epsilon^{IJKL}\widetilde{\alpha}^{cM}_I\ \widetilde{\alpha}^{(a}_{MJ}(D_c\widetilde{\alpha}^{b)})_{KL}\delta^3(x,y) \\
&+\ (\text{terms in } \phi^{ab})\ , && (11d)\\
\{V_a(x), V_b(y)\} =& - \text{tr } G\,\Omega_{ba}\ \delta^3(x,y)\ +\ (V_a(x)\ \partial^x_b\delta^3(x,y) \\
&- (x \leftrightarrow y))\ , && (11e)\\
\{V_a(x), G_{IJ}(y)\} =& 0\ , && (11f)\\
\{V_a(x), \phi^{bc}(y)\} =& \delta^b_a\ [\epsilon^{IJKL}\ G_{IJ}\ \widetilde{\alpha}^c_{KL}\delta^3(x,y) - \phi^{dc}\ (y)\ \partial^x_d\delta^3(x,y)] \\
&+ (b \leftrightarrow c) +\ (\phi^{bc}(x) + \phi^{bc}(y))\partial^x_a\ \delta^3(x,y)\ , && (11g)
\end{aligned}$$

$$\{G_{IJ}(x), G_{KL}(y)\} = -\tfrac{1}{2}((\eta_{IK}G_{JL} - \eta_{JK}G_{IL}) - (K \leftrightarrow L)) \,, \tag{11h}$$
$$\{G_{IJ}(x), \phi^{ab}(y)\} = 0 \,, \tag{11i}$$
$$\{\phi^{ab}(x), \phi^{cd}(y)\} = 0 \,. \tag{11j}$$

As before, D is the derivative operator defined by ω_a^{IJ}.

The right hand side of Eq.11a can be shown to be proportional to the vector constraint (Eq.6b) owing to the primary constraint Eq.9. (It is because we need Eq.9 to demonstrate this closure that the set of constraints (Eq.6) by itself is not closed.) Thus the only secondary constraint comes from Eq.11d; the right hand sides of all other Poisson brackets vanish weakly.[3] The secondary constraint can be expressed as:

$$\epsilon^{IJKL}\, \tilde{\alpha}^{c}{}_{I}{}^{M}\, \tilde{\alpha}^{(a}_{MJ} \left(D_c \tilde{\alpha}^{b)}\right)_{KL} \equiv \chi^{ab} \approx 0 \,. \tag{12}$$

There are no tertiary constraints. Thus, the full set of constraints consists of Eqs.6, 9 and 12 among which the last two constraints, Eqs. 9 and 12, are second class. *Note that all constraints are polynomial in the basic canonically conjugate variables* $\tilde{\alpha}^a_{IJ}$ *and* ω_b^{IJ}. The Poisson brackets between the second class constraints are:

$$\{\phi^{ab}(x), \chi^{cd}(y)\} \approx 4\,(\mathrm{tr}\ \tilde{\alpha}^a \tilde{\alpha}^b\ \ \mathrm{tr}\ \tilde{\alpha}^c \tilde{\alpha}^d - \mathrm{tr}\ \tilde{\alpha}^c \tilde{\alpha}^{(a}\ \ \mathrm{tr}\ \tilde{\alpha}^{b)} \tilde{\alpha}^d)\delta^3(x,y) \tag{13}$$

Finally, let us carry out the counting of degrees of freedom again. We have, as before, 18 configuration variables, together with 10 first class and $6+6=12$ second class constraints. Each first class constraint reduces the configuration degrees of freedom by 1 and each second class constraint reduces them by $\frac{1}{2}$. Therefore, now the degrees of freedom work out to be $18 - 10 - 6 = 2$. These are the degrees of freedom of general relativity.

---

3 Eq.11d needs some clarification. To begin with, one obtains

$$\begin{aligned}\{S_{\underset{\sim}{N}}, \phi^{ab}(y)\} &= -4\int d^3x\, D_c\left(\underset{\sim}{N}\tilde{\alpha}^c_{IN}\tilde{\alpha}^{(aN}{}_J\right)\epsilon^{IJKL}\tilde{\alpha}^{b)}_{KL}\delta^3(x,y)\\ &= -4\int d^3x\, D_c\left(\underset{\sim}{N}\tilde{\alpha}^c_{IN}\tilde{\alpha}^{(aN}{}_J\epsilon^{IJKL}\tilde{\alpha}^{b)}_{KL}\right)\delta^3(x,y)\\ &\quad +4\int d^3x\, \underset{\sim}{N}\tilde{\alpha}^c_{IN}\tilde{\alpha}^{(aN}{}_J\epsilon^{IJKL}D_c\left(\tilde{\alpha}^{b)}_{KL}\right)\delta^3(x,y),\end{aligned}$$

where the second line has been obtained by an integration by parts. By substituting the solution (Eq.(14′)) to Eqs.9 and 10, the first term in the second line can be shown to vanish on the constraint surface determined by $\phi^{ab} = 0$. Peeling off $\underset{\sim}{N}$, one now obtains Eq.11d. Since in the top line of the above equation the term in square brackets is a skew tensor density of weight $+1$, the expression is independent of the choice of the (torsion-free) extension of D to tensor indices made in its evaluation. The same is true of Eq.12 for χ_{ab}, on the constraint surface $\phi^{ab} = 0$.

The next step in the Dirac procedure is to solve the second class constraints. Since $\widetilde{\alpha}^a_{IJ}$ is a 2-form in the internal indices, we can decompose it into its "electric and magnetic parts". Fix a unit, time-like internal vector n^I. Then, we can decompose $\widetilde{\alpha}^a_{IJ}$ as:

$$\widetilde{\alpha}^a_{IJ} = \widetilde{E}^a_{[I}\, n_{J]} + \widetilde{B}^a_{IJ}, \tag{14}$$

where $\widetilde{B}^a_{IJ} = q_I^M\, q_J^N\, \widetilde{\alpha}^a_{MN}$, with $q_I^M = \delta_I^M + n_I\, n^M$ the projection operator in the (internal) 3-flat orthogonal to n^I. Note that, by construction, $\widetilde{E}^a_I$ and $\widetilde{B}^a_{IJ}$ are orthogonal to n^I. Since

$$\mathrm{tr}\ \widetilde{\alpha}^a\widetilde{\alpha}^b = -\frac{1}{2}n_I n^I \widetilde{E}^{aJ}\widetilde{E}^b_J - \widetilde{B}^a_{IJ}\widetilde{B}^{bIJ}, \tag{10'}$$

where the second term on the right is non-positive, using Eq.10 we conclude that n^I is time-like and $\widetilde{E}^a_I$ is non-degenerate. Substituting in Eq.9 we obtain $\epsilon^{IJKL} n_I \widetilde{E}^{(a}_J \widetilde{B}^{b)}_{KL} = 0$. Finally, one can use the 3-dimensional freedom in the initial choice of n^I to set $\epsilon^{IJKL}\, n_I\, \widetilde{E}^{[a}_J \widetilde{B}^{b]}_{KL} = 0$. In this "internal rest frame" then we have $\widetilde{B}^a_{KL} = 0$. Thus, as expected, the solution to Eqs.9 and 10 is precisely:

$$\widetilde{\alpha}^a_{IJ} = \widetilde{E}^a_{[I}\, n_{J]}, \tag{14'}$$

for *some* unit time-like internal vector n^I, with $\widetilde{E}^a_I n^I = 0$. Thus, in place of the eighteen components of $\widetilde{\alpha}^a_{IJ}$, we now have 12 components of $(n^I,\ \widetilde{E}^a_I)$ to work with. (Note, however, that we are still left with all eighteen components of ω_a^{IJ}.)

Next, we have to solve the remaining second class constraint, Eq.12. It turns out that the simplest way to solve it is by using part of the Gauss constraint Eq.6c to gauge fix the internal vector n^I. (This gauge fixing further reduces the momentum variables to just 9; $\widetilde{\alpha}^a_{IJ}$ is now fully determined by the nine components of $\widetilde{E}^a_I$.) However, since we wish to gauge fix the internal vector n^I, we must also solve the "boost part" of the Gauss constraint (relative to n^I), i.e., the equation

$$n^J (D_a \widetilde{\alpha}^a)_{IJ} \approx 0, \tag{15}$$

The $(SO(3))$ internal rotations generated by the remaining Gauss constraints will leave the gauge-fixed n^I invariant.

These 9 equations (Eqs.12 and 15) reduce the independent components in ω_a^{IJ} from 18 to 9. To solve them, let us first introduce ∇_a, the (torsion-free) derivative operator compatible with the triad $\widetilde{E}_I^a$ and define a field $K_{aI}{}^J$ via:

$$D_a\lambda_I =: \nabla_a\lambda_I + K_{aI}{}^J\lambda_J. \tag{16}$$

(Thus, ω_a^{IJ} is given by $\omega_a^{IJ} = \Gamma_a^{IJ} + K_a^{IJ}$, where Γ_a^{IJ} is the spin- connection of ∇_a.) Substituting for D_a in terms of ∇_a and $K_{aI}{}^J$ in Eqs.12 and 15, we obtain:

$$\chi^{ab} = -\frac{1}{4}\epsilon^{IJM}\,\widetilde{E}_L^{(a}\widetilde{E}_J^{b)}\,\widetilde{E}_M^c\,K_{cI}{}^L = 0 \tag{12'}$$

and

$$n^J D_a\widetilde{\alpha}_{IJ}^a = -\frac{1}{2}\widetilde{E}_N^a q_I^M K_{aM}{}^N = 0, \tag{15'}$$

where $q_I^J = \delta_I^J + n_I n^J$ is the internal projection operator. Let us decompose the 2-form indices of $K_{aI}{}^J$ in terms of their electric and magnetic parts: $K_a^{MN} = 2K_a^{[M}n^{N]} + \overline{K}_a^{MN}$, where $\overline{K}_a^{MN} = \overline{K}_a^{[MN]}$ and $K_a^M n_M = 0 = \overline{K}_a^{MN}n_M$. Substituting this decomposition in Eqs.12′ and 15′ one finds: $\overline{K}_a^{MN} = 0$.[4] Thus, Eqs.12 and 15 imply that the boost part of ω_a^{IJ} is free and is coordinatized by the field K_a^I while the rotation part is completely determined by the triad $\widetilde{E}_I^a$.

To summarize, we have now solved the 12 second class constraints Eqs.9 and 12 and eliminated the 3 first class constraints of Eq.15 by solving them and fixing the corresponding gauge. The resulting dynamical variables are $(\widetilde{E}_I^a, K_a^I)$. Since they are both orthogonal to n^I (which is gauge fixed), their internal indices effectively take only the values 1, 2, 3; to emphasize this fact, one may express these variables as $(\widetilde{E}_i^a, K_a^i)$. Having eliminated the second class constraints, our next task is to express the symplectic structure in terms of these reduced variables. Since the original symplectic structure in terms of $(\omega_a^{IJ}, \widetilde{\alpha}_{IJ}^a)$ is just $\int d^3x \mathrm{d\!I}\widetilde{\alpha}_{IJ}^a \wedge \mathrm{d\!I}\omega_a^{IJ}$, pulling it back to the surface defined by $\widetilde{\alpha}_{IJ}^a = \widetilde{E}_{[I}^a n_{J]}$ and Eq.15 (with our gauge choice for n^I), we obtain the symplectic structure on the phase space of reduced variables. It is given simply by: $\int d^3x\, \mathrm{d\!I}K_a^i \wedge \mathrm{d\!I}\widetilde{E}_i^a$. Thus, the reduced phase space is coordinatized by the canonically conjugate pairs $(\widetilde{E}_i^a, K_a^i)$, the only non-vanishing fundamental Poisson bracket being:

$$\{\widetilde{E}_i^a(x), K_b^j(y)\} = \delta_b^a\,\delta_i^j\,\delta^3(x,y)\,. \tag{17}$$

Our final task is to rewrite the 7 first class constraints – S, V_a and the rotation part of G_{MN} in terms of this new canonical pair. This task is straightforward.

4 To see this, define a field $\overline{K}^{IJ} := \widetilde{E}^{aI}\epsilon^{JKL}\overline{K}_{aKL}$. Now we find that Eqs.12′ and 15′ reduce respectively to $\overline{K}_M{}^M q^{IJ} - \overline{K}^{(IJ)} = 0$ and $\overline{K}^{[IJ]} = 0$, which, using the invertibility of the triad, is equivalent to $\overline{K}_a^{MN} = 0$.

However, as one might expect, the result is just the ADM form of the constraints for tetrad gravity:

$$q^{-\frac{1}{2}}(\widetilde{E}_i^b\widetilde{E}_j^a - \widetilde{E}_i^a\widetilde{E}_j^b)K_a^iK_b^j - q^{\frac{1}{2}}\mathcal{R} \equiv \mathcal{S}' \approx 0 \tag{18a}$$

$$D_b(K_a^i\widetilde{E}_i^b - K_c^i\widetilde{E}_i^c\delta_a^b) \equiv \mathcal{V}_a' \approx 0 \tag{18b}$$

$$\epsilon_{ijk}K_a^j\widetilde{E}^{ak} \equiv \mathcal{G}_i' \approx 0 \tag{18c}$$

where D is now the unique derivative operator (on vector *and* internal indices) compatible with the triad and $\mathcal{R}$ its scalar curvature. The field $q^{ab} = E_i^a\, E^{bi}$ is the spatial 3-metric and $K_{(a}{}^i\, E_{b)i}$ is the extrinsic curvature. Eqs.18 are non-polynomial in the canonical variables and the close relation to Yang-Mills theory is now lost.

So far, we have considered *real* general relativity. The situation for the complex theory is completely analogous. Here, one considers complex solutions g_{ab} to Einstein's equation on a real 4-manifold M. Now, the Palatini action (Eq.1) is a complex valued function of complex tetrads and connections. However, (since the underlying manifold M is real), the internal metric η_{IJ} continues to be a real Lorentzian metric and the internal group continues to remain $SO(3,1)$. The space-time metric g^{ab} can be complex because the tetrads are complex: $g^{ab} = \eta^{IJ}e_I^a\, e_J^b$. As in the real case (see section 3.2), the variation of the action with respect to the connection ω_a tells us that it is the internal connection of the derivative operator compatible with the tetrad. Extremizing the action with respect to the tetrad, and substituting the solution for the connection implies that the Einstein tensor of the complex metric $g^{ab} = e^{aI}e_J^b$, vanishes. To perform the Legendre transform, we introduce, as in the real case, a function t whose level surfaces provide a foliation of the real space-time manifold and a vector field t^a, transversal to the foliation, which is normalized such that $t^a\nabla_a t = 1$. The only difference is that now the lapse $\underset{\sim}{N}$, the shift vector N^a and other quantities are complex. The resulting constraints have the same form as Eq.6 and the Poisson brackets between $\widetilde{\alpha}_{IJ}^a$ and ω_b^{MN} again have the form of Eq.7. Since the action is complex and does not contain the complex conjugates of the tetrads and connections, we do not obtain *any* Poisson brackets involving the complex conjugates of $\widetilde{\alpha}_a^{IJ}$ and ω_b^{MN}. The constraint algebra is identical, yielding the secondary constraint of Eq.12, which together with Eq.9, forms a second class pair. In place of Eq.10, however, we only have the (non-holonomic) requirement that $\mathrm{tr}(\widetilde{\alpha}^a\widetilde{\alpha}^b)$ be non-degenerate. The solutions to these second class constraints are of the same form as in real general relativity. The only difference is that all canonical variables are now complex. Finally, rewriting the 7 (complex) first class constraints in terms of the new canonical pair (Eq.17) we again obtain the ADM framework but now for complex tetrad gravity. The simplicity of constraints is again lost –they are non-polynomial– and so is any resemblance to a theory of connections.

Thus, in the Palatini framework, regardless of whether we consider real or complex general relativity, it is while solving the second class constraints that we lose simplicity. We will see in the next section that the situation is quite different if we use the self dual action.

3 Self dual action

We saw in chapter 3 that using the self dual connection ${}^4A_a^{MN}$ in place of the real connection ${}^4\omega_a^{MN}$, one *can* obtain a canonical formulation for complex general relativity in which all equations are polynomial in the basic variables and the close relation to Yang-Mills theories is maintained. We now wish to analyze the Legendre transform in the self dual case following, as far as possible, the procedure used in section 2 to bring out the similarities and differences between the Palatini and the self dual frameworks.

Recall from section 3.3 that the self dual action for complex general relativity is given by:

$$S(e, {}^4A) = \int d^4x\,(e)\, e_I^a\, e_J^b\, {}^4F_{ab}^{IJ}\,, \tag{19}$$

where, as before, e_I^a is a *complex* tetrad and ${}^4F_{ab}{}^{IJ}$ is the curvature of the self dual connection 1-form ${}^4A_a^{IJ}$. Although in this discussion ${}^4A_a^{IJ}$ is an independent variable, to make contact with the Palatini framework, we note that it can be regarded as the self dual part of ${}^4\omega_a^{IJ}$:

$${}^4A_a^{IJ} \equiv \tfrac{1}{2}\Big({}^4\omega_a^{IJ} - \tfrac{i}{2}\epsilon^{IJ}{}_{KL}\,{}^4\omega_a^{KL}\Big). \tag{20}$$

Now we can repeat the steps followed in section 2 (as extended to complex general relativity) to perform the Legendre transform of the action $S(e, {}^4A)$. The canonical variables[5] satisfying the analog of Eq.7 are $(\widetilde{\Pi}_{IJ}^a, A_b^{IJ})$, while the constraints analogous to Eq.6 are given by

$$\begin{aligned} \mathcal{S} &\equiv \operatorname{tr}\, \widetilde{\Pi}^a\widetilde{\Pi}^b F_{ab} \approx 0\,, && (21a)\\ \mathcal{V}_b &\equiv \operatorname{tr}\, \widetilde{\Pi}^a F_{ab} \approx 0 && (21b)\\ \text{and}\quad \mathcal{G}_{IJ} &\equiv (\mathcal{D}_a\widetilde{\Pi}^a)_{IJ} \approx 0\,. && (21c) \end{aligned}$$

5 Note that $\widetilde{\Pi}_{IJ}^a := (\widetilde{\alpha}_{IJ}^a - \frac{i}{2}\epsilon_{IJ}{}^{KL}\widetilde{\alpha}_{KL}^a)$ is twice the self dual part of $\widetilde{\alpha}_{IJ}^a$. Since the commutator of self dual fields is also self dual, *all* fields we will consider henceforth will be self dual.

These equations are clearly polynomial in the basic canonical variables and in this form, closely resemble the constraint equations Eq.4 reported in chapter 2. The constraint algebra is straightforward to compute and the resulting equations are similar in form to the relevant Poisson brackets of Eqs.11. The key point now is the following: *Since $\widetilde{\Pi}^a_{IJ}$ is an arbitrary self dual field (in internal indices), there are* no *additional constraints on its form.* This is in striking contrast to what happened in the Palatini framework to the momentum $\widetilde{\alpha}^a_{IJ}$. Therefore, if we are concerned only with complex general relativity, our analysis would end here: by a simple algebraic manipulation, we have obtained the desired (self dual) canonical framework for complex general relativity. To confirm that there are no hidden constraints, let us count the *complex* degrees of freedom. Unlike the Palatini case, since the configuration variables A_a^{IJ} are *self dual*, we only have $3 \times 3 = 9$ configuration variables (per space point). There are $1 + 3 + 3 = 7$ first class constraints, leaving the expected 2 degrees of freedom of complex general relativity. As emphasized in chapter 3, however, to recover the canonical framework for *real* general relativity, additional work is needed.

To conclude this discussion, we will show in detail how the basic variables and the constraints of the self dual framework can be recast in the more familiar language used in chapters 2 and 3. The idea is to set up an isomorphism $I_{iM}{}^N$ between the self dual Lie-algebra and the Lie algebra of complexified $SO(3)$. One can choose the isomorphism such that the following equations hold:

$$[I_i, I_j] = \epsilon_{ij}{}^k I_k \qquad\qquad \mathrm{tr}\, I_i I_j = -\delta_{ij} \tag{23a}$$

and,

$$I_i^{MN}\, I^i{}_{IJ} = \frac{1}{2}(\delta^M_{[I}\, \delta^N_{J]} - \tfrac{i}{2}\epsilon_{IJ}{}^{MN}). \tag{23b}$$

(Such an isomorphism can be constructed "explicitly" by introducing basis in the self-dual Lie-algebra: $I_i^{MN} = -(\frac{1}{2}\epsilon_i{}^{MN} + q_i^{[M} n^{J]})$, where, as usual, $\epsilon_{iMN} = \epsilon_{iMNP}\, n^P$ and $q_{iM} = \eta_{iM} + n_i\, n_M$.) Let us further require that the isomorphism satisfy: $\partial_a I_i^{MN} = 0$. Then, any self dual $\Lambda_I{}^J$ can be expanded as :

$$\Lambda_I{}^J = \Lambda^i\, I_{iI}{}^J \tag{24a}$$

where the components Λ^i are given by:

$$\Lambda^i = -\,\mathrm{tr}\, \Lambda I^i. \tag{24b}$$

Then, we have $A_a^i = -\,\mathrm{tr}\, A_a I^i$ as well as $F_{ab}^i = -\,\mathrm{tr}\, F_{ab}\, I^i = 2\partial_{[a} A^i_{b]} + \epsilon^{ijk} A_{bj} A_{ck}$. Finally, let us define $\widetilde{E}_i^a$ via $\widetilde{E}_i^a = -i\, \mathrm{tr}\, \widetilde{\Pi}^a\, I_i$. Then it is possible to rewrite all our equations using A_a^i and $\widetilde{E}_i^a$ in place of $A_a{}^{MN}$ and $\widetilde{\Pi}^a_{MN}$.

To begin with, the canonically conjugate variables are the arbitrary complex fields $(\widetilde{E}_i^a, A_a^i)$, with the Poisson brackets:

$$\begin{aligned}\{\widetilde{E}_i^a(x),\, \widetilde{E}_j^b(y)\} = 0\,, \quad \{A_a^i(x),\, A_b^j(y)\} = 0\,, \\ \{\widetilde{E}_i^a(x),\, A_b^j(y)\} = -i\delta_b^a \delta_i^j\, \delta^3(x,\, y)\,.\end{aligned} \tag{25}$$

Let us next re-express the constraints Eq.21 in terms of these variables. The result is simply the constraint equations reported in chapter 2:

$$\epsilon^{ijk}\, \widetilde{E}_i^a\, \widetilde{E}_j^b\, F_{abk} = 0; \quad \widetilde{E}_i^a\, F_{ab}^i = 0; \text{ and } \mathcal{D}_a \widetilde{E}_i^a = 0\,. \tag{26}$$

Again, all the constraints are polynomial in the reduced variables. To recover real general relativity, we impose the (also polynomial) reality conditions:

$$(\widetilde{E}_i^a \widetilde{E}^{bi}) \quad \text{is real and positive, and} \quad (\widetilde{E}_i^a \widetilde{E}^{bi})^{\bullet} \quad \text{is real.} \tag{27}$$

Finally, let us briefly consider the situation in higher dimensional space-times. The Palatini analysis given in section 2 goes through in all dimensions. However, since the notion of self duality of 2-forms is meaningful only in four dimensions, it seems difficult to extend the analysis of section 3 to higher dimensions. Therefore, even the existence of the new variables is in question in higher dimensions. Perhaps the best avenue to analyze this issue is to use chiral spinors. However, since the dimension of the spin space grows rapidly in higher dimensions, it appears that one would be faced with an increasing number of primary constraints analogous to Eq.9 and the analysis of possible secondary constraints may well become cumbersome.

References

[1] J. Samuel, Pramana J. Phys. **28**, L429 (1987).

[2] T. Jacobson and L. Smolin, Phys. Lett. **B196**, 39 (1987); Class. & Quant. Grav. **5**, 583 (1988).

[3] A. Ashtekar, A.P. Balachandran and S. Jo, Int. J. Theo. Phys. **A4**, 1493 (1989).

5 FROM TRIADS TO $SU(2)$ SPINORS: A QUICK TRANSITION

In this chapter, I will show how one can go back and forth between the triads used in chapters 3 and 4 and the $SU(2)$ spinors which will be used in the remaining chapters in this Part. This chapter is "practically oriented" in the sense that the discussion will be focussed on calculations. A conceptual introduction to $SU(2)$ as well as $SL(2,\mathbb{C})$ spinors is given in appendix A.

The fact that $SU(2)$ spinors can be regarded as "square-roots" of 3-vectors must be familiar to most readers from non-relativistic quantum mechanics. The relation between the two is generally made explicit through Pauli matrices, $\tau^i{}_A{}^B$, where i, running over $1,2,3$, labels the matrix while the indices A, B refer to the elements of the matrix. Thus, $\tau^3{}_1{}^2$ is the entry in the first row and second column of the third Pauli matrix. Recall that $\tau^i{}_A{}^B$ are 2×2, traceless Hermitian matrices, satisfying the relation

$$\tau^i{}_A{}^B\,\tau^j{}_B{}^D \equiv (\tau^i\,\tau^j)_A{}^D = i\,\epsilon^{ijk}\,\tau_{kA}{}^D + \delta^{ij}\delta_A{}^D\,. \tag{1}$$

Given a real triad E^a_i, let us set

$$\sigma^a{}_A{}^B = -\tfrac{i}{\sqrt{2}}E^a_i\,\tau^i{}_A{}^B \tag{2}$$

At each point of the 3-manifold Σ, $\sigma^a{}_A{}^B$ provides an isomorphism between the 3-real dimensional tangent space at that point and the 3-real dimensional vector space H of 2×2 trace-free Hermitian matrices. Thus, the objects $\sigma^a{}_A{}^B$ "solder" the vector space H to the tangent space of each point of Σ. They are therefore called the *soldering forms.* (Note that the vector index in Eq.2 is contravariant; the word "form" is used in a non-technical sense here.) Due to Eq.1 we have:

$$\mathrm{tr}\,\sigma^a\sigma^b \equiv \sigma^a{}_A{}^B\sigma^b{}_B{}^A = -q^{ab}, \text{ and } [\sigma^a,\sigma^b]_A{}^B = \sqrt{2}\epsilon^{abc}\sigma_{cA}{}^B\,, \tag{3}$$

where $[.,.]$ stands for the matrix commutator. The numerical factor $\frac{-i}{\sqrt{2}}$ in front of Eq.2 has been adjusted precisely to yield Eq.3 (with its numerical factors), which are the standard relations in the relativity literature. The first of Eqs.3 tells us that the soldering form can be thought of as the square-root of the 3-metric while the second equation implies that once a soldering form has been chosen, an orientation can be fixed on Σ, i.e., a preferred choice of sign can be made for the alternating tensor ϵ^{abc} on (Σ, q_{ab}) (which, otherwise, is unique only up to sign).

Let us next turn to the connection 1-form A_a^i. The relation given above between the triads and the soldering forms is dictated by the Riemannian geometry of (Σ, q_{ab}). The fact that the group $SU(2)$ –generated by the Pauli matrices– is the double cover of the group $SO(3)$ acting on the tangent space at each point of Σ provides further structure which is interesting and extremely useful. However, this fact did not *directly* enter Eq.2. The situation is different for connections. Now, it is the fact that the Lie algebra of $SU(2)$ and $SO(3)$ are the same that governs the relation between the connections A_a^i considered so far and the connections $A_{aA}{}^B$ we now want to introduce. Let us, for a moment, ignore soldering forms and triads and concentrate only on connections (and their curvatures) which take values in the Lie algebra of $SO(3)$. Now, we shall use the Pauli matrices $\tau^i{}_A{}^B$ to work in the spin-$\frac{1}{2}$ representation of $SO(3)$. Let us set: $A_{aA}{}^B := kA_a^i\,\tau_{iA}{}^B$, for some constant k. Then, it follows that the curvatures

$$F_{ab}^i = 2\partial_{[a}A_{b]}^i + \epsilon^{ijk}\,A_{aj}A_{bk} \text{ and } F_{abA}{}^B := 2\partial_{[a}A_{b]A}{}^B + [A_a, A_b]_A{}^B \tag{4}$$

are related by $F_{abA}{}^B = k\,F_{ab}^i\,\tau_{iA}{}^B$ if and only if we set $k = -\frac{i}{2}$. Thus, we now have the recipe:

$$A_{aA}{}^B := -\tfrac{i}{2}A_a^i\tau_{iA}{}^B. \tag{5}$$

Thus, because we wanted to incorporate the spinorial conventions from Riemannian geometry as well as the standard conventions from Yang-Mills theory for the expressions of curvatures in terms of connections, we are led to use *different* numerical factors in Eqs.2 and 5.

In the spinorial language, the basic canonical variables are the density weighted soldering form $\widetilde{\sigma}^a{}_A{}^B := \sqrt{q}\sigma^a{}_A{}^B$, (where q, as usual, is the determinant of the covariant metric q_{ab} obtained from $\sigma^a{}_A{}^B$) and the connection $A_{aA}{}^B$. Using Eqs.2, 5 and the Poisson brackets between the canonical variables A_a^i and $\widetilde{E}_i^a$, it is straightforward to calculate the Poisson brackets between these spinorial canonical variables. The only non-vanishing Poisson bracket is:

$$\{\widetilde{\sigma}^a{}_{CD}(x)\,, A_b^{AB}(y)\,\} = -\tfrac{i}{\sqrt{2}}\,\delta_b^a\,\delta_C{}^{(A}\delta_D{}^{B)}\,\delta^3(x,y)\,. \tag{6}$$

This translation of the basic phase space structure provides the point of departure for the spinorial canonical formulation of general relativity.

To conclude, I would like to point out an important caveat in this procedure. While most of the algebraic calculations involved in the translation are straightforward, there is a potential source of confusion in what is meant by "Hermitian adjoint". There are in fact two distinct notions. The first, and perhaps the more familiar notion comes directly from matrices. Given a column vector (i.e., a ket), $\lambda^A = \begin{pmatrix} a \\ b \end{pmatrix}$, its matrix adjoint is a row vector (i.e., a bra) given by $(\lambda^\ddagger)_A = (\bar{a}, \bar{b})$. This operation is inconvenient to use in the spinor algebra because it maps vectors to co-vectors; it is ill-suited to an index notation. Recall, however, that the spin space is 2-dimensional and equipped with alternating tensors ϵ^{AB} and ϵ_{AB} (since we are dealing with $SU(2)$ rather than $U(2)$). Therefore, we can combine the operation '$\ddagger$' with contraction by ϵ^{AB} to define a new operation '$\dagger$' which maps column vectors to column vectors (and, similarly, row vectors to row vectors). Thus, we set:

$$(\alpha^\dagger)^A \equiv -(\alpha^A)^\dagger := \epsilon^{AB}\,(\alpha^\ddagger)_B \qquad \text{i.e.,} \qquad \begin{pmatrix} a \\ b \end{pmatrix}^\dagger = \begin{pmatrix} \bar{b} \\ -\bar{a} \end{pmatrix}, \tag{7}$$

where we have used

$$\epsilon^{AB} \equiv \begin{pmatrix} 0 & -1 \\ 1 & 0 \end{pmatrix}. \tag{8}$$

It is this operation $\dagger$ that is used in the following chapters. Note that, while the square of $\ddagger$ gives the identity transformation, the square of $\dagger$ gives *minus* the identity: $(\lambda_A^\dagger)^\dagger = -\lambda_A$ for any spinor λ_A.

Once we know its action on the row and column vectors, its action on arbitrary spinors $\alpha^{A\ldots B}{}_{D\ldots E}$ is defined simply by requiring that it interact in an obvious way with the algebraic operations available on the tensor algebra:

$$\begin{gathered}(\alpha^A + c\beta^A)^\dagger = (\alpha^A)^\dagger + \bar{c}(\beta^A)^\dagger \quad \text{and,} \\ (\alpha^{A\ldots B}{}_{D\ldots E}\beta^{M\ldots N}{}_{P\ldots Q})^\dagger = (\alpha^{A\ldots B}{}_{D\ldots E})^\dagger(\beta^{M\ldots N}{}_{P\ldots Q})^\dagger\,.\end{gathered} \tag{9}$$

The second of these conditions may seem surprising, because for matrices A and B, one normally sets $(AB)^\ddagger = (B^\ddagger)(A^\ddagger)$. However, that convention is adapted to the fact that the operation $\ddagger$ interchanges rows and columns. Thus, for example, if A is a $m \times n$ matrix and B is a $n \times p$ matrix, AB is $m \times p$, while $(AB)^\ddagger$ and $(B^\ddagger A^\ddagger)$ are both $p \times m$ matrices; the product $(A^\ddagger B^\ddagger)$ is not even defined. The operation $\dagger$, on the other hand, does *not* interchange rows and columns. Furthermore, because of the explicit presence of indices, there is *no* distinction between $(\alpha^{A\ldots B}{}_{D\ldots E}\beta^{M\ldots N}{}_{P\ldots Q})$ and $(\beta^{M\ldots N}{}_{P\ldots Q}\alpha^{A\ldots B}{}_{D\ldots E})$. An important consequence of these differences is the following. Consider a trace-free second rank spinor field $\alpha_A{}^B$ which is Hermitian, i.e. which satisfies $(\alpha_A{}^B)^\dagger = \alpha_A{}^B$. To express it as a matrix, let us introduce a basis $(e_1^A = \lambda^A, e_2^A = (\lambda^A)^\dagger)$, where the spinors are so normalized that $(\lambda^\dagger)^A\lambda_A = 1$. (The co-basis is then given by $(e^1_A = -(\lambda_A)^\dagger, e^2_A = \lambda_A)$.) In such a basis, the components of $\alpha_A{}^B$ are given by a matrix of the type:

$$\alpha_A{}^B = i \begin{pmatrix} a & b \\ \bar{b} & -a \end{pmatrix} \equiv -(\alpha_A{}^B)^\ddagger \tag{10}$$

Note that the matrix is *anti-Hermitian*! As a consequence, while $\tilde{\sigma}^a{}_A{}^B$, regarded as matrices, are anti-Hermitian with respect to the operation ‡, the fields themselves are Hermitian with respect to the operation †.

6 HAMILTONIAN FORMULATION

1 Introduction

In chapter 3 we arrived at a Hamiltonian formulation of Einstein's theory in terms of new variables, ($\widetilde{E}^a{}_i$ and $A_a{}^i$), starting from a Lagrangian. The Hamiltonian formulation was obtained for complex general relativity and we noted that the real theory can be recovered by restricting ourselves to the appropriate real section of the complex phase space. In this chapter, we will show that the Hamiltonian description of the real theory can be obtained by performing a canonical transformation directly on the traditional, Palatini phase space of section 4.2. While we used tetrads and triads in the previous discussion, in this chapter we shall present the Hamiltonian description using $SU(2)$ spinors; our canonical variables will be $(\widetilde{\sigma}^a{}_A{}^B,\ A_{aA}{}^B)$, introduced in chapter 5. It is clear from the overview presented in chapter 2 that the transition to spinors is *not* essential. We could have worked with the traditional phase-space based on triads and their conjugate momenta and performed a canonical transformation to arrive at $(\widetilde{E}^a{}_i, A_a{}^i)$.[1] My motivation here is pedagogical. I hope that the employment of spinors in this and subsequent chapters will make it easier for the reader to switch back and forth between the two frameworks. Indeed, while triads are probably more familiar to readers, many calculations are simpler with spinors since they are tailored to handle self dual fields. As an exercise, the reader is encouraged to translate the triad equations we obtained in chapters 3 and 4 to spinors using the dictionary given in chapter 5 and verify that they are the same as the equations we will now obtain directly in the spinorial language.

At first it may seem that re-deriving the constraints and the Hamiltonians via a canonical transformation is somewhat redundant since we have already obtained these expressions through a Legendre transform of the self dual action. However, there *are* a number of points that this approach will clarify, some of which play an important role in the strategy we will eventually adopt in the quantization program. In particular, the present treatment will make it clear that to obtain the new

1 This procedure was carried out independently by a number of researchers including Ashtekar, Friedman and Jack, Henneaux, Nelson and Schomblond and Lee [1].

canonical framework, it is *not* essential to first pass to complex general relativity; one can work throughout in the confines of the real theory and simply regard $A_{aA}{}^{B}$ as a complex function on the real phase space; it plays a role that is similar to the one played by the complex function $z = q - ip$ on the real phase space of a harmonic oscillator. Consequently, in this approach, much of the confusion surrounding the role of the "reality conditions" in classical and quantum theory is avoided. Recall also that the analysis presented in chapters 3 and 4 is incomplete because we did not show that the new constraints (Eqs.3.10) form a first class system. This gap will now be filled. In addition, we will display the relation between the new Hamiltonian formulation and the one traditionally used in geometrodynamics.[2] Finally, in the quantum theory, certain aspects of the relation between the self dual and the metric representations –as well as some subtleties of the self dual representation itself– can easily be missed unless one uses this Hamiltonian framework as the point of departure [3].

In the traditional Hamiltonian framework, the dynamical variables $(q_{ab}, \tilde{p}^{ab})$, are tensorial and the kinematical arena one needs to introduce them is just a 3-manifold, Σ. The dynamical variables we now wish to use, $(\tilde{\sigma}^{a}{}_{A}{}^{B}, A_{aA}{}^{B})$, on the other hand, are spinorial. Now, one normally thinks of spinors only in the context of manifolds equipped with metrics [4]. Unfortunately, in the phase space formulation of general relativity, the metric is not given *a priori*; it is a dynamical variable. Our first task then is to construct a kinematical arena that will enable us to introduce objects such as the soldering forms meaningfully, without committing ourselves to any specific metric. That this should be possible is clear from our brief treatment of spinors in chapter 5 based on triads. This task is carried out in section 2 following a somewhat different, although equivalent procedure which does not refer to triads. Using these ideas, in section 3 we enlarge the standard phase space of general relativity based on 3-metrics and their conjugate momenta to incorporate spinors. Such an enlargement is needed, in any case, to treat spinorial matter and has appeared in the literature, generally in the form of triad formulations (as outlined in chapter 5). New ideas appear only in section 4 which introduces the canonical transformation to new variables.

I have attempted to make this chapter sufficiently self-contained so as to be comprehensible without a prior knowledge of the Lagrangian formulation. Readers familiar with spinors can read it as a direct sequel to chapter 2. Further details can be found in [5,6].

2 We assume that the reader is familiar with the ADM phase space formulation. For details, see e.g., Part II of [2].

2 Kinematical arena

Let Σ be a 3-manifold which is either compact or asymptotically flat. Thus, either Σ is compact, or the complement of a compact set in Σ is diffeomorphic to the complement of a closed ball in $\mathbb{R}^3$. Since Σ is a manifold, the notion of tensor fields, $T^{a..b}{}_{c..d}$, on Σ is well-defined. Consider, in addition to tensor fields, objects such as $T^{a...b}{}_{c...d}{}^{A...B}{}_{C...D}$ with internal $SU(2)$ indices, $A \ldots B, C \ldots D$. Such fields naturally arise in the description of $SU(2)$ Yang-Mills fields; examples are the connection 1-forms $A_{aA}{}^B$ and the field strength $F_{abA}{}^B$. Formally, one can regard these fields as ***generalized tensors*** in the sense of Ref. [7].[3]

Fields $\lambda^{A...B}{}_{C...D}$ with only internal indices may be thought of as "Higgs scalars" of the hypothetical $SU(2)$ Yang-Mills system under consideration. The idea is to regard them as "potential $SU(2)$ spinors" which become actual $SU(2)$ spinors once a soldering form $\sigma^a{}_A{}^B$ is introduced. The $SU(2)$ structure provides us with a volume element ϵ_{AB}—with which we can raise and lower the internal indices—and a Hermitian conjugation operation '$\dagger$'. The following conventions/conditions hold (see chapter 5 and section 3 in appendix A):

$$\lambda^A = \epsilon^{AB}\lambda_B, \qquad \lambda_A = \lambda^B\epsilon_{BA}, \tag{1}$$

and

$$\begin{aligned} &(\lambda + c\mu)^\dagger_A = \lambda^\dagger_A + \bar{c}\mu^\dagger_A, \qquad (\lambda^\dagger)^\dagger_A = -\lambda_A \\ &(\lambda)^{\dagger A}\lambda_A \geq 0, \quad \text{with equality iff} \quad \lambda_A = 0, \\ &\epsilon^\dagger_{AB} = \epsilon_{AB}, \qquad (\lambda_A\mu_B)^\dagger := \lambda^\dagger_A\mu^\dagger_B. \end{aligned} \tag{2}$$

In the absence of a soldering form—which will eventually solder the tangent space at any point p of Σ to the space of second-rank, traceless, Hermitian Higgs scalars in the fiber over p—the internal indices, as noted above, can be treated exactly as in Yang-Mills theory. In particular, one can introduce connections which act on internal indices and/or tensor indices. This fact will play an important role

---

3 One can introduce these generalized tensor fields as follows. Consider a vector bundle **B** over Σ each of whose fibers is isomorphic with a 2-dimensional complex vector space W, equipped with a symplectic and a Hermitian structure. Denote membership in W by an upper case latin contravariant index, e.g., $\lambda^A \in W$. Since W is 2-dimensional, all non-degenerate 2-forms on it are proportional to one another. Pick one and call it ϵ_{AB}. A "Higgs scalar" λ^A is then a cross-section of the bundle **B**. More general "Higgs scalars", $\lambda^{A...B}{}_{C...D}$, are cross-sections of associated bundles where the fibers are the appropriate tensor products of W and its dual. Finally, generalized tensor fields, $T^{a...b}{}_{c...d}{}^{A...B}{}_{C...D}$, are cross-sections of associated bundles where the fibers are tensor products of a suitable number of copies of W, the tangent space at each point of Σ and their dual vector spaces.

in the new phase space formulation. Here, we simply note a few facts that will turn out to be useful later. Let us restrict ourselves to those connections, say D, which are *compatible with* ϵ_{AB}, i.e., which satisfy $D_a\epsilon_{AB} = 0$. It then follows that the curvature 2-forms, $F_{abA}{}^B$ are all traceless in the internal indices. Secondly, any two connections D and D' on internal indices are related by a generalized tensor field $C_{aA}{}^B$:

$$(D'_a - D_a)\lambda_A = C_{aA}{}^B\lambda_B, \tag{3a}$$

where $C_{aA}{}^B$ satisfies:

$$C_a{}^{[AB]} = 0. \tag{3b}$$

All this structure is to be regarded as "kinematical"; it is fixed once and for all prior to the introduction of the dynamical variables. It will provide us with the tools required in the next section to extend the phase space appropriately.

3 Extended phase space

Let us begin with a basic definition. An $SU(2)$ *soldering form* $\sigma^a{}_A{}^B$ is a general isomorphism from the space of trace-free, Hermitian Higgs scalars $\lambda_B{}^A$ to the space of vectors λ^a tangent to the 3-manifold Σ: $\lambda^a := -\sigma^a{}_A{}^B\lambda_B{}^A \equiv -\operatorname{tr}(\sigma^a\lambda)$. We shall denote the inverse mapping by $\sigma_{aA}{}^B$. The conventions fixed in section 2 imply that

$$q_{ab} := -\operatorname{tr}(\sigma_a\sigma_b) \tag{4}$$

is a positive definite metric on Σ. Thus, *given a soldering form* $\sigma^a{}_A{}^B$, we are back to the spinorial scenario discussed in [4]: $\sigma^a{}_A{}^B$ is the soldering form of q_{ab} which ties the abstractly defined internal indices $A, B, \ldots$ to the tangent space of the 3-manifold Σ thereby making them spinorial indices. In particular, we have the result that each soldering form $\sigma^a{}_A{}^B$ picks out a unique torsion-free connection D acting on *both* internal and tensor indices.

The idea now is to regard $\sigma^a{}_A{}^B$ as the basic dynamical variable and q_{ab} as a derived, secondary object. The new (extended) configuration space $\mathcal{C}$ can therefore be defined as follows. Fix a soldering form $\mathring{\sigma}^a{}_A{}^B$ whose 3-metric e_{ab} is flat outside a compact set and consider soldering forms $\sigma^a{}_A{}^B$ such that

$$\sigma^a{}_A{}^B = \left(1 + \frac{M(\theta,\phi)}{r}\right)^2 \mathring{\sigma}^a{}_A{}^B + \mathcal{O}(1/r^2), \tag{5}$$

for some function M of the angles, where r is a radial coordinate of e_{ab}. The new configuration space $\mathcal{C}$ is the space of all soldering forms $\sigma^a{}_A{}^B$ satisfying the boundary condition Eq.5. (We shall see in the next section that it is important to use

$\sigma^a{}_A{}^B$ rather than $\sigma_{aA}{}^B$ as the new configuration variable.) The natural mapping ψ: $\psi(\sigma^a{}_A{}^B) := q_{ab}$ then sends the new configuration space $\mathcal{C}$ to the traditional configuration space C of asymptotically flat, positive-definite 3-metrics q_{ab}. Let ${}^1\sigma^a{}_A{}^B$ and ${}^2\sigma^a{}_A{}^B$ be two elements of $\mathcal{C}$ which project down to the same 3-metric q_{ab}. Then it follows from Eq.4 that the two soldering forms are related by a local $SU(2)$ transformation. Thus, the enlargement of the configuration space from C to $\mathcal{C}$ has been brought about because of the freedom to perform internal $SU(2)$ rotations. Indeed, while q_{ab} has six components per point of Σ, $\sigma^a{}_A{}^B$ has nine; the three new degrees of freedom correspond precisely to the three $SU(2)$ rotations.

The momentum conjugate to $\sigma^a{}_A{}^B$ is a density of weight one, $\widetilde{M}_{aA}{}^B$, whose index structure is opposite to that of $\sigma^a{}_A{}^B$ and whose fall-off is given by:

$$\mathrm{tr}\,(\widetilde{M}_a\sigma^a) = \mathcal{O}(1/r^3), \qquad \widetilde{M}_{aA}{}^B + \tfrac{1}{3}(\mathrm{tr}\,\widetilde{M}_m\sigma^m)\,\sigma_{aA}{}^B = \mathcal{O}(1/r^2). \tag{6}$$

The action of the cotangent vector $\widetilde{M}_{aA}{}^B$ on any tangent vector $(\delta\sigma)^a{}_A{}^B$ at a point $\sigma^a{}_A{}^B$ of $\mathcal{C}$ is given by

$$\widetilde{M}(\delta\sigma) := -\int_\Sigma d^3x\,\mathrm{tr}\,(\widetilde{M}_a\delta\sigma^a). \tag{7}$$

Note that the fall-off (Eq.6) is precisely such that the integral on the right hand side of Eq.7 converges.

The *extended phase space* Γ is the cotangent bundle over $\mathcal{C}$. Thus, a point of Γ is represented by a pair $(\sigma^a{}_A{}^B, \widetilde{M}_{aA}{}^B)$. The natural symplectic structure on Γ is given by $\Omega = \int d^3x \mathbb{d}\widetilde{M}_a{}^{AB} \wedge \mathbb{d}\sigma^a{}_{AB}$, or:

$$\{\sigma^a{}_{AB}\,,\, M_b{}^{CD}\} = \delta^b_a\,\delta^C_{(A}\delta^D_{B)}\delta^3(x,y) \tag{8}$$

where $\mathbb{d}$ denotes the infinite dimensional exterior derivative and $\{\ \ ,\ \ \}$, the Poisson bracket. Consequently, the Hamiltonian vector field X_f generated by any observable f is

$$X_f = \int_\Sigma d^3x\,\mathrm{tr}\left(\frac{\delta f}{\delta\sigma^a}\frac{\delta}{\delta\widetilde{M}_a} - \frac{\delta f}{\delta\widetilde{M}_a}\frac{\delta}{\delta\sigma^a}\right), \tag{9a}$$

and the Poisson bracket between any two observables f and g is

$$\{f,g\} = \int_\Sigma d^3x\,\mathrm{tr}\left(\frac{\delta f}{\delta\widetilde{M}_a}\frac{\delta g}{\delta\sigma^a} - \frac{\delta f}{\delta\sigma^a}\frac{\delta g}{\delta\widetilde{M}_a}\right). \tag{9b}$$

Next, let us examine the constraints. In the transition from C to $\mathcal{C}$, we have added three degrees of freedom to the configuration variables. Since the physical degrees haven't changed—we are still dealing with vacuum general relativity—we have three new constraints. From a Lagrangian viewpoint, these arise because (after 3+1-decomposition) the Lagrangian depends on $\sigma^a{}_A{}^B$ (or, equivalently, the triad) only through q_{ab}. From a Hamiltonian viewpoint, one expects the situation to be analogous to that in the Yang-Mills theory: since the internal rotations are "pure gauge", their generating functionals must vanish. This expectation turns out to be correct. The new constraints are:

$$\widetilde{C}_{ab} := -\,\mathrm{tr}\,(\widetilde{M}_{[a}\sigma_{b]}) \equiv \widetilde{M}_{[ab]} = 0, \tag{10a}$$

or

$$\widetilde{C}^{AB} := \sigma^a{}_C{}^{(A}\widetilde{M}_{aD}{}^{B)}\epsilon^{CD} = 0. \tag{10b}$$

To see this, let us compute the canonical transformation generated by these constraints. Given any trace-free, anti-Hermitian spinor field $\Lambda_A{}^B$ we can define a constraint function on Γ,

$$C_\Lambda(\sigma,\widetilde{M}) := \int_\Sigma d^3x\, \Lambda_A{}^B C_B{}^A. \tag{11}$$

This function can generate a canonical transformation only if it is differentiable on Γ. This differentiability requirement imposes a fall-off condition on the smearing field $\Lambda_A{}^B$: it has to fall off faster than $1/r$ at infinity. When this condition is met, the Hamiltonian vector field is given by:

$$X_\Lambda = \tfrac{1}{2}\int_\Sigma d^3x\,\mathrm{tr}\left([\Lambda,\sigma^a]\frac{\delta}{\delta\sigma^a} + [\Lambda,\widetilde{M}_a]\frac{\delta}{\delta\widetilde{M}_a}\right). \tag{12}$$

Thus, the infinitesimal changes in $\sigma^a{}_A{}^B$ and $\widetilde{M}_{aA}{}^B$ caused by the canonical transformation are precisely the (infinitesimal) rotations of the internal indices by $\Lambda_A{}^B$. Algebraic symmetries of $\Lambda_A{}^B$ imply that it is a generator of an $SU(2)$ transformation. Hence, the new constraints generate *small* $SU(2)$ transformations—i.e., transformations which are asymptotically identity and which belong to the connected component of identity of the group of local $SU(2)$ rotations—of the basic dynamical variables.

It follows, in particular, that, as in $SU(2)$ Yang-Mills theory, these three constraints form a first class set. We can therefore pass to the reduced phase space following the procedure given in appendix B. Recall that each point on the reduced phase space is an equivalence class of points on *the constraint surface*, where two points are regarded as equivalent if they differ by a canonical transformation generated by constraints. To see the structure of the resulting reduced phase space, it is convenient to define $\tilde{p}^{ab}$ by

$$\tilde{p}^{ab} := \widetilde{M}^{(ab)}, \tag{13}$$

so that the constraint surface is characterized by the equation $\tilde{p}^{ab} = \widetilde{M}^{ab}$. Then it follows that the reduced phase space can be coordinatized precisely by the (unconstrained) pairs $(q_{ab}, \tilde{p}^{ab})$ satisfying the canonical Poisson bracket relations. The definitions (Eqs.4 and 13) of q_{ab} and $\tilde{p}^{ab}$ imply that we can identify them with the 3-metric and its canonical conjugate momentum. Thus, the reduction of the phase space Γ *with respect to the new constraints* yields precisely the standard phase space T^*C, the cotangent bundle over C. This result is a succinct expression of the fact that the enlargement of the phase space in the transition from T^*C to Γ is brought about simply by introducing the freedom to perform internal, $SU(2)$ rotations.

In the standard Hamiltonian formulation, (see, e.g., Part II of [2]), the phase space T^*C of pairs $(q_{ab}, \tilde{p}^{ab})$ carries a vector and a scalar constraint. It is straightforward to "lift" these constraints to Γ since they are written in terms of q_{ab} and $\tilde{p}^{ab}$ which themselves admit lifts (given by Eqs.4 and 13). Hence, the remaining (or "old") constraints are

$$C_a(\sigma, \widetilde{M}) := -2\, q_{am} D_n \tilde{p}^{mn} = 0, \tag{14}$$

and

$$C(\sigma, \widetilde{M}) = -q^{1/2} R + q^{-1/2}(\tilde{p}^{ab}\tilde{p}_{ab} - \tfrac{1}{2}\tilde{p}^2) = 0, \tag{15}$$

where, however, q_{ab} and $\tilde{p}^{ab}$ are now regarded as secondary variables, defined in terms of $\sigma^a{}_A{}^B$ and $\widetilde{M}_{aA}{}^B$by Eqs.4 and 13 respectively. Thus, the constraint of Eq.10 reduces the new phase space to the old one, and Eqs.14 and 15 constrain those degrees of freedom in $(\sigma^a{}_A{}^B,\ \widetilde{M}_{aA}{}^B)$ which are insensitive to internal rotations so as to satisfy the initial value equations of Einstein's theory. The canonical transformations generated by the constraints in Eqs.14 and 15 continue to retain their interpretation. Thus, the motions generated on the phase space by the vector constraint (Eq.14) correspond precisely to changes in $(q_{ab}, \tilde{p}^{ab})$ resulting from spatial diffeomorphisms on Σ while those generated by by the scalar constraint (Eq.15) correspond to "time-evolution" of the initial data $(q_{ab}, \tilde{p}^{ab})$. In all we now have $3+3+1 = 7$ constraints. The configuration variable $\sigma^a{}_A{}^B$ has nine components per space point. Thus, we have two degrees of freedom per space point. Indeed, it is straightforward to verify that the phase space obtained by reduction of $\mathbf{T}$ with respect to *all* constraints is *naturally* isomorphic to that obtained by reduction of T^*C with respect to Eqs.14 and 15. Thus, our enlarged phase space description is indeed equivalent to the standard one.

To summarize, the extended phase space Γ consists of pairs $(\sigma^a{}_A{}^B, \widetilde{M}_{aA}{}^B)$ satisfying the boundary conditions of Eqs.5 and 6. The Poisson brackets are given by Eq.9. There are seven constraints. Six of them, Eqs.10 and 14, are linear in momenta while the seventh, Eq.15, is quadratic.

We conclude this section with two remarks.

i) Although on the phase space $\widetilde{\Gamma}$, obtained by the reduction with respect only to the new constraints Eq.10, q_{ab} and $\widetilde{p}^{ab}$ *are* canonically conjugate, their Poisson brackets on Γ are more complicated away from the surface defined by Eq.10. At a general point $(\sigma^a{}_A{}^B, \widetilde{M}_{aA}{}^B)$ of Γ, they are given by:

$$\begin{aligned} \{q_{ab}(x), q_{cd}(y)\} &= 0, \\ \{q_{ab}(x), \widetilde{p}^{cd}(y)\} &= -2\delta_a{}^{(c}\delta_b{}^{d)}\,\delta^3(x,y), \\ \{\widetilde{p}^{ab}(x), \widetilde{p}^{cd}(y)\} &= \tfrac{1}{2}\delta^3(x,y)\left(\widetilde{M}^{[ca]}q^{bd} + \widetilde{M}^{[da]}q^{bc} + \widetilde{M}^{[cb]}q^{ad} + \widetilde{M}^{[db]}q^{ac}\right). \end{aligned} \tag{16}$$

It is only when Eq.10 is satisfied that the $\widetilde{p}^{ab}$ commute with each other. (The minus sign in the second Poisson bracket arises because σ^a –rather than σ_a– is now the configuration variable.)

ii) In the above enlargement, we first extended the configuration space from C to $\mathcal{C}$ by introducing the freedom to perform internal $SU(2)$ rotations and then removed this gauge freedom by imposing the new constraints (Eq.10). The passage from q_{ab} to $\sigma^a{}_A{}^B$ is essential for the introduction of the new variables. However, could we not have avoided the introduction and the subsequent elimination of the new degrees of freedom by a "gauge fixing procedure" which associates to each q_{ab} a canonical $\sigma^a{}_A{}^B$? Recall that there is a natural projection mapping ψ from $\mathcal{C}$ to C which sends each soldering form $\sigma^a{}_A{}^B$ to a 3-metric. Thus, $\mathcal{C}$ may be regarded as a fiber bundle over C, each fiber providing a natural realization of $SU(2)$. The question therefore is: Does $\mathcal{C}$ admit natural horizontal cross-sections? If it does, we could have used such a cross-section for the new configuration space. This space would be isomorphic to C so that there would be no additional gauge. At the same time, being a subspace of $\mathcal{C}$, it would provide soldering forms and not just metrics. To analyze this issue, let us first consider tangent vectors $\delta\sigma^a{}_A{}^B$ at any point $\sigma^a{}_A{}^B$ of $\mathcal{C}$. Using the "background" $\sigma^a{}_A{}^B$, we can convert $\delta\sigma^a{}_A{}^B$ to a tensor field $\delta\sigma^{ab}$. The symmetric part of this tensor field gives us the variation of the metric q_{ab} caused by $\delta\sigma^a{}_A{}^B$, while the skew symmetric part is "pure gauge". Thus, we can divide the tangent vector $\delta\sigma^{ab}$ into a horizontal part $\delta\sigma^{(ab)}$ and a vertical part $\delta\sigma^{[ab]}$. We are therefore led to ask: Are the horizontal subspaces integrable? If so, their integral manifolds would provide the required cross-sections. Unfortunately, however, the integrability condition fails to hold and we are forced to take the route of first enlarging the configuration space and then eliminating the unwanted degrees of freedom by introducing additional constraints.

4 New variables

The constraints, Eqs.14 and 15, have remained intact in the transition from T^*C to Γ; the addition of new degrees of freedom does not, by itself, simplify the constraints. The key step in the simplification is the next one: introduction of new canonical variables. The extension to Γ is, however, necessary because these variables cannot be defined on T^*C. For simplicity of presentation, we shall first state the main result and then discuss the steps leading to it.

Fix a point $(\sigma^a{}_A{}^B, \widetilde{M}_{aA}{}^B)$ of Γ. Just as in Riemannian geometry the metric selects a unique torsion-free connection on tensors, now $\sigma^a{}_A{}^B$ provides us with a unique (torsion-free) connection D that acts on both tensor and internal indices. We now introduce a new derivative operator, $\mathcal{D}$, in which the information about both the soldering form $\sigma^a{}_A{}^B$ and the momentum $\widetilde{M}_{aA}{}^B$ is coded just in the right way so that the constraints, Eqs.10, 14 and 15, can be expressed in terms of it in a simple way. The definition was motivated by a certain connection that Sen [8] had introduced to simplify the discussion of the initial value formulation of Einstein's theory. (The Sen connection is discussed in appendix A.) Set:

$$\mathcal{D}_a\lambda_{bM} := D_a\lambda_{bM} - \tfrac{i}{\sqrt{2}}\,\Pi_{aM}{}^N\lambda_{bN}, \tag{17}$$

where $\Pi_{aM}{}^N$ is given by:

$$\Pi_{aM}{}^N := q^{-1/2}\Big(\widetilde{M}_{aM}{}^N + \tfrac{1}{2}\,(\mathrm{tr}\,\widetilde{M}_b\sigma^b)\,\sigma_{aM}{}^N\Big), \tag{18a}$$

or

$$\widetilde{M}_{aM}{}^N := q^{1/2}\Big(\Pi_{aM}{}^N + (\mathrm{tr}\,\Pi_b\sigma^b)\,\sigma_{aM}{}^N\Big). \tag{18b}$$

Thus, $\mathcal{D}$ and D have the same action on tensors. Their action on internal indices is related by Eq.3, the role of $C_{aA}{}^B$ being played by $\Pi_{aA}{}^B$. Note that $\Pi_{aM}{}^N$ is related to $\widetilde{M}_{aA}{}^B$ in the same way that the extrinsic curvature K^{ab} is related to $\widetilde{p}^{ab}$. In fact, when the Eq.10 is satisfied, Π^{ab} reduces to the extrinsic curvature K^{ab}.

Since $\Pi_{ab} := -\,\mathrm{tr}\,(\Pi_a\sigma_b)$ is not symmetric in a and b—except on the constraint surface given by Eq.10—the connection defined above *is not the same* as the Sen connection introduced in [8]. Why do we not simply use the symmetric part of Π_{ab} in Eq.17. While this strategy seems attractive at first, it ruins certain crucial Poisson bracket relations. Consequently, to construct a useful Hamiltonian description and especially for the passage to quantum theory to be manageable, it is important that Eq.17 be used as is, *without* symmetrizing the second term on the RHS.

As in gauge theories, it is convenient to work with connection 1-forms $A_{aA}{}^{B}$ in place of the derivative operators. Let us therefore fix a fiducial derivative operator ∂. For simplicity, we shall assume that ∂ commutes with the Hermitian conjugation operation, $\partial_a \lambda^\dagger_B = (\partial_a \lambda_B)^\dagger$, and has zero internal curvature, $\partial_{[a}\partial_{b]}\lambda_A = 0$. (These assumptions are not essential for the formalism to go through; they will however, simplify the explicit calculations that will lead us to the final results.) Then, we can set

$$\mathcal{D}_a \lambda_M = \partial_a \lambda_M + A_{aM}{}^{N} \lambda_N. \tag{19a}$$

If we denote the spin connection 1-form of D by $\Gamma_{aM}{}^{N}$, we can rewrite Eq.17 as:

$$A_{aM}{}^{N} = \Gamma_{aM}{}^{N} - \tfrac{i}{\sqrt{2}} \Pi_{aM}{}^{N}. \tag{19b}$$

Our new variables will be $A_{aA}{}^{B}$ *and* $\tilde{\sigma}^a{}_A{}^B := q^{1/2}\sigma^a{}_A{}^B$. Since $A_{aA}{}^{B}$ has density weight zero, and since one of the canonically conjugate pair of variables has to be a density, it is not surprising that we had to "densitize" $\sigma^a{}_A{}^B$ to obtain the second variable. Results on the Sen connection [8] (see also Appendix A)suggest that the constraints, Eqs.10, 14 and 15 will simplify when expressed in terms of $\tilde{\sigma}^a{}_A{}^B$ and $A_{aA}{}^{B}$.

To see that these variables are viable, let us compute the Poisson brackets between them. The first Poisson bracket follows trivially from the expression of the bracket (see Eq.9*b*):

$$\{\tilde{\sigma}^a{}_A{}^B(x), \tilde{\sigma}^b{}_M{}^N(y)\} = 0. \tag{20a}$$

Next, a straightforward calculation shows that $\tilde{\sigma}^a{}_A{}^B(x)$ and $\Pi_{bM}{}^N(y)$ are canonically conjugate. Since $\Gamma_{bM}{}^N$ is a function only of $\sigma^a{}_A{}^B$, it follows that the Poisson bracket between $\tilde{\sigma}^a{}_A{}^B$ and $A_{aA}{}^B$ is the same as that between $\tilde{\sigma}^a{}_A{}^B$ and $\frac{i}{\sqrt{2}}\Pi_{bM}{}^N$. Thus, we have:

$$\{A_a{}^{AB}(x), \tilde{\sigma}^m{}_{MN}(y)\} = \tfrac{i}{\sqrt{2}}\, \delta_M{}^{(A}\delta_N{}^{B)}\delta_a{}^m\, \delta^3(x,y). \tag{20b}$$

(Note that $\sigma^a{}_A{}^B$ does not have a simple, c-number Poisson bracket with $\Pi_{bM}{}^N$. Hence the bracket between $\sigma^a{}_A{}^B$ and $A_{aA}{}^B$ also fails to be a c-number. It is because of this that we had to "densitize" $\sigma^a{}_A{}^B$.)

Finally, we have to compute the bracket between $A_{aA}{}^B(x)$ and $A_{aM}{}^N(y)$. For this, we note that $(\sigma^a{}_A{}^B,\ \widetilde{M}_{aA}{}^B) \mapsto (\widetilde{\sigma}^a{}_A{}^B,\ \widetilde{\Pi}_{aA}{}^B)$ is a canonical transformation, and that the transformation $(\widetilde{\sigma}^a{}_A{}^B,\ \widetilde{\Pi}_{aA}{}^B) \mapsto (\widetilde{\sigma}^a{}_A{}^B,\ A_{aA}{}^B)$ is of the type $(q^{\underline{i}}, p_{\underline{i}}) \mapsto (q^{\underline{i}}, p'_{\underline{i}} := -ip_{\underline{i}} + f_{\underline{i}}(q))$. It is easy to verify that the last of these transformations is a canonical transformation—one only needs to check that $p'_{\underline{i}}$ and $p'_{\underline{j}}$ commute—if and only if $f_{\underline{i}}$ is (locally) a gradient. Thus, to show that the $A_{aA}{}^B$ commute, it suffices to show that $\Gamma_{aA}{}^B$ is a gradient of a generating function G of $\widetilde{\sigma}^a{}_A{}^B$.

To find the required function $G(\widetilde{\sigma}^a)$, following [1], let us consider first the variation $\delta\Gamma_{aA}{}^B$ of $\Gamma_{aA}{}^B$ resulting from an infinitesimal variation of $\widetilde{\sigma}^a{}_A{}^B$. Let $\underset{\leftarrow}{\nabla}_a$ be the torsion free derivative operator which is compatible with the 3-metric q_{ab} and has the same action as ∂_a on spinors. Then, the spin connection of $\widetilde{\sigma}^a{}_A{}^B$ is given by $\Gamma_{aA}{}^B = -\frac{1}{2}\underset{\sim}{\sigma}_b{}^{BC}\underset{\leftarrow}{\nabla}_a\widetilde{\sigma}^b{}_{AC}$. It follows that, under a change $\delta\widetilde{\sigma}^a$ of the soldering form, the spin-connection changes by:

$$\delta\Gamma_{aA}{}^B = -\frac{1}{2}\Big[-\underset{\sim}{\sigma}_b{}^{MN}\underset{\sim}{\sigma}_d{}^{BC}(\delta\widetilde{\sigma}^d{}_{MN})\underset{\leftarrow}{\nabla}_a\widetilde{\sigma}^b{}_{AC} + \underset{\sim}{\sigma}_b{}^{BC}(\delta\Gamma_{ac}{}^c\widetilde{\sigma}^b{}_{AC} - \delta\Gamma_{ac}{}^b\widetilde{\sigma}^c{}_{AC} + \underset{\leftarrow}{\nabla}_a\delta\widetilde{\sigma}^b{}_{AC})\Big], \tag{21}$$

where $\Gamma_{ab}{}^c$ is the connection 1-form of $\underset{\leftarrow}{\nabla}_a$. Now, using the identity (Eq.3, chapter 5) satisfied by $SU(2)$ soldering forms,

$$\underset{\sim}{\sigma}_a{}^{M(A}\widetilde{\sigma}^b{}_M{}^{B)} = \frac{1}{\sqrt{2}}q_{ad}\widetilde{\eta}^{dbc}\underset{\sim}{\sigma}_c{}^{AB}, \tag{22}$$

we find that

$$\mathrm{tr}(\delta\Gamma_a\widetilde{\sigma}^a) = \frac{1}{2\sqrt{2}}\partial_a\Big(\widetilde{\eta}^{ade}q_{be}\,\mathrm{tr}(\underset{\sim}{\sigma}_d\,\delta\widetilde{\sigma}^b)\Big), \tag{21b}$$

which is a total divergence of a vector density. Therefore, if we set $G(\widetilde{\sigma}^a)$ to be

$$G(\widetilde{\sigma}^a) := \int_\Sigma d^3x\,\mathrm{tr}\,(\Gamma_a\widetilde{\sigma}^a), \tag{23}$$

we have, as required,

$$\frac{\delta G}{\delta\widetilde{\sigma}^a{}_A{}^B} = \Gamma_{aB}{}^A. \tag{24}$$

Consequently, we have the desired Poisson bracket relation

$$\{A_{aA}{}^B(x), A_{bM}{}^N(y)\} = 0. \tag{20c}$$

This relation would *not* have held if we had used the Sen connection, i.e., if we had used the symmetric part of Π_{ab} in the definition Eq.17 of $\mathcal{D}$.

Let us summarize. First, we make a straightforward canonical transformation $(\sigma^a{}_A{}^B,\ \widetilde{M}_{aA}{}^B) \mapsto (\widetilde{\sigma}^a{}_A{}^B,\ \Pi_{aA}{}^B)$. All this does is to give a density weight to $\sigma^a{}_A{}^B$ and remove an appropriate trace factor from $\widetilde{M}_{aA}{}^B$. In a second step, we mix up the configuration and momentum variables genuinely. This is achieved by leaving the configuration variable $\widetilde{\sigma}^a{}_A{}^B$ as it is, and adding to the momentum variable, $\Pi_{aA}{}^B$, the gradient of the function $G(\widetilde{\sigma})$. The resulting pair, satisfying the canonical commutation relations (Eqs.20), is $(\widetilde{\sigma}^a{}_A{}^B,\ A_{aA}{}^B)$. Thus, we have arrived at the new variables by composing two canonical transformations.[4]

At this stage, it is useful to return to the analogy with the harmonic oscillator. Consider the phase space of the N-dimensional harmonic oscillator, coordinatized by $(q^{\underline{i}}, p_{\underline{i}})$. We are to think of these as analogs of the real variables $(\widetilde{\sigma}^a{}_A{}^B,\ \Pi_{aA}{}^B)$ on Γ. One often introduces a complex coordinate $z_{\underline{i}} := (m\omega)^{1/2} q_{\underline{i}} - i\,(m\omega)^{-1/2} p_{\underline{i}}$, where m and ω are the mass and the frequency of the oscillator. $A_{aA}{}^B$ is analogous to $z_{\underline{i}}$. In the case of the oscillator, the parameters m and ω enable us to form dimensionally meaningful linear combinations of the real variables $q^{\underline{i}}$ and $p_{\underline{i}}$. In the gravitational case, the only available constant, G (which has been set equal to one here), does not enable us to form a dimensionally meaningful linear combination of $\sigma^a{}_A{}^B$ and $\widetilde{M}_{aA}{}^B$. However, since $\Pi_{aA}{}^B$ has the same dimensions as the derivatives of $\widetilde{\sigma}^a{}_A{}^B$, it *is* possible to add $\Pi_{aA}{}^B$ and the spin connection forms $\Gamma_{aA}{}^B$ of $\widetilde{\sigma}^a{}_A{}^B$. In this sense, $(\widetilde{\sigma}^a{}_A{}^B,\ A_{aA}{}^B)$ is as close to the pair $(q^{\underline{i}}, z_{\underline{i}})$ as is dimensionally possible. This analogy also brings out the fact that, although $(\widetilde{\sigma}^a{}_A{}^B,\ A_{aA}{}^B)$ have canonical Poisson bracket relations, they are not quite "canonically conjugate" in the traditional sense, since $\widetilde{\sigma}^a{}_A{}^B$ is "real", while $A_{aA}{}^B$ is "complex". We wish to emphasize, however, that it *is* a viable pair of basic variables: in the case of the harmonic oscillator, for example, one *can* carry out the entire Hamiltonian dynamics as well as the passage to quantum theory in terms of $(q^{\underline{i}}, z_{\underline{i}})$. (For details, see chapters 8 and 10). The key reason for choosing $(\widetilde{\sigma}^a{}_A{}^B,\ A_{aA}{}^B)$ as the basic variables is, as we will see, that the field equations are particularly simple in terms of them.

To conclude this section, let us recast the constraints in terms of the new variables. Let us begin with the new constraint, Eq.10. Using the definition Eq.17 of the connection $\mathcal{D}$, it is easy to show that:

$$\mathcal{D}_a \widetilde{\sigma}^a{}_A{}^B = \sqrt{2}i\, M_{[ab]} \sigma^a{}_A{}^M \sigma^b{}_M{}^B. \tag{25}$$

Hence, Eq.10 is completely equivalent to

$$\mathcal{D}_a \widetilde{\sigma}^a{}_A{}^B = 0. \tag{10'}$$

4 Note that the boundary conditions imposed on $\sigma^a{}_A{}^B$ and $\widetilde{M}_{aA}{}^B$ imply that the pair $(\widetilde{\sigma}^a{}_A{}^B,\ A_{aA}{}^B)$ satisfies the analogs of Eqs.5 and 6, where the only change consists of replacement of $\sigma^a{}_A{}^B$ by $\widetilde{\sigma}^a{}_A{}^B$ and of $\widetilde{M}_{aA}{}^B$ by $A_{aA}{}^B$.

Note that, since the divergence of a vector density of weight one is independent of the choice of the derivative operator, one can expand Eq.10′ knowing only the action

$$\mathcal{D}_a\lambda_M = \partial_a\lambda_M + A_{aM}{}^N\lambda_N \tag{26}$$

of $\mathcal{D}$ on internal indices. We have:

$$\mathcal{D}_a\tilde{\sigma}^a{}_A{}^B = \partial_a\tilde{\sigma}^a{}_A{}^B + [A_a, \tilde{\sigma}^a]_A{}^B = 0. \tag{10''}$$

Thus, the Gauss constraint, Eq.10, has been re-expressed in terms of $(\tilde{\sigma}^a{}_A{}^B,\ A_{aA}{}^B)$ only. To re-express Eqs.14 and 15 let us first calculate the curvature $F_{abM}{}^N$ of $\mathcal{D}$ in terms of the curvature of q_{ab} and the extrinsic curvature K^{ab}. Using the definition (Eq.17) of $\mathcal{D}$ in terms of D and $\Pi_{aM}{}^N$ it is straightforward to show that

$$\begin{aligned} F_{abM}{}^N\lambda_N &:= 2\,\mathcal{D}_{[a}\mathcal{D}_{b]}\lambda_M \\ &= \left(R_{abM}{}^N - \Pi_{[aM}{}^P\Pi_{b]P}{}^N - \sqrt{2}i\, D_{[a}(\Pi_{b]M}{}^N)\right)\ \lambda_N, \end{aligned} \tag{27}$$

where $R_{abM}{}^N$ is the internal or spinorial curvature of D. Let us use $\sigma^a{}_A{}^B$ (and its inverse) to convert spinorial indices to tensorial ones. Then, we have :

$$\begin{aligned} \mathrm{tr}\,(\sigma^a F_{ab}) &= \tfrac{1}{2\sqrt{2}}\,(\Pi_{am}\Pi_{bn} - \Pi_{bm}\Pi_{an})\,\epsilon^{mna} + \tfrac{i}{\sqrt{2}}\,D^a(\Pi_{ba} - \Pi q_{ba}) \\ &\approx \tfrac{i}{\sqrt{2}}\,D^a(K_{ab} - Kq_{ab}), \end{aligned} \tag{28}$$

and

$$\begin{aligned} \mathrm{tr}\,(\sigma^a\sigma^b F_{ab}) &= \tfrac{1}{2}\,(R + \Pi^2 - \Pi_{ab}\Pi^{ba}) + i\epsilon^{abc}D_a\Pi_{bc} \\ &\approx \tfrac{1}{2}\,(R + K^2 - K_{ab}K^{ab}), \end{aligned} \tag{29}$$

where $\approx$ *stands for equality modulo the Gauss constraint*, Eq.10. Thus, the constraints given by Eqs.14 and 15 can be respectively written as:

$$C_a(\sigma, M) = -2\sqrt{2}i\,\mathrm{tr}\,(\tilde{\sigma}^b F_{ab}) = 0, \tag{14'}$$

and

$$C(\sigma, M) = -2q^{-1/2}\,\mathrm{tr}\,(\tilde{\sigma}^a\tilde{\sigma}^b F_{ab}) = 0. \tag{15'}$$

Again, note that the right hand sides of these equations are functionals of $\tilde{\sigma}^a{}_A{}^B$ and $A_{aA}{}^B$ only. Thus, we have re-expressed all constraints in terms of the new variables.

To summarize, as reported in chapter 2, the set of Einstein constraints on Γ becomes simply:

$$\mathcal{D}_a\tilde{\sigma}^a{}_A{}^B = 0; \qquad \mathrm{tr}\,(\tilde{\sigma}^a F_{ab}) = 0; \qquad \text{and} \qquad \mathrm{tr}\,(\tilde{\sigma}^a\tilde{\sigma}^b F_{ab}) = 0. \tag{30}$$

These are precisely the spinorial version of the triad constraints we obtained in chapters 3 and 4 by performing the Legendre transform of the self dual action.

A number of remarks are in order.

i) Note that, in their final form, the constraints involve only our basic variables, $\tilde{\sigma}^a{}_A{}^B$ and $A_{aA}{}^B$ and their ∂ derivatives. In particular, as pointed out in chapter 3, we do not need to raise or lower the tensor indices on these fields; the inverse of $\tilde{\sigma}^a{}_A{}^B$ never enters the constraints. In fact, the constraints are at worst quadratic in each of the basic variables. This is a significant improvement because the dependence of the constraints on q_{ab} and $\tilde{p}^{ab}$ was rather complicated and in fact *non-polynomial.* Note, in particular, that the non-polynomial scalar curvature term in Eq.15 —which has been at the root of many a problem in canonical quantization—has disappeared in the transition to Eq.15′. This came about because the connection $A_{aA}{}^B$ contains just the right combination of the metric and its momentum.

ii) The form of the constraints also brings out a reason behind our choice of $\tilde{\sigma}^a{}_A{}^B$ as the configuration variable rather than its inverse, which is more closely related to the usual configuration variable q_{ab}. A second reason is that, since the connection 1-forms $A_{aA}{}^B$ naturally occur as covectors, the conjugate variable should have a *contravariant* vector index.

iii) An important feature of Eqs.10′, 14′ and 15′ is that they involve the action of $\mathcal{D}$ *only on internal indices.* As noted above, since $\tilde{\sigma}^a{}_A{}^B$ is a vector density of weight one, the divergence of $\tilde{\sigma}^a{}_A{}^B$ does *not* involve the Christoffel symbols of q_{ab}. Similarly, the definition of $F_{abM}{}^N$ involves the action of $\mathcal{D}$ *only* on internal indices. Thus, although initially $\mathcal{D}$ was defined *both* on tensor and internal indices (Eq.17), in the final analysis, it is only the action on internal indices that matters; to operate on tensor indices, we can choose *any* torsion-free extension of Eq.19 to tensors.

iv) In terms of q_{ab} and $\tilde{p}^{ab}$, the vector constraint is linear in momentum, while the scalar constraint is quadratic. If we regard $A_{aA}{}^B$ as the configuration variable and $\tilde{\sigma}^a{}_A{}^B$ as its momentum, we see that the new vector constraint is again linear in the new momentum and the new scalar constraint, quadratic. It is curious that this general structure has remained intact in spite of the fact that the momentum has "flipped"; the new momentum $\tilde{\sigma}^a{}_A{}^B$ is geometrically analogous to the old configuration variable q_{ab}.

v) Note, however, the absence in the new scalar constraint (Eq.15′) of the analog of the of the scalar curvature –or, the "potential"– term we had in the old scalar constraint (Eq.15). Thus, the new scalar constraint has only a "kinetic" term and resembles, in its form, the strong coupling limit of the old scalar constraint. This is a significant simplification. There is, however, a price. The "supermetric" –i.e., the coefficient multiplying the quadratic combination of momenta– in the old variables is *ultralocal* in the sense that it depends only on the configuration variables q_{ab} and not on their derivatives. The analogous supermetric in the new scalar constraint, on the other hand, is only local; since it is given by the curvature, $F_{abA}{}^{B}$, it depends not only on the new configuration variable $A_{aA}{}^{B}$ but also its derivatives. In homogeneous cosmologies, however, all spatial derivatives can be traded for structure constants of the spatial isometry group and the new supermetric again becomes ultralocal. Thus, in these mini-superspaces, the new scalar constraint has all the nice features of the old scalar constraint without any of its drawbacks. Therefore, I expect that the use of new variables will considerably simplify the discussion of both classical and quantum homogeneous cosmologies.

vi) One can in fact combine the scalar and the vector constraints into a single equation $(\tilde{\sigma}^a\tilde{\sigma}^b\, F_{ab})_A{}^B = 0$. The trace of this equation gives us the scalar constraint while (assuming that the soldering form is non-degenerate, the symmetric part in the spinorial indices gives us the vector constraint. Furthermore, since free indices of *all* constraints are now "internal"; the canonical transformations they generate operate only in the internal space. One might hope that this fact would simplify the issue of time and dynamics in general relativity. Unfortunately it does not because the Poisson structure of these combined constraints is complicated. (This issue was investigated in detail by Joseph Romano and Peter Thomi.)

vii) It was recently pointed out that under an assumption of genericity, a general solution to the vector and the scalar constraints can be given [9]. Let us suppose that $\tilde{B}^a{}_A{}^B := \tilde{\eta}^{abc} F_{bcA}{}^B$ is a non-degenerate mapping from the space of vector densities to the space of trace-free Hermitian spinors. Then, we can express the soldering form $\tilde{\sigma}^a{}_{AB}$ as a linear combination, $\tilde{\sigma}^a{}_{AB} = M_{AB}{}^{MN}\tilde{B}^a{}_{MN}$ of $\tilde{B}^a_{MN}$ and substitute this expression in the scalar and the vector constraints. These constraints then impose only an algebraic relation on $M_{AB}{}^{MN}$ which can be solved easily.

viii) Finally, let us ask: What is the geometrical interpretation of the new variables? $\tilde{\sigma}^a{}_A{}^B$ is of course the square root of the (density-weighted) 3-metric. The interpretation of $A_{aA}{}^B$ is less immediate. It is not difficult to show [5,8] (see also section 5 in appendix A) that, in any solution to the field equations,

$A_{aA}{}^B$ is the natural potential for the self dual part of the 4-dimensional Weyl curvature. More precisely, we have the following:

$$\mathrm{tr}\,(\widetilde{\sigma}^c F_{ab})\widetilde{\eta}^{abd} = -\sqrt{2}q\,(E^{cd} - iB^{cd}), \tag{31}$$

where $\widetilde{\eta}^{abc}$ is the natural (metric-independent) totally skew density of weight one, and where E^{ab} and B^{ab} are, respectively, the electric and the magnetic parts of the Weyl tensor relative to Σ. Therefore the use of $(\widetilde{\sigma}^a{}_A{}^B, A_{aA}{}^B)$ simplifies the analysis of half-flat 4-metrics considerably [10].

References

[1] A. Ashtekar, Talk given at the *Eighth Workshop on Grand Unification* (1988); J.L. Friedman and I. Jack, Phys. Rev. **D37**, 3495 (1988); H. Henneaux, J.E. Nelson and C. Schomblond, Phys. Rev. **D39**, 437 (1989).

[2] A.Ashtekar (with invited contributions), *New Perspectives in Canonical Gravity* (Bibliopolis, Naples, 1988).

[3] H. Kodama, Phys. Rev. **D42**, 2548 (1990).

[4] R. Penrose and W. Rindler, *Spinors and space-time*, vol. 1 (Cambridge University Press 1984).

[5] A. Ashtekar, Phys. Rev. Lett. **57**, 2244 (1986); Phys. Rev. **D36**, 1587 (1987).

[6] A. Ashtekar, P. Mazur and C.G. Torre, Phys. Rev. **D36**, 2955 (1987).

[7] A. Ashtekar, G.T. Horowitz and A. Magnon, Gen. Rel. & Grav. **14**, 411 (1982).

[8] A. Sen, J. Math. Phys. **22**, 1781 (1981); Phys. Lett. **119B**, 89 (1982).

[9] R. Capovilla, T. Jacobson and J. Dell, Phys. Rev. Lett. **63**, 2325 (1989).

[10] A. Ashtekar, T. Jacobson and L. Smolin, Comm. Math. Phys. **115**, 631 (1988); L. Mason and E.T. Mason, Comm. Math. Phys. **121**, 659 (1989).

7 CONSTRAINT ALGEBRA, HAMILTONIANS AND DYNAMICS

1 Introduction

In this chapter, we continue the discussion of the Hamiltonian structure of the theory. First we consider the canonical transformations generated by constraints. Transformations generated by six of the seven constraints have a direct geometrical interpretation which trivializes the task of calculating their Poisson algebra. A direct calculation is needed only for the evaluation of the Poisson bracket between two scalar constraints. Here, the use of new variables simplifies the task since these constraints are at worst quadratic in each of the basic canonical variables $\tilde{\sigma}^a{}_A{}^B$ and $A_{aA}{}^B$. Next, we investigate dynamics. The Hamiltonian generating asymptotic time-translations is obtained by just adding a surface term to the volume integral of the scalar constraint. On physical states, constraints are satisfied, the volume term vanishes and the numerical value of the Hamiltonian is given just by the surface term. This is the ADM energy.

In this chapter, we continue to use the convention, notation and boundary conditions that were introduced in chapter 6.

2 Constraint algebra

Let us begin by analyzing the canonical transformations generated by the constraints. We have seen that, apart from an overall factor, $M_{[ab]}$ equals $\mathcal{D}_a\tilde{\sigma}^a{}_A{}^B$. Since the canonical transformations generated by $M_{[ab]}$ were seen to be local $SU(2)$ rotations on internal indices, it follows that these are also the canonical transformations generated by $\mathcal{D}_a\tilde{\sigma}^a{}_A{}^B$. In fact, this interpretation has become more transparent in the transition from Eq.6.10 to Eq.6.10′. For, if one regards $\tilde{\sigma}^a{}_A{}^B$ as the momentum conjugate to $A_{aA}{}^B$ one sees that Eq.6.10′ has exactly the same form as the Gauss constraint of Yang-Mills theory, $\mathbf{D}_a\mathbf{E}^a = 0$, the well-known generator of local gauge rotations (see ,e.g., appendix B). Therefore, Eq.6.10′ will be referred to as the *Gauss constraint of general relativity.* For later convenience, let us introduce some notation. Given any $SU(2)$ generator $N_A{}^B$, i.e., an anti-Hermitian traceless field $N_A{}^B$, let us set

$$C_N(\tilde{\sigma}, A) = -\sqrt{2}i \int_\Sigma d^3x \, \mathrm{tr}\,(N \mathcal{D}_a \tilde{\sigma}^a). \tag{1}$$

One can check that this is a differentiable function on Γ if and only if the smearing function $N_A{}^B$ tends to zero at infinity faster than $1/r$.[1] If it is differentiable, it can generate canonical transformations. A simple calculation yields the Poisson brackets

$$\begin{aligned} \{C_N, \tilde{\sigma}^m{}_A{}^B\} &= [N, \tilde{\sigma}^m]_A{}^B \\ \{C_N, A_{mA}{}^B\} &= -\mathcal{D}_m N_A{}^B, \end{aligned} \tag{2}$$

confirming the interpretation of Eq.6.10. Since the generating function $N_A{}^B$ must vanish at infinity, it follows that, as in Yang-Mills theory, the Gauss law constraint of general relativity generates "small gauge transformations" ; i.e., elements of the connected component of the group of asymptotically identity local gauge transformations.

Next, let us consider the vector constraint, Eq.6.14′. Since $-i\sqrt{2}\,\mathrm{tr}\,(\tilde{\sigma}^b F_{ab})$ equals $q_{ac} D_b \tilde{p}^{bc}$ *only* when Eq.6.10′ is satisfied, one does not expect Eq.6.14′ by itself to generate spatial diffeomorphisms. This expectation is correct. To obtain the generator of spatial diffeomorphisms, one has to add to Eq.6.14′ a multiple of Eq.6.10′ (with a "q-number coefficient"). Given a shift field N^a, let us therefore set

$$C_{\vec{N}}(\tilde{\sigma}, A) := -\sqrt{2}i \int_\Sigma d^3x \, N^a \, \mathrm{tr}\,(\tilde{\sigma}^b F_{ab} - A_a \mathcal{D}_b \tilde{\sigma}^b). \tag{3}$$

As in the case of the Gauss law constraint, this function, $C_{\vec{N}}(\tilde{\sigma}, A)$, is differentiable on Γ if and only if N^a tends to zero at infinity as $1/r$. Given such a shift field, one can use the fundamental Poisson bracket relations (Eq.6.20) and the definition of $F_{abA}{}^B$ to show that :

$$\begin{aligned} \{C_{\vec{N}}, \tilde{\sigma}^a{}_A{}^B\} &= \mathcal{L}_{\vec{N}} \tilde{\sigma}^a{}_A{}^B, \\ \{C_{\vec{N}}, A_{aA}{}^B\} &= \mathcal{L}_{\vec{N}} A_{aA}{}^B, \end{aligned} \tag{4}$$

where $\mathcal{L}_{\vec{N}}$ is the Lie derivative, with respect to the vector field N^a, which treats internal indices as scalars. (For example, we have: $\mathcal{L}_{\vec{N}} A_{aA}{}^B = N^m \partial_m A_{aA}{}^B +$

---

1 While taking the functional derivative of $C_N(\tilde{\sigma}, A)$, with respect to $\tilde{\sigma}^a{}_A{}^B$,one encounters a surface term. Unless this term is zero, the functional derivative yields a genuine distribution on Σ. Cotangent vectors to the configuration space $\mathcal{C}$, on the other hand, are *not* distributions. Thus, unless the surface-term vanishes, the function $C_N(\tilde{\sigma}, A)$ is *not* differentiable on Γ. The fall-off of $N_A{}^B$ is needed to ensure that the surface term vanishes.

$A_{mA}{}^B \partial_a N^m$.) Thus, $C_{\vec{N}}$ does indeed generate diffeomorphisms along the vector field N^a. Since the vector field N^a is required to vanish at infinity, the diffeomorphisms it generates are all asymptotically identity. Thus, $C_{\vec{N}}$ generates 'small diffeomorphisms'. We shall refer to $C_{\vec{N}}$ as the *diffeomorphism constraint* thereby distinguishing it from the left side of Eq.6.14′ which is referred to as the *vector constraint.* Note, however, that $C_{\vec{N}}$ is *not* a (internal) gauge-invariant function on Γ. We shall see that this reflects the fact that the action of the diffeomorphism group on the 3-manifold Σ cannot be unambiguously lifted to generalized tensors; given a lifting, one can obtain another by composing it with a local rotation on internal indices.

Finally, let us consider the scalar constraint, Eq.6.15′. Set

$$C_{\underset{\sim}{N}}(\widetilde{\sigma}, A) := -\int_\Sigma d^3x \underset{\sim}{N} \operatorname{tr}(\widetilde{\sigma}^a \widetilde{\sigma}^b F_{ab}), \tag{5}$$

where, $\underset{\sim}{N}$—the lapse—is now a scalar density of weight -1. (We can recover the lapse *function* N from $\underset{\sim}{N}$ by setting $N = q^{\frac{1}{2}} \underset{\sim}{N}$. Note that $\underset{\sim}{N}$ is considered as the fundamental, metric independent field.) Again, $C_{\underset{\sim}{N}}$ is differentiable on Γ if and only if $\underset{\sim}{N}$ tends to zero at infinity as $1/r$. In that case, $C_{\underset{\sim}{N}}$ has the following Poisson bracket relations:

$$\begin{aligned} \{C_{\underset{\sim}{N}}, \widetilde{\sigma}^a{}_A{}^B\} &= \sqrt{2} i\, \mathcal{D}_b(\underset{\sim}{N} \widetilde{\sigma}^{[a} \widetilde{\sigma}^{b]})_A{}^B, \\ \{C_{\underset{\sim}{N}}, A_{aA}{}^B\} &= \tfrac{i}{\sqrt{2}} [\underset{\sim}{N} \widetilde{\sigma}^b, F_{ba}]_A{}^B. \end{aligned} \tag{6}$$

As in geometrodynamics, the geometrical meaning of the canonical transformation generated by the scalar constraint is clear only when restricted to the constraint surface of Γ. Indeed, when Eqs.6.10′, 6.14′ and 6.15′ are all satisfied, the right hand sides of Eqs.6 give us the "time rate of change" of $\widetilde{\sigma}^a{}_A{}^B$ and $A_{aA}{}^B$ respectively with respect to the time parameter defined by the lapse function $\underset{\sim}{N}$. Again, since the lapse is required to vanish at infinity, these are all "bubble-time evolutions" which are asymptotically identity. This interpretation will become clearer in the next section, on dynamics and Hamiltonians.

We can now compute the Poisson bracket algebra. The brackets of the Gauss law constraint and the diffeomorphism constraint with other constraints can be evaluated almost by inspection because, as we saw, the canonical transformations that they generate (Eqs.2 and 4) on the basic canonical variables have simple geometric meaning. Using this interpretation and integrating once by parts, one readily obtains:

$$\begin{aligned}
\{C_N, C_M\} &= -C_{[N,M]}\\
\{C_{\vec{N}}, C_M\} &= -C_{\mathcal{L}_{\vec{N}}M}\\
\{C_{\vec{N}}, C_{\vec{M}}\} &= -C_{[\vec{N},\vec{M}]}\\
\{C_N, C_{\underset{\sim}{M}}\} &= 0\\
\{C_{\vec{N}}, C_{\underset{\sim}{M}}\} &= -C_{\mathcal{L}_{\vec{N}}\underset{\sim}{M}}
\end{aligned} \tag{7a}$$

It only remains to evaluate the Poisson bracket between two scalar constraints. This evaluation has to be made by a direct computation; there is, unfortunately, no general argument that one can now use. However, even this calculation is significantly simpler than the one in geometrodynamics because the constraints are now polynomial in the basic "canonically conjugate" variables $\widetilde{\sigma}^a{}_A{}^B$ and $A_{aA}{}^B$. Using Eq.6 and performing an integration by parts, one obtains:

$$\begin{aligned}
\{C_{\underset{\sim}{N}}, C_{\underset{\sim}{M}}\} &= -i\sqrt{2}\int d^3x(\underset{\sim}{N}\partial_a\underset{\sim}{M} - \underset{\sim}{M}\partial_a\underset{\sim}{N})(\widetilde{\sigma}^{[a}\widetilde{\sigma}^{b]})_A{}^B[\widetilde{\sigma}^c, F_{bc}]_B{}^A\\
&= -i\sqrt{2}\int d^3x\,(\underset{\sim}{N}\partial_a\underset{\sim}{M} - \underset{\sim}{M}\partial_a\underset{\sim}{N})\,\mathrm{tr}(\widetilde{\sigma}^c\widetilde{\sigma}^{(b}\widetilde{\sigma}^{a)}F_{bc} + \widetilde{\sigma}^{(c}\widetilde{\sigma}^{a)}\widetilde{\sigma}^b F_{cb}).
\end{aligned}$$

Now, since the $SU(2)$ soldering forms are trace-free, they satisfy the identity: $(\widetilde{\sigma}^{(a}\widetilde{\sigma}^{b)})_{AB} = \frac{1}{2}\,\mathrm{tr}(\widetilde{\sigma}^a\widetilde{\sigma}^b)\epsilon_{AB}$. Substitution of this identity in the last equation implies that

$$\begin{aligned}
\{C_{\underset{\sim}{N}}, C_{\underset{\sim}{M}}\} &= -i\sqrt{2}\int d^3x\,(\underset{\sim}{N}\partial_a\underset{\sim}{M} - \underset{\sim}{M}\partial_a\underset{\sim}{N})\ \mathrm{tr}(\widetilde{\sigma}^a\widetilde{\sigma}^b)\ \mathrm{tr}(\widetilde{\sigma}^c F_{bc})\\
&= C_{\vec{K}} + C_{A_m K^m},
\end{aligned} \tag{7b}$$

where $\vec{K}$ is given by

$$K^a := (\underset{\sim}{N}\partial_b\underset{\sim}{M} - \underset{\sim}{M}\partial_b\underset{\sim}{N})\,\mathrm{tr}\,(\widetilde{\sigma}^a\widetilde{\sigma}^b).$$

Thus, *the constraints do form a first class system.* Note however, that the constraint algebra is *not* a Lie algebra: inspection of the very last Poisson bracket shows that we are faced with structure functions rather than structure constants. In the BRST terminology, we have here an *open* algebra. An explicit computation of the second order structure functions shows that they all vanish [2]. Thus, the BRST rank of this system of constraints is 1. Note also that nowhere on this calculation, or in any of the previous ones, did we have to assume the non-degeneracy of $\widetilde{\sigma}^a{}_A{}^B$. This is an important point to which we will return in section 9.8.

A number of remarks are in order.

i) In the above formulation, the lapse naturally arises as a density of weight minus one because the scalar constraint is a density of weight two. The integrands of all constraint functions, Eqs.1, 3 and 5, are densities of weight one; the integration can be carried out without reference to a specific volume element.

ii) The constraint Eqs.6.10′ and 6.14′, linear in the new momenta $\tilde{\sigma}^a{}_A{}^B$, can be regarded as 'kinematical', since the canonical transformations they generate have a simple geometrical meaning in terms of the 3-manifold Σ itself. Furthermore, since their Poisson brackets involve only *structure constants*, they *do* generate a group. To unravel the structure of this group, let us consider the bundle **B** over Σ each of whose fibers is a copy of the 2-dimensional complex vector space W, $\lambda^A \in W$, equipped with the symplectic form ϵ_{AB} and a Hermitian metric $G_{AA'}$ (see appendix A). Roughly, automorphisms of **B** are the "kinematical" symmetries of our framework. Denote the group of these automorphisms by G. Since G maps entire fibers to entire fibers, each element of G can be projected down to Σ unambiguously to give us a diffeomorphism on Σ. The subgroup of G that projects down to identity is the group $SU(2)_{loc}$ of local $SU(2)$ transformations. It is easy to check that $SU(2)_{loc}$ is a normal subgroup of G. The quotient, $G/SU(2)_{loc}$ is naturally isomorphic to $Diff(\Sigma)$. Thus, G is a semi-direct product of $SU(2)_{loc}$ with $Diff(\Sigma)$. It is easy to see that the Lie algebra generated by the constraints C_N and $C_{\vec{N}}$ is isomorphic to the Lie algebra of the subgroup G_0 of G consisting of those elements which are asymptotically identity. Thus, the canonical transformations generated by these two constraints span the connected component of the identity of G_0. This is the precise ***kinematical symmetry group*** of the extended Hamiltonian framework.

iii) Although the constraint algebra continues to be open in the new framework, the structure functions now depend only ***polynomially*** on the basic canonical variables. From a computational point of view, (particularly in the BRST framework, see, e.g.,[2]) this is a significant improvement over the situation one is faced with if one uses the geometrodynamical variables.

iv) The phase space of general relativity introduced here is identical to that of complex Yang-Mills theory with internal group $SU(2)$. The Gauss constraint of general relativity, Eq.1, is identical in form to the Gauss constraint of Yang-Mills theory. In addition, general relativity has the vector and scalar constraints, Eqs.3 and 5. Thus, as discussed in detail in section 2.2, the constraint surface of general relativity is naturally imbedded in the Yang-Mills constraint surface.

Considered from *within* Yang-Mills theory, Eqs.3 and 5 are the simplest gauge and diffeomorphism invariant expressions one could write down in the absence of any background structure, such as a metric on the space-time. Thus from the point of view of gauge theories, one could consider the following question: Take Yang-Mills theory with an *arbitrary* internal group, and impose on it, *ad hoc*, the additional constraints given by Eqs.3 and 5 – is the resulting theory an interesting one? In general the answer will be "no"; precisely because the Poisson bracket between two scalar constraints, $\{C_{\underset{\sim}{N}}, C_{\underset{\sim}{M}}\}$, will *not* yield a linear combination of constraints, and the set of constraints will fail to close. The general conditions sufficient for an interesting theory to result are not known. (For a partial answer, see [3]).

3 Dynamics and Hamiltonians

We can now discuss dynamics. As we just saw in the previous section, $C_{\underset{\sim}{N}}$ is not differentiable on **T**—and hence cannot generate a well-defined canonical transformation—unless $\underset{\sim}{N}$ tends to zero at infinity as $1/r$. To obtain dynamical evolution, on the other hand, we need the lapse to go to constant values (i.e., multiples of $e^{-\frac{1}{2}}$, where e is the determinant of the Euclidean metric e_{ab}) at infinity. Consequently, to obtain a Hamiltonian generating a time-translation, one must add suitable surface terms to the constraint function $C_{\underset{\sim}{N}}$.

Let $\underset{\sim}{T}$ be a scalar density of weight -1 on Σ which equals $q^{-\frac{1}{2}}$ outside a compact set. Our experience with geometrodynamics suggests that to obtain the Hamiltonian generating the corresponding time translation, we should subtract from $C_{\underset{\sim}{T}}$ that surface term which we would obtain if we performed an integration by parts in $C_{\underset{\sim}{T}}$ to remove terms involving derivatives of $A_{aA}{}^{B}$.(For details, see, e.g., chapter II.2 of [1].) We shall now see that this expectation turns out to be correct: the Hamiltonian so obtained *is* differentiable on Γ and generates the desired canonical transformation. Let us set

$$\begin{aligned} H_{\underset{\sim}{T}}(\tilde{\sigma}, A) :&= \lim_{S\to\Sigma}\left(\int_S d^3x\, \underset{\sim}{T}\, \mathrm{tr}\,(\tilde{\sigma}^a\tilde{\sigma}^b F_{ab}) - 2\oint_{\partial S} dS_a\, \underset{\sim}{T}\ \mathrm{tr}\,(\tilde{\sigma}^{[a}\tilde{\sigma}^{b]} A_b)\right) \\ &= -2\int_\Sigma d^3x\, \mathrm{tr}\left((\partial_a \underset{\sim}{T}\, \tilde{\sigma}^{[a}\tilde{\sigma}^{b]}) A_b - \underset{\sim}{T}\, \tilde{\sigma}^{[a}\tilde{\sigma}^{b]} A_a A_b\right), \end{aligned} \tag{8}$$

where the integral in the first step is first performed over a finite portion S of Σ

and the limit of the result is then taken as S expands out to fill out all of Σ.[2] The right hand side of the first step brings out the relation between the Hamiltonian $H_{\underset{\sim}{T}}$ and the constraint $C_{\underset{\sim}{T}}$ while the right hand side of the second step shows that the Hamiltonian is well-defined; since each term of the integrand falls off as $1/r^4$, the integral is manifestly convergent. Note that the Hamiltonian is manifestly finite. The overall numerical factor has been chosen to reproduce the standard evolution equations for Cauchy data.

To obtain the evolution equations, we take the Poisson bracket of the basic variables with the Hamiltonian. The result is:

$$\dot{\tilde{\sigma}}^a := \{\tilde{\sigma}^a, H_{\underset{\sim}{T}}\} = -\sqrt{2}i\, \mathcal{D}_b(\underset{\sim}{T}\tilde{\sigma}^{[b}\tilde{\sigma}^{a]}), \tag{9a}$$

and

$$\dot{A}_a := \{A_a, H_{\underset{\sim}{T}}\} = -\frac{i}{\sqrt{2}}[\underset{\sim}{T}\tilde{\sigma}^b, F_{ab}]. \tag{9b}$$

Using the definition of the 3-metric q_{ab} and the extrinsic curvature K^{ab} in terms of $\tilde{\sigma}^a{}_A{}^B$ and $A_{aA}{}^B$ (see Eqs.6.4 and 6.17, and note that when Eq.6.10′ is satisfied, $\Pi^{ab} = K^{ab}$) one can now obtain the evolution equations for q_{ab} and K^{ab}. The result is *precisely* the equations

$$\begin{aligned} \dot{q}_{ab} &= 2NK_{ab} \\ \dot{K}_{ab} &= (Nq_a{}^m q_b{}^n {}^4R_{mn} - NR_{ab} + 2NK_a{}^m K_{bm} - NKK_{ab} + D_aD_bN) \end{aligned} \tag{10}$$

one normally has in the initial value formulation with lapse $N = q^{1/2}\underset{\sim}{T}$ and shift $N^a = 0$. (Here R_{ab} and D are respectively the Ricci tensor and the derivative operator compatible with the 3-metric q_{ab}. This confirms our interpretation of $H_{\underset{\sim}{T}}$ as the Hamiltonian generating "time translations".)

On the constraint hypersurface, the numerical value of the Hamiltonian is given by just the surface term in Eq.8. Even though the integrand is in general complex

---

2 This subtlety arises because our boundary conditions guarantee that, in general, $\underset{\sim}{T}\,\mathrm{tr}\,\tilde{\sigma}^a\tilde{\sigma}^b F_{ab}$ falls-off only as $\frac{1}{r^3}$. Therefore, there is no *a priori* guarantee that, without the limiting procedure, the volume integral in the first step would converge. However, if we first perform the integral over a finite portion S of the 3-manifold Σ, the integral is obviously convergent and the integration by parts needed to pass on to the second step in (8) is well-defined. We now notice that the integrand of the resulting integral falls-off as $\frac{1}{r^4}$ so that the limit as S tends to Σ is well-defined and is in fact equal to the value of the integral over all of Σ. Thus, care is needed because whereas the limit of integrals in the first step is well-defined, without the limiting procedure, the volume integral may be ill-behaved. This is analogous to the fact that while $Lim_{r_0\to\infty}\{\int_{r\le r_0} d^3x\, r\sin\varphi\}$ is well-behaved (in fact zero), the integral $\int d^3x\, r\sin\varphi$ over $\mathbb{R}^3$ is ill-defined.

on Γ since $A_{aA}{}^{B}$ is complex, it turns out that, on the constraint surface, it is in fact real. Substituting the expression (Eq.6.19) for $A_{aA}{}^{B}$ in this integrand, and using the boundary conditions, one obtains :

$$H_{\underset{\sim}{T}}(\tilde{\sigma}, A) \approx -\tfrac{1}{2} \oint_{\partial\Sigma} dS^c\, T\, (\partial_b q_{ac} - \partial_c q_{ab})\, e^{ab}, \tag{11}$$

where, as usual, $\approx$ stands for "equals on the constraint surface", e_{ab} is the Euclidean 3-metric at infinity, and ∂ its derivative operator. Thus, except for overall factors, the numerical value of the Hamiltonian on physical states is precisely the ADM energy! Note that, if one regards the phase space as a complex manifold coordinatized by $A_{aA}{}^{B}$, the boundary term is in fact a holomorphic function on Γ. However, this holomorphic function takes on real values on the constraint surface and this coincides precisely with the ADM energy.

Let us conclude this section with some remarks.

i) At a general point of the phase space the Hamiltonian is a complex function. Therefore, even on the constraint surface, the time evolution generated by $H_{\underset{\sim}{T}}$ fails to preserve the "reality" of of $\tilde{\sigma}^a{}_A{}^B$. However, the "imaginary part" picked up by $\tilde{\sigma}^a{}_A{}^B$ is pure gauge, whence the evolution of q_{ab} and K^{ab} does preserve their reality. It is nonetheless desirable to have a Hamiltonian which preserves the reality of $\sigma^a{}_A{}^B$ itself, i.e., whose Hamiltonian vector field is real on **T**. To obtain such a Hamiltonian, one has only to add to $H_{\underset{\sim}{T}}$ a suitably weighted Gauss constraint functional (Eq.1):

$$H'_{\underset{\sim}{T}} := H_{\underset{\sim}{T}} - \int_\Sigma d^3x\, (D_a \underset{\sim}{T})\, \mathrm{tr}\, (\tilde{\sigma}^a \mathcal{D}_b \tilde{\sigma}^b) \tag{12}$$

is the Hamiltonian which is real on the entire phase space and which therefore generates real evolution:

$$\dot{\tilde{\sigma}}^a = -\, i\sqrt{2}\left(\mathcal{D}_b(\underset{\sim}{T}\, \tilde{\sigma}^{[b}\tilde{\sigma}^{a]}) - \tfrac{1}{2}\, (D_b \underset{\sim}{T})\, [\tilde{\sigma}^b, \tilde{\sigma}^a]\right) \tag{13a}$$

$$\text{and} \quad \dot{A}_a = -\, \frac{i}{\sqrt{2}}\left(\underset{\sim}{T}\, [\tilde{\sigma}^b, F_{ab}] + \mathcal{D}_a((D_b \underset{\sim}{T})\, \tilde{\sigma}^b) - (D_a \underset{\sim}{T}) \mathcal{D}_c \tilde{\sigma}^c\right). \tag{13b}$$

ii) Note that the asymptotic time translation discussed above (as well as space translations which can be treated in a completely analogous way) leave the background $\mathring{\sigma}^a{}_A{}^B$ invariant and preserve the boundary conditions on the phase space variables ($\tilde{\sigma}^a{}_A{}^B$, $A_{aA}{}^B$). This property is *not* shared by an internal rotation $T_A{}^B$ which approaches a non-zero constant value at infinity. That is, "global rotations" are incompatible with our boundary conditions. Hence, in contrast to the Yang-Mills theory, we do not have internal (color) charges on our phase space. This is to be expected on physical grounds because such charges—unlike energy-momentum—have no role in general relativity.

iii) The Hamiltonian simplifies considerably if we set $\underset{\sim}{T} = q^{-1/2}$ everywhere on Γ. First of all, in this case Eq.8 equals Eq.12, so that reality is automatically preserved. Secondly, the expression of the Hamiltonian reduces to :

$$H_{\underset{\sim}{T}}(\widetilde{\sigma}, A) = -\int_{\Sigma} d^3x \left(\overline{A_{ab}}A^{ba} - \overline{A}A\right) q^{\frac{1}{2}}, \tag{14}$$

where $A_{ab} = -\operatorname{tr}(A_a\sigma_b)$, overbar denotes complex conjugation and $A = -\operatorname{tr}(A_a\sigma^a)$. This form of the Hamiltonian is useful especially in the weak field and strong coupling limits.[3]

iv) For simplicity, we have set the shift equal to zero in this discussion. The Hamiltonian for a lapse-shift pair representing a general asymptotic translation can be worked out by extending the discussion of this section in a straightforward way (see Ref. [4]). The final result is:

$$\begin{aligned} H(\widetilde{\sigma}^a, A_a) = & \int_{\Sigma} d^3x \operatorname{tr}\left(\underset{\sim}{T}\widetilde{\sigma}^a\widetilde{\sigma}^b F_{ab} + i\sqrt{2}T^a\widetilde{\sigma}^b F_{ab}\right) \\ & - 2\oint_{\partial\Sigma} d^2S_a \operatorname{tr}\left(\widetilde{T}\sigma^{[a}\widetilde{\sigma}^{b]}A_b + i\sqrt{2}T^{[a}\widetilde{\sigma}^{b]}A_b\right). \end{aligned} \tag{15}$$

v) In the Lagrangian formulation discussed in chapters 3 and 4, we worked with tetrads. For completeness, let now us summarize the results of the present Hamiltonian description in terms of triads $\widetilde{E}^a_i$ and the conjugate connections A^i_a. The connection defines a derivative operator $\mathcal{D}$ on fields with internal indices via: $\mathcal{D}_a\lambda^i := \partial_a\lambda^i + \epsilon^{ijk} A_{aj}\lambda_k$. The curvature of $\mathcal{D}$, defined by $2\mathcal{D}_{[a}\mathcal{D}_{b]}\lambda^i = \epsilon^{ijk} F_{abj}\lambda_k$, can be expressed in terms of the connection A^i_a: $F^i_{ab} = 2\partial_{[a}A^i_{b]} + \epsilon^{ijk} A_{aj}A_{bk}$. The fundamental non-vanishing Poisson brackets

3 This observation is due to Joohan Lee. He also pointed out an error in Ref. [4] which suggested that one may be able to obtain a new proof of positivity of energy by going to a gauge in which A_{ab} is symmetric and trace-free. This strategy does work in the linearized case (See chapter III.3 in Ref. [1]). However, in the full non-linear case, there is a problem: if A_{ab} is symmetric everywhere on Σ, the surface term in (8)—i.e., energy—vanishes identically! A more subtle gauge choice is therefore needed. For example, one may be able to gauge away all but the "leading, $1/r^2$-part" of $A_{[ab]}$.

are:

$$\{A_a^i(x)\ ,\ \widetilde{E}_j^b(y)\} = i\delta_a^b\,\delta_j^i\,\delta^3(x,y) \tag{16}$$

In terms of them, the constraints (Eqs.6.10′, 6.14′ and 6.15′) can be expressed as:

$$\mathcal{D}_a\widetilde{E}_i^a = 0, \qquad \widetilde{E}_i^b F_{ab}^i = 0, \qquad \widetilde{E}_i^a\widetilde{E}_j^b F_{abk}\,\epsilon^{ijk} = 0, \tag{17}$$

and the Hamiltonian with lapse $\underset{\sim}{T}$, shift T^a is given by :

$$H(\widetilde{E},A) = -\int_\Sigma d^3x\,\left(\tfrac{1}{2}\underset{\sim}{T}\widetilde{E}_i^a\widetilde{E}_j^b F_{abk}\,\epsilon^{ijk} + iT^a\widetilde{E}_i^b F_{ab}^i\right) + \oint_{\partial\Sigma} d^2S_a\,\left(\underset{\sim}{T}\widetilde{E}_i^a\widetilde{E}_j^b A_{bk}\,\epsilon^{ijk} + 2iT^{[a}\widetilde{E}_i^{b]}A_b^i\right). \tag{18}$$

The dynamical equations for the basic variables, $\dot{f} = \{f,H\}$, analogous to Eq.9 are:

$$\begin{aligned} \dot{A}_b^i &= i\underset{\sim}{T}\widetilde{E}_j^a F_{abk}\epsilon^{ijk} + T^a F_{ab}^i, \\ \dot{\widetilde{E}}_i^b &= -i\mathcal{D}_a(\underset{\sim}{T}\widetilde{E}_j^a\widetilde{E}_k^b)\epsilon_i{}^{jk} + 2\mathcal{D}_a(T^{[a}\widetilde{E}_i^{b]}). \end{aligned} \tag{19}$$

Finally, the reality conditions which ensure that the complex-valued connection A_a^i is in fact obtained by a canonical transformation on the real phase space now take the form:

$$\widetilde{E}_i^a \text{ be real} \quad \text{and} \quad i(A_a^i - \Gamma_a^i) \text{ be imaginary}, \tag{20}$$

where Γ_a^i is the spin-connection of the triad $\widetilde{E}_i^a$.

4 Discussion

Let us begin with a summary of the results obtained in the previous two chapters. The canonical framework constructed here is the same as the one obtained in chapters 3 and 4. However, whereas in those chapters we started from a Lagrangian formulation of complex relativity, performed a Legendre transform and then restricted ourselves to the real section of the resulting complex phase space by imposing the reality conditions "by hand", in chapters 6 and 7 we arrived at this Hamiltonian description by working entirely in the confines of real general relativity. In this summary, I will discuss the relation between the two approaches.

Consider then *complex* general relativity on a real 4-manifold M. Thus, g_{ab} is a complex, symmetric, non-degenerate tensor field on M satisfying $R_{ab} = 0$.[4] The idea is to regard real general relativity, the physical theory of interest, as a "sub-theory" of complex general relativity, obtained by imposing a "reality condition" at the end. In terms of the new variables, the Hamiltonian description of the complex theory is rather simple. The phase space, $\Gamma_{\mathbb{C}}$, consists of pairs, $(\tilde{\sigma}^a{}_A{}^B,\ A_{aA}{}^B)$, defined on a real 3-manifold Σ, both trace-free in internal indices, with $\tilde{\sigma}^a{}_A{}^B$ satisfying an additional non-degeneracy condition. (Thus, $\tilde{\sigma}^a{}_A{}^B$ is an isomorphism between the 3-complex dimensional vector space consisting of trace-free fields $\lambda_A{}^B$ and vector densities, $\tilde{\lambda}^a$, in the *complexified* tangent space to Σ. Note that, now, there is no Hermiticity condition on $\tilde{\sigma}^a{}_A{}^B$.) These fields are subject to canonical commutation relations. Note also that $(\tilde{\sigma}^a{}_A{}^B,\ A_{aA}{}^B)$ are now genuinely canonically conjugate since they are *both* "complex". However, unlike the text book treatments of, say, the complex scalar fields, here, the symplectic structure is *holomorphic*; complex conjugates of $\tilde{\sigma}^a{}_A{}^B$ and of $A_{aA}{}^B$ never enter the discussion. (Thus, the symplectic structure is non-degenerate only in the holomorphic sense.) The constraint surface is given by:

$$\mathcal{D}_a \tilde{\sigma}^a{}_A{}^B = 0, \qquad \operatorname{tr}(\tilde{\sigma}^a F_{ab}) = 0, \qquad \text{and} \qquad \operatorname{tr}(\tilde{\sigma}^a \tilde{\sigma}^b F_{ab}) = 0. \tag{21}$$

The Poisson brackets between these constraints are again given by Eq.7 where, however, the smearing fields are now allowed to be complex. Thus, we have a first class system of constraints, where, moreover, all constraints are *polynomial* in the basic canonical variables. (Note that the Gauss constraint now generates $SL(2,\mathbb{C})$—rather than $SU(2)$—rotations on the internal indices.) The Hamiltonian is again given by Eq.8; it is also polynomial in the basic variables. Finally, to recover real, Lorentzian general relativity from this framework, we impose the "reality conditions". The most straightforward way of writing these conditions is to use the relation (Eq.6.19) between the new variable $A_{aA}{}^B$ and the old variables $\sigma^a{}_A{}^B$ and $\Pi_{aA}{}^B$. Let us begin by introducing, a Hermitian conjugation operation $\dagger$ on fields with internal indices satisfying Eq.6.2. Then Eq.6.19 tells us that the "real section" Γ of $\Gamma_{\mathbb{C}}$ is given by:

$$(\tilde{\sigma}^a)^\dagger = \tilde{\sigma}^a, \qquad \text{and} \qquad (A_{aA}{}^B - \Gamma_{aA}{}^B)^\dagger = -(A_{aA}{}^B - \Gamma_{aA}{}^B). \tag{21}$$

These conditions are preserved by dynamics and project out real, Lorentzian general relativity.[5] As we noted in chapter 4, this procedure has a parallel in the phase

4 To see that this is a meaningful and interesting theory, see, eg., section 2 of [4].

5 Note that, in Euclidean general relativity, $A_{aA}{}^B$ is given just by $A_a = \Gamma_a + \frac{1}{\sqrt{2}}\Pi_a$. Therefore, the reality conditions are just that both $\tilde{\sigma}^a{}_A{}^B$ and $A_{aA}{}^B$ be Hermitian.

space description of a simple harmonic oscillator. There, one may similarly begin with a *complex* phase space consisting of pairs of complex numbers (q, z), with $z = q - ip$ and the symplectic structure $\Omega = i\, dq \wedge dz$, and work out dynamics using the Hamiltonian $H(q, z) := \frac{z}{2}(2q - z)$. One may then restrict oneself to the "real slice" of the complex phase space by imposing the "reality conditions": $q = \overline{q}$, and $\overline{(z - q)} = -(z - q)$. Since these conditions are left invariant by dynamics , we recover, after their imposition, the standard Hamiltonian description of the simple harmonic oscillator in terms of $(q,\ p = i(z-q))$. (For further details, see chapters 8 and 10.)

The second of the reality conditions (Eq.21) has a non-polynomial dependence on the basic canonical variables. However, we shall see in the next chapter that it can be replaced by an equivalent condition that *is* polynomial. The reason we chose to first present the reality conditions in the form Eq.21 is that, in this form, their origin in the Hamiltonian formulation is transparent. We know from the standard spinorial description of general relativity that $\tilde{\sigma}^a{}_A{}^B$–and hence, also $\Gamma_{aA}{}^B$ – as well as $\Pi_{aA}{}^B$ are both Hermitian on the real section of the phase-space of the theory and this is precisely the content of Eq.21. The problem is that we have reverted to the old variables in this procedure and this, as we shall see, is what leads to the *apparent* non-polynomial dependence.

Finally, we note that it is useful, in the transition to quantum theory, to distinguish between first class constraints and reality conditions. The former are to be imposed as operator constraints on the wave functions while the latter are to be used in a second step to determine an appropriate inner product. It is because of this that the viewpoint adopted in this section, of dealing first with the constraints and then with the reality conditions, is well-suited for quantum gravity. The standard Dirac quantization procedure is usually applied to systems with "trivial" reality conditions, and thus their possible role in the quantum theory is not explicit. The new element in our quantization procedure is the realization that the freedom in the choice of the inner product can be used to incorporate the reality conditions as Hermitian adjointness relations on certain operators, *and*, that this step need not be carried out until *after* the constraints have been solved. The quantum analogs of the constraint Eqs.14 and 15 of chapter 6 are difficult to solve. In the new procedure, we can divide the problem into two parts. First, we can solve the quantum versions of the easier constraints Eqs.6.10′, 6.14′ and 6.15′ and then we can use the reality conditions to determine the inner product.

References

[1] A.Ashtekar (with invited contributions), *New Perspectives in Canonical Gravity* (Bibliopolis, Naples, 1988).

[2] A. Ashtekar, P. Mazur and C.G. Torre, Phys. Rev. **D36**, 2955 (1987).

[3] I. Bengtsson, Phys. Lett.**B220**, 51 (1989).

[4] A. Ashtekar, Phys. Rev. Lett. **57**, 2244 (1986); Phys. Rev. **D36**, 1587 (1987).

[5] A. Ashtekar, T. Jacobson and L. Smolin, Comm. Math. Phys. **115**, 631 (1988).

8 REALITY CONDITIONS

1 Introduction

Since there has been considerable confusion on the subject of reality conditions, at the risk of some repetition, I will first summarize the overall situation and postpone the main result of the chapter –the polynomial character of the reality conditions– until the next section.

As is clear from chapters 3 and 4-7, there are two approaches to the new canonical variables for (real) general relativity. In the first – discussed in chapter 3 – one begins with *complex* general relativity on a real 4-manifold, uses the self dual action and performs the Legendre transform. Then, to recover the real theory, one simply restricts oneself to the real section of the complex phase-space. The last step is carried out by hand: one simply imposes the appropriate reality conditions which single out the desired real section. However, in this approach, one does not, strictly speaking, obtain a canonical description of the real theory. In particular, it is not even obvious that the pull-back of the symplectic structure to the real section, so defined, is real. Nonetheless, as far as the classical theory is concerned, the resulting description is both complete and useful. To investigate any issue in classical general relativity using this framework, we can first formulate it in the complex theory, analyze it and then see if further simplifications occur by going to the real section. This approach has been useful, in particular, in the investigation of Bianchi models. In type I or II models, for example, one can explicitly integrate the equations of motion in the full complex theory and then restrict the solutions to the real section. The result is a new mathematical description of these models which provides some new insights.

The second approach, followed in chapters 4-7, is more closely related to the traditional canonical framework. Here one works entirely with real general relativity. One knows that the symplectic structure is real. In fact, one has manifestly real (i.e. Hermitian) canonical variables $(\tilde{\sigma}^a_{AB}, \Pi_a^{AB})$ on the real phase space. However, they are inconvenient to use; in particular, in the quantum theory constraints are unmanageable in terms of them. One can introduce new canonical variables,

$(\tilde{\sigma}^a{}_A{}^B,\ A_{aA}{}^B)$, in terms of which the constraints simplify. The price paid is that while $\tilde{\sigma}^a{}_A{}^B$is Hermitian, $A_{aA}{}^B$is not. Thus, $(\tilde{\sigma}^a{}_A{}^B,\ A_{aA}{}^B)$ do *not* provide a chart on the real phase space. There are "hidden" relations between them. These are the reality conditions. However, as far as the Poisson brackets are concerned, $\tilde{\sigma}^a{}_A{}^B$and $A_{aA}{}^B$ *can be* considered as canonically conjugate variables. Furthermore, both the constraint and the evolution equations are polynomial in terms of them. Therefore, it is extremely convenient to regard them as the basic variables. A bonus, then, is that we have a close relation between general relativity and Yang-Mills theory. All equations of interest can be now expressed just in terms of $(\tilde{\sigma}^a{}_A{}^B,\ A_{aA}{}^B)$. However, to recover predictions of general relativity, one must also include the reality conditions in the calculations. Although this description is closely related to the traditional canonical framework, the idea of using hybrid canonical variables is somewhat unconventional. There is also a new strategy involved: One first ignores the reality conditions, analyzes physical problem and only at the end recovers predictions of general relativity by imposing the reality conditions.

One's first reaction to these two descriptions may be that while they are acceptable in the classical theory, since neither is a proper canonical description (*a la* Dirac) of general relativity, they would not serve as stepping stones to quantum theory. This is a legitimate worry. However, it turns out that one *can* construct a meaningful quantization program based on these somewhat unusual frameworks. In simple models, the program can be carried out to completion and provides the correct quantum description.

Let me illustrate the main ideas with a simple harmonic oscillator with unit mass and unit spring constant. For brevity, I will discuss the second of the above two strategies in some detail and then comment briefly on the first. The phase space Γ of this system is of 2 real dimensions. It can be coordinatized by the usual variables (q,p). As discussed at the end of chapter 7, one can introduce on Γ a complex function $z = q - ip$, and note that the pair (q,z) satisfies the Poisson bracket relations $\{q,q\} = 0, \{z,z\} = 0$ and $\{z,q\} = i$. One can therefore regard (q,z) as being "canonically conjugate" even though they do not constitute a chart on the 2-real dimensional phase space. Further, given any function $f(q,p)$ on Γ, one can obtain a function $\tilde{f}(q,z) := f(q, i(z-q))$ of the new "canonical variables" (z,q) and compute the Poisson brackets between these functions using the fundamental Poisson brackets between z and q given above. Thus, for example, the equations of motion for q and z are obtained by first noting that the Hamiltonian can be written as $\tilde{H}(q,z) = \frac{1}{2}(q^2 - (z-q)^2) = zq - \frac{1}{2}z^2$, whence $\dot{q} = \{q, \tilde{H}\} = iz - iq$, and, similarly, $\dot{z} = \{z, \tilde{H}\} = iz$. To recover the equations of motion in terms of q and p, we first define $p := i(z-q)$ and then restrict ourselves to the part of the abstractly defined (z,q) space on which p, so defined, is real. That is, the original harmonic oscillator can be recovered by imposing the "reality condition" that $(z - q)$ be pure imaginary.

The quantization program can be carried out in a completely analogous fashion; one can base it on the pair (q, z) rather than (q, p). Here we will only sketch the procedure; details are given in chapter 10 on the quantization program. To do so, let us introduce an abstract $\star$-algebra $\mathcal{A}$, generated by operators $\hat{q}$ and $\hat{z}$, subject to the "canonical commutation relations" $[\hat{q}, \hat{z}] = \hbar$ and the $\star$-relations $\hat{q}^\star = \hat{q}$ and $(\hat{z})^\star = -\hat{z} + 2\hat{q}$, which are motivated dictated by the classical Poisson brackets and reality conditions respectively. It is straightforward to check that this $\star$ algebra is naturally isomorphic to the standard one generated by the usual canonically conjugate operators $(\hat{q}, \hat{p})$. The next problem is that of finding its appropriate $\star$-representation. The idea, as in the classical description, is to first ignore the $\star$-relations. Let us work in the z representation. Then, the wave functions are holomorphic functions $\Psi(z)$ of z. We can represent $\hat{z}$ by the multiplication operator, $\hat{z} \cdot \Psi(z) := z\Psi(z)$, and $\hat{q}$ by a derivative operator, $\hat{q} \cdot \Psi(z) := \hbar(d\Psi/dz)$, so that the canonical commutation relations are satisfied. The next step is to incorporate the $\star$-relations. The idea is to regard these relations as conditions on the inner product. That is, we seek, on the space of holomorphic functions, an inner product which will make the concrete representation, defined above, of the operator $\hat{q}$ self adjoint and $(-\hat{z} + 2\hat{q})$ the adjoint of $\hat{z}$. The key point is that such an inner product exists and is unique: $\langle \Psi(z) \mid \Phi(z) \rangle = i \int dz \wedge d\bar{z}\, \overline{\Psi(z)}\, \Phi(z) \exp\left(-\frac{1}{4\hbar}(z + \bar{z})^2\right)$. The normalizable functions are thus of the type $\Psi(z) = f(z) \exp[\frac{z^2}{4\hbar}]$, where $f(z)$ are polynomials. Using this fact, it is easy to show that the quantum description obtained here is the same as that given in the standard Bargmann representation of a harmonic oscillator. As we shall see in Part III, the idea is to repeat this program in general relativity. This program *is* somewhat different from the standard Dirac quantization of constrained systems. The new element is that one first recognizes that in the quantization procedure, there is some freedom in the choice of the Hermitian inner-product on the space of states and then one uses this freedom to simplify the problem of solving quantum constraints. The program has been completed in a number of simpler model systems including certain Bianchi models, 2+1 gravity, Maxwell theory in Minkowski space and linearized gravity.

Finally, in the case of the harmonic oscillator, it is also possible to adopt the viewpoint of chapter 3. We can begin with the *complex* oscillator, construct the $\star$-algebra and its representation and finally impose the reality conditions to recover the quantum description of the real oscillator. These steps have been worked out in detail by Ranjeet Tate and will appear in his doctoral dissertation at Syracuse University.

2 Polynomial form

Let us now turn to the main result of this chapter. We shall show that the reality conditions obtained in chapters 3 and 7 for source-free general relativity are in fact polynomial in the new variables.[1] In chapter 9, where we discuss matter couplings, we shall present the modifications to these conditions needed in the presence of sources.

As we saw in chapter 7, the most straightforward way to state the reality conditions is to use the relation between the new and the old variables and directly translate the well-known reality conditions on the old variables to those on $\tilde{\sigma}^a{}_A{}^B$ and $A_{aA}{}^B$. However, this procedure leads one to conditions (Eq.7.21) which are non-polynomial in $\tilde{\sigma}^a{}_A{}^B$ and $A_{aA}{}^B$. This is why in the early literature on new variables, it was often stated that these conditions are non-polynomial in the basic variables. We shall now show that, while there was no computational error, this conclusion is nevertheless incorrect: One *can* give an alternate but equivalent formulation of the reality conditions in which all expressions are polynomial in $\tilde{\sigma}^a{}_A{}^B$ and $A_{aA}{}^B$. It is in order to emphasize this point that I decided to separate this issue from those discussed in the last two chapters and present it in a separate chapter.

To see the origin of reality conditions, let us return, for a moment, to the geometrodynamical variables, the 3-metric q_{ab} and the extrinsic curvature K_{ab}. One can, if one so desires, begin with complex-valued fields (q_{ab}, K_{ab}), set up the constraint and evolution equations and, at the end, take the real section of the complex phase-space by imposing the reality condition: $q_{ab} = \bar{q}_{ab}$ and $K_{ab} = \overline{K}_{ab}$ (or, equivalently, $\dot{q}_{ab} = \dot{\bar{q}}_{ab}$). These conditions are automatically preserved under time-evolution. We want to impose a similar restriction on the phase-space spanned by our pairs $(A_{aA}{}^B, \tilde{\sigma}^a{}_A{}^B)$. Hermiticity of $\tilde{\sigma}^a{}_A{}^B$ guarantees the reality of the (densitized) 3-metric $\tilde{\tilde{q}}^{ab} := -\operatorname{tr}\tilde{\sigma}^a\tilde{\sigma}^b$. The additional condition we need is that its time derivative, $(\tilde{\tilde{q}}^{ab})^\bullet$, also be real. Using Eq.9*a* of chapter 7, we have:

$$(\tilde{\tilde{q}}^{ab})^\bullet = \sqrt{2}i\underset{\sim}{T}\Big(\operatorname{tr} \mathcal{D}_m(\tilde{\sigma}^{[m}\tilde{\sigma}^{a]})\tilde{\sigma}^b + \operatorname{tr} \mathcal{D}_m(\tilde{\sigma}^{[m}\tilde{\sigma}^{b]})\tilde{\sigma}^a\Big) \qquad (1)$$

Since $\underset{\sim}{T}$ appears as an overall multiplicative factor in this equation and since the Poisson bracket of $H_{\vec{T}}$ with $\tilde{\tilde{q}}^{ab}$ gives just the Lie derivative of $\tilde{\tilde{q}}^{ab}$, the reality of $\tilde{\tilde{q}}^{ab}$ and of its time-derivative under *arbitrary* real lapse and shift evolutions is ensured by requiring:

---

1 This result first appeared in [1]

$$\begin{aligned} \overline{\operatorname{tr}\tilde{\sigma}^a\tilde{\sigma}^b} &= \operatorname{tr}\tilde{\sigma}^a\tilde{\sigma}^b \\ \overline{(\operatorname{tr}\mathcal{D}_m(\tilde{\sigma}^{[m}\tilde{\sigma}^{a]})\tilde{\sigma}^b + a \leftrightarrow b)} &= -(\operatorname{tr}\mathcal{D}_m(\tilde{\sigma}^{[m}\tilde{\sigma}^{a]})\tilde{\sigma}^b + a \leftrightarrow b) \end{aligned} \tag{2}$$

This is the required reality condition in the source-free case. It is clearly polynomial in the basic canonical variables. Indeed, $\tilde{\tilde{q}}^{ab}$ is just quadratic, and, its time-derivative, being its Poisson bracket with the Hamiltonian which is at worst quartic, is also at worst quartic in $\tilde{\sigma}^a{}_A{}^B$ and $A_{aA}{}^B$. Again, if the reality condition is imposed initially on a pair $(A_{aA}{}^B, \tilde{\sigma}^a{}_A{}^B)$ satisfying constraints, under the Hamiltonian flow, it is preserved in time. Why then was it previously stated that these conditions are non-polynomial? It is because, in the earlier work, one insisted on expressing them in terms of the geometrodynamical variables, the metric and the extrinsic curvature, and the non-polynomial dependence entered in the transition from the new to the old variables. More precisely, in the notation used in the previous chapter, the situation is the following. Let $\tilde{\sigma}^a{}_A{}^B$ be non-degenerate and let D denote the unique torsion-free derivative operator which annihilates $\tilde{\sigma}^a{}_A{}^B$. Recall that the difference between $\mathcal{D}$ and D is captured in the field $\Pi_{aA}{}^B$: $(\mathcal{D}_a - D_a)\lambda_A = -\frac{i}{\sqrt{2}}\Pi_{aA}{}^B\lambda_B$. Then, using the evolution equation for $\tilde{\tilde{q}}^{ab}$, it is easy to check that $\dot{q}_{ab} = -2\underset{\sim}{T}\operatorname{tr}\Pi_{(a}\tilde{\sigma}_{b)}$. (Thus, $K_{ab} = -\operatorname{tr}\Pi_{(a}\sigma_{b)}$.) Hence, the second of the reality conditions may be replaced by Hermiticity of $\Pi_{aA}{}^B$, i.e., of $i(A_{aA}{}^B - \Gamma_{aA}{}^B)$, where $\Gamma_{aA}{}^B$ is the spin-connection of D. However, since $\Gamma_{aA}{}^B$ is non-polynomial in $\tilde{\sigma}^a{}_A{}^B$, this last condition is also non-polynomial. Put differently, the extrinsic curvature K_{ab} is a non-polynomial function of the basic phase-space variables so that the condition that K_{ab} be real is also non-polynomial. However, $(\tilde{\tilde{q}}^{ab})^{\cdot} = (det q)(K^{ab} - Kq^{ab})$ *is* polynomial and so is the condition that it be real. Since there is no reason to revert to the geometrodynamical variables in the formulation of the reality conditions, we shall just use Eq.2 as the reality conditions in the source-free case.

References

[1] A. Ashtekar, J.D. Romano, R.S. Tate, Phys. Rev. **D40**, 2572 (1989).

9 INCLUSION OF MATTER

1 Introduction

So far, we have restricted ourselves to source-free general relativity for simplicity. We saw that the use of new variables simplifies Hamiltonian general relativity and, in addition, enables one to cast general relativity in the language normally used in Yang-Mills theory. While the first of these features gives rise to technical simplifications in both classical and quantum gravity, the second, as we shall see in Part III, provides new conceptual tools in the passage to quantum gravity. A natural question now is whether these attractive features of the new framework survive the introduction of matter sources. In this chapter, we shall answer the question affirmatively.

As explained in the previous chapters in this Part, the general framework involving new canonical variables was first obtained using Hamiltonian methods [1-4]. However, a manifestly covariant Lagrangian formulation was soon given independently by Samuel [5] and by Jacobson and Smolin [6]. This formulation is better suited for inclusion of matter terms. For, in the Lagrangian formulation, one has only to choose the basic dynamical variables and their couplings. Given the source-free Lagrangian formulation and general facts about gravitational coupling, this is a relatively easy choice to make. In the Hamiltonian formulation, on the other hand, one has to guess the appropriate canonical transformation and this becomes rather complicated once Dirac fields are brought in. Therefore, we shall use the work of Samuel, Jacobson, and Smolin as our point of departure. Our purpose here is to extend their framework by including matter sources and to analyze the resulting algebra of constraints and Hamiltonians.[1] We shall see that the constraint functionals, the Hamiltonians, and the reality conditions continue to retain their polynomial form in the basic canonical variables for matter and gravity and that

1 Our presentation follows Ref. [7] closely. Coupling to Klein-Gordon fields was also discussed by Koshti [8] and to Dirac fields by Jacobson [9]. Our treatment will be more exhaustive than theirs.

one can continue to use $\tilde{\sigma}^a{}_A{}^B$ and $A_{aA}{}^B$ as the basic variables for gravity. Consequently, one can continue to borrow techniques from Yang-Mills theory and QCD to analyze physical predictions of general relativity and quantum gravity [10,11].

We begin, in section 2, by introducing the basic fields and the Lagrangian on a 4-manifold $\mathcal{M}$, topologically $\Sigma \times \mathbb{R}$. As in the source-free case, we use a first order formalism for the gravitational part of the action. However, we will now use spinors rather than tetrads. The basic fields will thus be the 4-dimensional $SL(2, \mathbb{C})$ soldering form, $\sigma^a{}_A{}^{A'}$, the unprimed spin (or "internal") connection, ${}^4A_{aA}{}^B$, and a possible cosmological constant Λ. As in Ref. [6], we restrict the soldering form to be anti-Hermitian from the beginning, i.e. $\overline{\sigma}^a{}_A{}^{A'} = -\sigma^a{}_A{}^{A'}$, so that *the space-time metric will always be real* (with signature $(-+++)$).[2] The matter fields consist of the minimally coupled, massive Klein-Gordon field ϕ, the massive Dirac field $(\xi^A, \eta_{A'})$, and a Yang-Mills connection 1-form ${}^4\mathbf{A}_a$ (with an arbitrary internal gauge group.) It is easy to extend the framework to include suitable interaction terms between Klein-Gordon, Yang-Mills, and Dirac fields and to allow several distinct fields of any given spin. In particular, one could replace the real Klein-Gordon field by a Higgs multiplet and/or let the Dirac field have an additional, Yang-Mills internal index. As in the source-free case, our total action is *complex*. One would *a priori* expect that this would lead to *twice* as many equations as one wants. In the source-free case, the extra equation turned out to be simply the Bianchi identity. We shall show that the situation remains unaltered despite the fact that the Dirac field now couples to the *complex-valued* self dual connection ${}^4A_{aA}{}^B$. Thus, the main result of section 2 is that, even though our action depends only on the self dual part of the spin connection, its variation does not lead to spurious equations of motion.

In section 3, we carry out the variation explicitly and write down the resulting equations of motion in a space-time language. We find that while the Klein-Gordon and Yang-Mills equations are the expected ones, the Dirac equation contains a cubic term in the spin-$\frac{1}{2}$ fields. This is not surprising: since we are using a first order formalism, one would suspect that the outcome would resemble the Einstein-Cartan theory rather than the minimally coupled Einstein-Dirac. However, it is not *a priori* clear that our theory is equivalent to the Einstein-Cartan theory since, unlike in the usual Palatini formulation, our Lagrangian depends only on the self dual part of the spin connection. Given any solution to the field equations, however, we can extend the connection to primed spinors *and* vectors by requiring that it annihilate the

2 With our conventions, the anti-Hermiticity of the 4-dimensional soldering form $\sigma^a{}_A{}^{A'}$ will ensure that the spatial soldering form $\sigma^a{}_A{}^B$ is Hermitian. The connection on the other hand will be complex throughout. Thus, inspite of the conceptual problems discussed in chapters 3 and 4, we are adopting here the strategy used in [5, 6]; in this chapter, the emphasis is on computational rather than conceptual issues.

metric. Then, the connection develops the standard torsion terms of the Einstein-Cartan theory. That is, as far as solutions to the classical equations of motion are concerned, our Lagrangian is equivalent to the standard first order Einstein-Cartan Lagrangian. Now, it is well-known (see, e.g. [12]) that the Einstein-Dirac theory can be recovered in a first-order framework by adding a quartic term to the Dirac Lagrangian. We show that this is also the case for our self dual Lagrangian by exhibiting the required quartic term.

One often uses Grassmann-valued fermion fields in path integrals. Therefore, it is desirable to let our Dirac fields be either complex or Grassmann-valued in the discussion of the action. This is straightforward to achieve: the discussion of sections 2 and 3 is completely insensitive to the choice. In the Grassmann case, one just has to be careful with orderings and use consistently the ordering of spin-$\frac{1}{2}$ fields specified in this chapter.

In section 4, we pass to a Hamiltonian formulation through a Legendre transform following the procedure used in chapters 3 and 4. All the canonically conjugate variables are now fields on the 3-manifold Σ. These are subject to four types of constraints: *the gravitational Gauss law, the Yang-Mills Gauss law, the vector and the scalar constraint.* We exhibit these constraints in terms of the canonically conjugate variables. They are all polynomial. For the convenience of the reader, we list them in the form of a table.

In section 5, we examine the Poisson-bracket algebra generated by these constraints. As in the source-free case, the gravitational Gauss constraint generates internal rotations on (unprimed) spinor indices and provides a representation of the Lie algebra of local $SU(2)$ transformations. However, now the expression of the functional contains, in addition to $\mathcal{D}_a\tilde{\sigma}^a{}_A{}^B$, contributions from Dirac fields and the $SU(2)$ rotation it generates acts also on these spin-$\frac{1}{2}$ fields. The Yang-Mills Gauss constraint provides, as usual, a representation of the Lie algebra of local gauge transformations. A linear combination (with "q-number coefficients") of the vector and the gravitational Gauss constraint generates diffeomorphisms on the 3-manifold Σ. This combination will be referred to as the *diffeomorphism constraint.* The Poisson bracket algebra of diffeomorphism constraints provides us with a representation of the Lie algebra of the diffeomorphism group of Σ. Thus, as in the source-free case, the Gauss and diffeomorphism (or vector) constraints generate a closed Poisson-bracket algebra which mirrors the "kinematic symmetries" associated with (suitable bundles constructed on) the 3-manifold Σ. The geometrical interpretation of these constraints again trivializes the task of computing their Poisson brackets with scalar constraints. Thus, the only non-trivial computation left is that of the Poisson brackets between scalar constraints themselves. As expected, the bracket is a vector constraint whose coefficient is a "q-number". Thus, although the constraints are of first class, their algebra is open in the sense of Becchi-Rouet-Stora-Tyutin (BRST).

Section 6 discusses Hamiltonians. As in the source-free case, if Σ is compact, the Hamiltonian consists only of a linear combination of constraints, basically because we do not have a background space-time geometry. A more interesting situation occurs in the asymptotically flat case. Therefore, in this section, we restrict ourselves to non-compact 3-manifolds Σ and canonically conjugate fields thereon satisfying suitable boundary conditions. As in the source-free case, constraints only generate those canonical transformations which are asymptotically identity. In particular, they do *not* generate space-time translations. Thus, the Hamiltonians are again distinct from constraints. To obtain Hamiltonians, one must add suitable surface terms to constraint functionals. As one might expect, these are precisely the surface integrals that define the Arnowitt-Deser-Misner energy-momentum. Finally, we consider the reality conditions. Since the source-free case has already been treated in chapter 8, we discuss here only the modifications caused by the presence of matter sources.

It is somewhat surprising that, although our action is not polynomial in the basic fields, in the resulting Hamiltonian formulation, *all* field equations are polynomial in the canonical variables. To understand this at least partially, in section 7 we present an alternate but equivalent action which *is* polynomial. The reason we did not use this action in the main discussion is that it requires the introduction of additional fields –it is of first order in matter variables as well– which makes the framework less transparent.

Our conventions are as follows. Throughout, we use Penrose's abstract index notation [13,14]. The signature of the space-time metric is $(-+++)$. Usually, the stem letters of various fields and/or their index structure are used to distinguish between space-time, "4-dimensional" fields on $\mathcal{M}$ and spatial, "3-dimensional" fields on Σ. Thus, while g_{ab} and $\sigma^a{}_A{}^{A'}$ are the space-time metric and the corresponding $SL(2,\mathbb{C})$ soldering form on $\mathcal{M}$, q_{ab} and $\sigma^a{}_A{}^B$ are the spatial metric and the $SU(2)$ soldering form on Σ. When there is possibility of an ambiguity, the "four-dimensional" fields carry an explicit prefix '4'. Thus, ${}^4A_{aA}{}^B$ is the space-time connection while $A_{aA}{}^B$ is its pull-back to Σ. The torsion-free derivative operator compatible with $\sigma^a{}_A{}^{A'}$ is denoted by ∇, while the one compatible with $\sigma^a{}_A{}^B$ is denoted by by D. The derivative operator on unprimed spinors, defined by the gravitational self dual connection ${}^4A_{aA}{}^B$ is denoted by ${}^4\mathcal{D}$ while that defined by the Yang-Mills connection ${}^4\mathbf{A}$ is denoted by ${}^4\mathbf{D}$. The relation between the "4-dimensional" $SL(2,\mathbb{C})$ spinors and the " 3-dimensional" $SU(2)$ spinors is as follows. Recall first (see, e.g., appendix A) that, given a Hermitian metric, $G_{AA'}(=\overline{G}_{A'A})$, the primed $SL(2,\mathbb{C})$ spinors define a Hermitian conjugation operation on the unprimed ones via $(\xi^\dagger)_A := G_A{}^{A'}\bar{\xi}_{A'}$. Given a space-like submanifold Σ with a unit normal field $n^{AA'} := \sigma_a^{AA'}n^a$, we will choose the required Hermitian metric to be $G^{AA'} = -i\sqrt{2}n^{AA'}$. Consequently, the "dagger" operation will satisfy: $(\xi^{\dagger})^A\xi_A \geq 0$; and, $[(\xi^{\dagger})^{\dagger}]_A = -\xi_A$. Finally, while we have set G=c=1 throughout these notes, at the end of section 3, we briefly indicate how to restore factors of G. In the present framework, they appear in rather unexpected places so that care is needed in taking the weak field or strong coupling limits.[3]

2 Self dual action and reality of equations

Fix a 4-manifold $\mathcal{M}$, with topology $\Sigma \times \mathbb{R}$ for some 3-manifold Σ. The total Lagrangian density, $\mathcal{L}_T$, on $\mathcal{M}$ will be a sum of several pieces, each of which is a scalar density of weight 1:

$$\mathcal{L}_T = \mathcal{L}_E + \mathcal{L}_C + \mathcal{L}_{KG} + \mathcal{L}_D + \mathcal{L}_{YM}. \tag{1}$$

Here $\mathcal{L}_E$ is the gravitational part of the Lagrangian density; $\mathcal{L}_C$ is the cosmological constant term; and $\mathcal{L}_{KG}$, $\mathcal{L}_D$, and $\mathcal{L}_{YM}$ are the matter Lagrangian densities for the spin-0 Klein-Gordon field, the spin-$\frac{1}{2}$ Dirac field, and the spin-1 Yang-Mills field, respectively. As in the source-free case, we use a first order framework for the gravitational part. Thus, $\mathcal{L}_E \equiv \mathcal{L}_E(\sigma^a{}_A{}^{A'}, {}^4A_{aA}{}^B)$ is a functional of an anti-Hermitian soldering form $\sigma^a{}_A{}^{A'}$ and a connection ${}^4A_{aA}{}^B$ which acts only on the unprimed spinor (or, rather, *internal*) indices. (Note that, being anti-Hermitian, $\sigma^a{}_{AA'}$ satisfies $\bar{\sigma}^a{}_{AA'} = -\sigma^a{}_{A'A}$.) It is given by:

$$\mathcal{L}_E(\sigma^a{}_A{}^{A'}, {}^4F_{abA}{}^B) = -({}^4\!\sigma)\sigma^a{}_A{}^{A'}\sigma^b{}_{BA'}{}^4F_{ab}{}^{AB}, \tag{2a}$$

where $({}^4\!\sigma)$ is the determinant of the inverse soldering form and ${}^4F_{abA}{}^B$ is the curvature tensor of ${}^4A_{aA}{}^B$. The soldering form $\sigma^a{}_A{}^{A'}$ defines a real 4-metric g^{ab} of signature $(-+++)$ via $g^{ab} = \sigma^a{}_{AA'}\sigma^{bAA'}$, and a (unique) torsion-free derivative operator ∇ which acts on unprimed and primed spinor as well as tensor indices. The connection ${}^4\mathcal{D}$ defined by ${}^4A_{aA}{}^B$ via ${}^4\mathcal{D}_a\lambda_A := \partial_a\lambda_A + {}^4A_{aA}{}^B\lambda_B$ bears no relation to ∇ at this stage. Indeed, ${}^4\mathcal{D}$ can only act on unprimed spinors. It is a somewhat remarkable fact that none of the essential equations of the theory require the availability of a specific extension of ${}^4\mathcal{D}$ to tensors or primed spinors. Thus, although for

3 On the whole, the conventions used in this chapter are pretty much the same as those used in [1]. However, there were a few sign errors in [1] particularly in the discussion of the relation between the $SL(2,\mathbb{C})$ and $SU(2)$ spinors. These have been corrected. Readers who are unfamiliar with $SL(2,\mathbb{C})$ spinors and their relation to $SU(2)$ spinors should consult, e.g., [13] or appendix A.

computational convenience we may occasionally extend ${}^4\mathcal{D}$ in an appropriate way, it should be borne in mind that the framework does *not* require the knowledge of *any* derivative operator on tensors or primed spinors. The cosmological constant Λ contributes via:

$$\mathcal{L}_C(\sigma^a{}_A{}^{A'}) := -\Lambda({}^4\!\sigma). \tag{2b}$$

The Klein-Gordon and Yang-Mills Lagrangian densities are the standard ones:

$$\mathcal{L}_{KG}(\sigma^a{}_A{}^{A'}, \phi) := -4\pi({}^4\!\sigma)(g^{ab}\partial_a\phi\partial_b\phi + \mu^2\phi^2), \tag{2c}$$

$$\mathcal{L}_{YM}(\sigma^a{}_A{}^{A'}, {}^4\mathbf{A}_a) := -\tfrac{1}{2}({}^4\!\sigma)g^{ac}g^{bd}\,\mathrm{tr}\,{}^4\mathbf{F}_{ab}{}^4\mathbf{F}_{cd}, \tag{2d}$$

where 'tr' denotes the trace over the (suppressed) Yang-Mills internal indices. Finally, we shall take the Dirac Lagrangian density to be

$$\begin{aligned}\mathcal{L}_D(\sigma^a{}_A{}^{A'}, {}^4A_{aA}{}^B, \xi^A, \bar{\xi}^{A'}, \bar{\eta}^A, \eta^{A'})\\ := -\sqrt{2}({}^4\!\sigma)[\sigma^a{}_{AA'}(\bar{\xi}^{A'}\,{}^4\mathcal{D}_a\xi^A - ({}^4\mathcal{D}_a\bar{\eta}^A)\eta^{A'}) - \frac{im}{\sqrt{2}}(\bar{\eta}_A\xi^A - \bar{\xi}^{A'}\eta_{A'})]\end{aligned} \tag{2e}$$

where the spin-$\frac{1}{2}$ fields ξ^A and $\bar{\eta}^A$, can be either complex or Grassmann-valued. Note that $\mathcal{L}_D$ contains derivatives only of unprimed spinors and that these fields are minimally coupled to gravity. This coupling is the same as in Yang-Mills theory except that the "Higgs scalars" are now spin-$\frac{1}{2}$ fields of one chirality; the Yang-Mills internal index is replaced by the unprimed spinor index. Note also that while all matter fields couple to $\sigma^a{}_A{}^{A'}$, only the Dirac field couples to ${}^4A_{aA}{}^B$. Because of this, much of the conceptual and computational non-triviality of our analysis lies only in the gravitational and Dirac parts, $\mathcal{L}_E$ and $\mathcal{L}_D$, of the total Lagrangian density $\mathcal{L}_T$.

Note first that while $\mathcal{L}_C$, $\mathcal{L}_{KG}$, and $\mathcal{L}_{YM}$ are manifestly real, $\mathcal{L}_E + \mathcal{L}_D$ is not, owing to the absence of the primed spin connection ${}^4\bar{A}_{aA'}{}^{B'}$, the complex conjugate of ${}^4A_{aA}{}^B$. This gives rise to the possibility that equations of motion may not be *real*, i.e., the Lagrangian density considered here may not lead to the familiar equations for the gravitational and Dirac fields. Indeed, *a priori*, it is quite possible that the variation of $\mathcal{L}_E + \mathcal{L}_D$ may give rise to spurious equations which may even lead to inconsistencies. Therefore, before proceeding with a detailed analysis of the theory, let us make a detour to ensure that the complex nature of $\mathcal{L}_E + \mathcal{L}_D$ does not give rise to such unforseen problems.

To analyze this issue, it is particularly convenient to use a general fact about variational principles mentioned in chapter 3. Let $S(x^i, y^\alpha)$ be an action depending on two types of dynamical variables, x^i and y^α. Solutions to dynamical equations are extrema of S with respect to both x^i and y^α. Let us suppose that the equations

$\partial S/\partial y^\alpha = 0$ admit *unique* solutions $y_0^\alpha(x)$ for each choice of x^i so that the surface in the $x - y$ space representing the solutions to this equation can be coordinatized by x^i. Then the pull-back, $\overline{S}(x) := S(x^i, y_0^\alpha(x))$, of the action to the solution set has the property that its extrema are precisely the extrema of the full action $S(x^i, y^\alpha)$. As in chapter 3, the role of y^α will be played by the connections ${}^4A_{aA}{}^B$ while the role of x^i will be now played by *all* the remaining variables, $\sigma^a{}_A{}^{A'}$, ξ^A, $\bar{\xi}^{A'}$, $\bar{\eta}^A$, $\eta^{A'}$, ϕ, and ${}^4\mathbf{A}_a$, that occur in $\mathcal{L}_T$. We shall first show that the extrema of $\mathcal{L}_T$ with respect to variations in ${}^4A_{aA}{}^B$ –i.e. solutions to the equations of motion of ${}^4A_{aA}{}^B$ – exist and are unique for each choice of the remaining variables. Therefore, equations of motion for $\sigma^a{}_A{}^{A'}$ and for the matter fields can be obtained by extremizing the pull-back, $\overline{S}_T(\sigma^a{}_{AA'}, \xi^A, \cdots)$, of the full action to the space of solutions ${}^4A_{aA}{}^B$. We shall show that this pull-back is real, ensuring that equations of motion for $\sigma^a{}_A{}^{A'}$, ϕ and ${}^4\mathbf{A}_a$ are real and those for ξ_A and $\bar{\eta}_A$ are the complex conjugates of the equations for $\bar{\xi}^{A'}$ and $\eta^{A'}$.

Let ∇ denote the unique torsion-free connection compatible with $\sigma^a{}_A{}^{A'}$. Then $({}^4\mathcal{D}_a - \nabla_a)\lambda_A = {}^4C_{aA}{}^B\lambda_B$ for some ${}^4C_{aA}{}^B$. Hence, varying the total action,

$$S_T := \int d^4x\ \mathcal{L}_T(\sigma, {}^4A, \phi, {}^4\mathbf{A}, \xi, \bar{\xi}, \bar{\eta}, \eta) \tag{3}$$

with respect to ${}^4A_{aA}{}^B$ keeping all other fields (including $\sigma^a{}_A{}^{A'}$) fixed is equivalent to varying S_T with respect to ${}^4C_{aA}{}^B$. The corresponding equation of motion for ${}^4C_{aA}{}^B$ (or ${}^4A_{aA}{}^B$), $\delta S_T/\delta\, {}^4C_{aA}{}^B = 0$, yields:

$$2\sigma^{[m}{}_M{}^{A'}\sigma^{a]}{}_{A'A}{}^4C_{aN}{}^A + 2\sigma^{[m}{}_N{}^{A'}\sigma^{a]}{}_{A'A}{}^4C_{aM}{}^A = -\frac{i}{2}(\sigma^m{}_{MA'}k_N{}^{A'} + \sigma^m{}_{NA'}k_M{}^{A'}) \tag{4a}$$

with

$$k^{AA'} := -i\sqrt{2}(\bar{\xi}^{A'}\xi^A - \bar{\eta}^A\eta^{A'}). \tag{4b}$$

This algebraic equation has the unique solution

$${}^4C_a{}^{AB} = \frac{i}{4}k^{(A}{}_{D'}\sigma_a{}^{B)D'}. \tag{5}$$

Thus, given any choice of field configurations $\sigma^a{}_A{}^{A'}$, ξ^A, and $\bar{\eta}^{A'}$, the connection ${}^4\mathcal{D}$ (or ${}^4A_{aA}{}^B$) is uniquely determined. Therefore, to obtain equations of motion for the remaining fields, we need only consider the reduced action $\overline{S}_T$. obtained by substituting the solution, Eq.5, for ${}^4A_{aA}{}^B$ in S_T. Simplifications occur only in the gravitational and the Dirac parts of S_T. The curvature tensor ${}^4F_{abA}{}^B$ in $\mathcal{L}_E$ can now be expressed in terms of the curvature tensor of $\sigma^a{}_A{}^{A'}$ and the field $k^{AA'}$

defined above, while the connection ${}^4\mathcal{D}$ in $\mathcal{L}_D$ can be replaced by the torsion-free connection ∇ compatible with $\sigma^a{}_A{}^{A'}$ and $k^{AA'}$. The reduced gravitational action is given by:

$$\overline{\mathcal{L}}_E = ({}^4\sigma)[\frac{1}{2}{}^4R + \frac{3i}{4}\nabla_a k^a + \frac{3}{16}k^a k_a] \tag{6a}$$

where ${}^4R$ is the scalar curvature of the 4-metric $g^{ab} = \sigma^a{}_{AA'}\sigma^{bAA'}$ and $k^a := \sigma^a{}_{AA'}k^{AA'}$, and the reduced Dirac action, by:

$$\overline{\mathcal{L}}_D = -\frac{1}{\sqrt{2}}({}^4\sigma)\sigma^a{}_{AA'}[\bar{\xi}^{A'}\nabla_a\xi^A - (\nabla_a\bar{\xi}^{A'})\xi^A m + \bar{\eta}^A\nabla_a\eta^{A'} - (\nabla_a\bar{\eta}^A)\eta^{A'}]$$
$$-({}^4\sigma)im(\bar{\eta}_A\xi^A - \bar{\xi}^{A'}\eta_{A'}) - \frac{3}{8}({}^4\sigma)k^a k_a - \frac{i}{2}({}^4\sigma)\nabla_a k^a. \tag{6b}$$

Note that $\overline{\mathcal{L}}_E$ now depends not only on the gravitational variable $\sigma^a{}_A{}^{A'}$ but also on spin-$\frac{1}{2}$ fields because we have solved for ${}^4A_{aA}{}^B$ in terms of the soldering form and Dirac fields. Since $\mathcal{L}_C$, $\mathcal{L}_{KG}$, and $\mathcal{L}_{YM}$ do not involve ${}^4A_{aA}{}^B$, the reduction has no effect on these terms. Combining all terms, we now have:

$$\overline{S}_T = \int d^4x\,[\overline{\mathcal{L}}_E + \overline{\mathcal{L}}_D + \mathcal{L}_C + \mathcal{L}_{KG} + \mathcal{L}_{YM}]. \tag{6c}$$

Since the vector field k^a is real by definition, each term in the integrand is manifestly real, except for a total divergence, $\frac{i}{4}({}^4\sigma)\nabla_a k^a$. Since total divergences do not affect equations of motion, we conclude that equations of motion arising from Eq.6c – and hence, from Eq.3 – for g_{ab}, ϕ, and ${}^4\mathbf{A}_a$ are all real and those for the spin-1/2 fields ξ^A and $\bar{\eta}^A$ are complex conjugates of the equations for $\bar{\xi}^{A'}$ and $\eta^{A'}$. Thus, even though S_T is complex, it does *not* give rise to any spurious equations of motion.

3 Euler-Lagrange equations

Let us now vary the total action S_T to obtain the Euler-Lagrange equations of motion for the soldering form and matter fields.

The variation with respect to the scalar field ϕ and the Yang-Mills potential $\mathbf{A}_a$ is straightforward and yields the expected equations for these fields:

$$g^{ab}\nabla_a\nabla_b\phi - \mu^2\phi = 0, \tag{7}$$

and,

$${}^4\mathbf{D}_a{}^4\widetilde{\mathbf{F}}^{ab} \equiv \nabla_a{}^4\widetilde{\mathbf{F}}^{ab} + [{}^4\mathbf{A}_a, {}^4\widetilde{\mathbf{F}}^{ab}] = 0, \tag{8}$$

where, as before, ∇ is the unique torsion-free derivative operator compatible with the metric g^{ab}. Variations with respect to the Dirac fields yield:

$$\begin{aligned}
\widetilde{\sigma}^a{}_{AA'}{}^4\mathcal{D}_a\xi^A - \frac{im}{\sqrt{2}}({}^4\sigma)\eta_{A'} &= 0, \\
{}^4\mathcal{D}_a(\widetilde{\sigma}^a{}_{AA'}\eta^{A'}) - \frac{im}{\sqrt{2}}({}^4\sigma)\xi_A &= 0, \\
{}^4\mathcal{D}_a(\widetilde{\sigma}^a{}_{AA'}\bar{\xi}^{A'}) - \frac{im}{\sqrt{2}}({}^4\sigma)\bar{\eta}_A &= 0 \\
\text{and}\quad \widetilde{\sigma}^a{}_{AA'}{}^4\mathcal{D}_a\bar{\eta}^A - \frac{im}{\sqrt{2}}({}^4\sigma)\bar{\xi}_{A'} &= 0.
\end{aligned} \tag{9}$$

To compare these equations with the standard ones, let us express ${}^4\mathcal{D}$ in terms of ∇ and the spinor fields using the equation of ${}^4\mathcal{D}$ obtained in the previous section. We then obtain:

$$\begin{aligned}
\sigma^a{}_{AA'}(\nabla_a - \frac{3i}{8}k_a)\xi^A &= \frac{im}{\sqrt{2}}\eta_{A'}, \\
\sigma^a{}_{AA'}(\nabla_a + \frac{3i}{8}k_a)\eta^{A'} &= \frac{im}{\sqrt{2}}\xi_A,
\end{aligned} \tag{10}$$

and the complex conjugate equations for $\bar{\xi}^{A'}$ and $\bar{\eta}^A$, where as before, $k_a = -i\sqrt{2}\sigma_a{}^{AA'} \times(\bar{\xi}_{A'}\xi_A - \bar{\eta}_A\eta_{A'})$. These are *not* the standard minimally coupled spin-$\frac{1}{2}$ equations owing to the presence of terms involving k_a. Since these terms are cubic in spin-$\frac{1}{2}$ fields, Eqs.10 are not even linear in Dirac fields. One can see the origin of the cubic term (in the equations of motion) in the reduced action (Eq.6), which has a term quartic in spin-$\frac{1}{2}$ fields. We will return to this point at the end of this section.

To obtain the gravitational field equations, we have to vary S_T with respect to $\sigma^a{}_{AA'}$. Since each piece in S_T depends on $\sigma^a{}_{AA'}$, let us carry out the variation term by term. We have:

$$\frac{\delta S_E}{\delta\sigma^a{}_{AA'}} = ({}^4\sigma)[2\sigma^b{}_B{}^{A'}{}^4F_{ab}{}^{AB} + \sigma^d{}_D{}^{D'}\sigma^b{}_{BD'}{}^4F_{db}{}^{DB}\sigma_a{}^{AA'}], \tag{11a}$$

$$\frac{\delta S_C}{\delta\sigma^a{}_{AA'}} = \Lambda({}^4\sigma)\sigma_a{}^{AA'}, \tag{11b}$$

$$\frac{\delta S_{KG}}{\delta\sigma^a{}_{AA'}} = -8\pi({}^4\sigma)[\sigma^{bAA'}\partial_a\phi\partial_b\phi - \frac{1}{2}\sigma_a{}^{AA'}(g^{bd}\partial_b\phi\partial_d\phi + \mu^2\phi^2)], \tag{11c}$$

$$\frac{\delta S_{YM}}{\delta\sigma^a{}_{AA'}} = -2({}^4\sigma)\,\mathrm{tr}[\sigma^{bAA'}g^{cd}\,\mathrm{tr}\,{}^4\mathbf{F}_{ac}{}^4\mathbf{F}_{bd} - \frac{1}{4}\sigma_a{}^{AA'}g^{cd}g^{mn}\,\mathrm{tr}\,{}^4\mathbf{F}_{cm}{}^4\mathbf{F}_{dn}], \tag{11d}$$

$$\frac{\delta S_D}{\delta\sigma^a{}_{AA'}} = -\sqrt{2}({}^4\sigma)[\bar{\xi}^{A'}{}^4\mathcal{D}_a\xi^A - ({}^4\mathcal{D}_a\bar{\eta}^A)\eta^{A'} - \frac{i}{2\sqrt{2}}\sigma_a{}^{AA'}\nabla_b k^b]. \tag{11e}$$

(Here, in the last equation, we have used the equation of motion (Eq.10), for spin-$\frac{1}{2}$ fields.) Thus, the Euler-Lagrange equations for $\sigma^a{}_A{}^{A'}$ say that the sum of right sides of these equations should vanish. The numerical factors in front of the matter terms have been chosen to agree with the conventions for stress-energy in the literature [15]. Set

$$H_{ab} = ({}^4\!\sigma)^{-1}\sigma_{bAA'}\frac{\delta S_E}{\delta \sigma^a{}_{AA'}}, \tag{12a}$$

and

$$E_{ab}(\text{matter}) = -(8\pi({}^4\!\sigma))^{-1}\sigma_{bAA'}\frac{\delta S_{\text{matter}}}{\delta \sigma^a{}_{AA'}}. \tag{12b}$$

Note that $E_{ab}(KG)$ and $E_{ab}(YM)$ are the standard stress-energy tensors of the Klein-Gordon and Yang-Mills fields. The field equation for $\sigma^a{}_A{}^{A'}$ now becomes:

$$H_{ab} + \Lambda g_{ab} = 8\pi E_{ab}(\text{matter}). \tag{13}$$

Let us now focus on the gravity-spin-$\frac{1}{2}$ part, $S_E + S_D$, of the total action S_T. We can simplify Eq.13 further by substituting for ${}^4\mathcal{D}$ and ${}^4F_{abA}{}^B$ in terms of $\sigma^a{}_A{}^{A'}$ and Dirac fields. We have:

$$H_{ab} = G_{ab} - \frac{1}{16}g_{ab}(k^c k_c + 8i\nabla_c k^c) + \frac{i}{2}\nabla_a k_b + \frac{1}{4}\epsilon_{ab}{}^{cd}\nabla_c k_d - \frac{1}{8}k_a k_b, \tag{14a}$$

and,

$$8\pi E_{ab}(D) = \sqrt{2}\sigma_{bAA'}(\bar{\xi}^{A'}\nabla_a\xi^A - (\nabla_a\bar{\eta}^A)\eta^{A'}) + \frac{1}{8}g_{ab}(k^c k_c - 4i\nabla_c k^c) - \frac{1}{8}k_a k_b, \tag{14b}$$

where G_{ab} is the Einstein tensor of the metric g_{ab}. Hence, if we had only the Dirac field as the matter source, the field equation for g^{ab} would have been:

$$\begin{aligned} G_{ab} = {} & \frac{1}{\sqrt{2}}\sigma_{bAA'}(\bar{\xi}^{A'}\nabla_a\xi^A - (\nabla_a\bar{\xi}^{A'})\xi^A + \bar{\eta}^A\nabla_a\eta^{A'} - (\nabla_a\bar{\eta}^A)\eta^{A'}) \\ & + \frac{3}{16}g_{ab}k^c k_c - \frac{1}{4}\epsilon_{ab}{}^{cd}\nabla_c k_d, \end{aligned} \tag{15}$$

where we have expressed the term $\frac{i}{2}\nabla_a k_b$ in terms of Dirac fields. We see that, as expected from our discussion in the previous section, the right hand side is manifestly real (recall that $\sigma^a{}_A{}^{A'}$is *anti*-Hermitian). Unfortunately, it is not manifestly symmetric. However, since the equations of motion obtained from S_T are the same as those obtained from the reduced action $\overline{S}_T$ (Eq.6), modulo equations of motion for matter fields, Eq.15 is just the result of setting the variation of $\overline{S}_T$ with respect to g^{ab} equal to zero. Thus, equations of motion for matter ensure that the RHS of Eq.15 is in fact symmetric in indices a and b.

Since the equations of motion (Eqs.10) for spin-$\frac{1}{2}$ fields contain a cubic term, and since the effective stress-energy term for spin-$\frac{1}{2}$ fields (i.e. the RHS of Eq.15) contains a term quartic in these fields, the equations of motion obtained above are *not* the standard Einstein-Dirac equations. This is not surprising since we are using a first order formalism and since the usual Palatini first order formalism leads to Einstein-Cartan theory rather than the standard Einstein-Dirac theory. In the Palatini framework, one can just add a quartic term to the action to recover the Einstein-Dirac equations. What is the situation in the present case? Let us return to the reduced action $\overline{S}_T$ of Eq.6. $\overline{S}_T$ contains precisely one term, $-\frac{3}{16}k^a k_a$, involving Dirac fields, which fails to be quadratic in these fields. Therefore, one would expect that the removal of this term from S_T would lead us to the Einstein-Dirac theory. This expectation is correct. Set

$$S'_T = S_T + \frac{3}{16}\int d^4x\, ({}^4\sigma) k^a k_a. \tag{16}$$

Since the added term does not involve ${}^4A_{aA}{}^B$, the solution (Eq.5) to the equation of motion of ${}^4A_{aA}{}^B$ remains unaffected. Therefore, $\overline{S}'_T$ is obtained from $\overline{S}_T$ simply by removing quartic terms, $k^a k_a$, from $\overline{S}_T$. The new reduced action, $\overline{S}'_T$, is precisely the usual Einstein-Dirac action. The new equations of motion for spin-$\frac{1}{2}$ fields are simply:

$$\sigma^a{}_{AA'}\nabla_a \xi^A = \frac{im}{\sqrt{2}}\eta_{A'}, \qquad \text{and} \qquad \sigma^a{}_{AA'}\nabla_a \eta^{A'} = \frac{im}{\sqrt{2}}\xi_A \tag{10$'$}$$

and their complex conjugates, while the new equations for g^{ab} are

$$\begin{aligned} G_{ab} + \Lambda g_{ab} = {} & \frac{1}{\sqrt{2}}\sigma_{bAA'}(\bar{\xi}^{A'}\nabla_a \xi^A - (\nabla_a \bar{\xi}^{A'})\xi^A + \bar{\eta}^A \nabla_a \eta^{A'} - (\nabla_a \bar{\eta}^A)\eta^{A'}) \\ & -\frac{1}{4}\epsilon_{ab}{}^{cd}\nabla_c k_d + 8\pi E_{ab}(KG) + 8\pi E_{ab}(YM) \end{aligned} \tag{15$'$}$$

where, as before, $E_{ab}(KG)$ and $E_{ab}(YM)$ are the standard [15] stress-energy tensors for the Klein-Gordon and Yang-Mills fields. Thus, if we had used S'_T as our total action, we would have obtained the standard equations of motion for all fields, including, in particular, spin-$\frac{1}{2}$ fields. Note however, that it is only the action S_T that admits a direct extension to supergravity [16].

A number of remarks are in order.

i) As we have seen, the action $S_E + S_D$ leads to the Einstein-Cartan theory. Why then did we not encounter torsion terms in the derivative operator? It is because we have formulated the theory using *only* the self dual connection $^4\mathcal{D}$ which does not act on tensors. Nonetheless, our description is complete and there is no compelling reason to extend its action to tensors or unprimed spinors. However, let us suppose that we do extend the action by requiring that it be Hermitian and annihilate g_{ab}. Then, the resulting connection does admit torsion. The torsion tensor is, however, governed entirely by the spinor fields: it is given by $T_{ab}{}^c = \frac{1}{2}\epsilon_{mab}{}^c k^m$, where ϵ_{abcd} is the unique alternating tensor determined by the metric g_{ab}, and k^a (as before) is the vector field determined by the soldering form and the spinor fields.

ii) For simplicity, we have set $G = 1$ through out these notes. However, since factors of G are reshuffled in the passage to new variables, let us discuss briefly how G would have entered various expressions had it not been set equal to one. In this framework, G plays the role of the Yang-Mills coupling constant. (However, since it is dimensionful, dimensions of ${}^4A_{aA}{}^B$and ${}^4F_{abA}{}^B$in the gravitational case are different from those in the Yang-Mills theory.) Thus, $^4\mathcal{D}$ has the expression ${}^4\mathcal{D}_a\xi_A = \partial_a\xi_A + G\ {}^4A_{aA}{}^B\xi_B$ and ${}^4F_{abA}{}^B$ is given by ${}^4F_{abA}{}^B = 2\partial_{[a}{}^4A_{a]A}{}^B + G[{}^4A_a, {}^4A_b]_A{}^B$. As a result, ${}^4F_{abA}{}^B$ has the dimensions of Lagrangian density. Since $\sigma^a{}_A{}^{A'}$ is dimensionless, the expression of the gravitational action does not have a multiplicative factor of G. Since matter fields have conventional dimensions, there is no factor of G whatsoever in the matter action, except through $^4\mathcal{D}$ (which appears only in the Dirac Lagrangian). Thus, the only term in the action which has an explicit G dependence is the cosmological term, which has to be multiplied by an overall factor of $\frac{1}{G}$. In the absence of a cosmological constant term, the only G dependence in the entire action comes through the expressions of ${}^4F_{abA}{}^B$ (in S_E) and $^4\mathcal{D}$ (in S_D) in terms of ${}^4A_{aA}{}^B$.

iii) Why is there no surface term in the expression of the gravitational action S_E? It is because we are using a first order formalism. Thus, in the expression of S_E, we have treated $\sigma^a{}_A{}^{A'}$ and ${}^4A_{aA}{}^B$ as independent variables. In particular, while deriving the equation of motion for ${}^4A_{aA}{}^B$ we keep ${}^4A_{aA}{}^B$ fixed on the boundary (and $\sigma^a{}_A{}^{A'}$ fixed throughout the volume.) Therefore, there is no need to add a surface term to obtain the correct equations of motion from variations. In the second order formalism a surface term is essential: since the soldering form $\sigma^a{}_A{}^{A'}$ is the only dynamical variable in this case, one is allowed to keep only $\sigma^a{}_A{}^{A'}$ –rather than both $\sigma^a{}_A{}^{A'}$ and its derivatives– fixed on the boundary while performing the variations. How can we then reconcile the fact that, on solving for ${}^4A_{aA}{}^B$, we obtained a reduced second-order action $\overline{S}_T$ without any surface-term? The answer is that the reduction procedure comes with a prescription on how to do variations. For paths which lie entirely in the solution set of $\delta S_T/\delta^4 A_a = 0$, fixing $\sigma^a{}_A{}^{A'}$ on the boundary automatically fixes certain derivatives of $\sigma^a{}_A{}^{A'}$ on the boundary as well, so that a surface term is not needed in the action for the variation to give the correct equation of motion.

4 (3+1)-decomposition

In this section, we introduce a foliation in the space-time manifold $\mathcal{M}$ and carry out a (3+1)-decomposition of the action to pass on to the Hamiltonian framework.

As in chapters 3 and 4, we begin by introducing on $\mathcal{M}$ a smooth function t whose gradient is nowhere vanishing and whose level surfaces Σ_t are each diffeomorphic to Σ, and a smooth vector field t^a with affine parameter t; i.e. which satisfies $t^a\nabla_a t = +1$. Given a soldering form $\sigma^a{}_{AA'}$ on $\mathcal{M}$ with respect to which each level surface Σ_t is spacelike, denote by n^a the future-directed, unit, timelike vector field everywhere orthogonal to Σ_t. Denote the induced, positive-definite metric on Σ_t by q_{ab} ($= g_{ab} + n_a n_b$), and obtain the lapse and shift fields N and N^a by projecting t^a into and orthogonal to Σ_t ; $t^a = Nn^a + N^a$. Using $G_{AA'} := -i\sqrt{2}n_{AA'} \equiv -i\sqrt{2}n^a\sigma_{aAA'}$ as the Hermitian metric for $SL(2,\mathbb{C})$ spinors, we can identify unprimed $SL(2,\mathbb{C})$ spinors on $\mathcal{M}$ with $SU(2)$ spinors on Σ_t and, at the same time, introduce a dagger operation on these spinors: we set $G_A{}^{A'}\bar{\xi}_{A'} = (\xi^\dagger)_A$, the Hermitian conjugate of ξ_A. Using this identification, we can now introduce a soldering form $\sigma^a{}_A{}^B$ on $SU(2)$ spinors:

$$\sigma^a{}_{AB} := i\sqrt{2}\sigma^a{}_{(A}{}^{A'}n_{B)A'}. \tag{17a}$$

This form is automatically Hermitian, $(\sigma^a{}_{AB})^\dagger = \sigma^a{}_{AB}$, and, trace-free, $\sigma^a{}_A{}^A = 0$. It is also non-degenerate; it defines an isomorphism between the space of second rank, trace- free Hermitian spinors ξ_{AB} at any point of Σ_t and the tangent space to Σ_t at that point. Finally, it serves as the square-root of the 3-metric q_{ab} on Σ_t : $q^{ab} = -\operatorname{tr}\sigma^a\sigma^b$. We can "invert" Eq.17$a$ to express $\sigma^a{}_A{}^{A'}$ in terms of $\sigma^a{}_A{}^B$ and $n^{AA'}$:

$$\sigma^a{}_{AA'} = i\sqrt{2}\sigma^a{}_{AB}n^B{}_{A'} - n^a n_{AA'} \tag{17b}$$

Finally, we note a useful identity

$$n_{AA'}n_B{}^{A'} = -\frac{1}{2}\epsilon_{AB} \tag{18}$$

which, in particular, implies that $\sigma^a{}_{AB}$ defined by Eq.17a is already projected into the 3-surfaces Σ_t.

We can now obtain a $(3+1)$-decomposition of the action S_T. Let us begin with the gravitational part S_E. Using Eq.17*b* in the gravitational Lagrangian density $\mathcal{L}_E$ and substituting $N^{-1}(t^a - N^a)$ for n^a we get:

$$-({}^4\sigma)\sigma^a{}_A{}^{A'}\sigma^b{}_{BA'}{}^4F_{ab}{}^{AB} = -({}^4\sigma)\,\mathrm{tr}(i\sqrt{2}n^a\sigma^{b4}F_{ab} - \sigma^a\sigma^{b4}F_{ab})$$

$$= ({}^4\sigma)\,\mathrm{tr}(-i\sqrt{2}N^{-1}\sigma^b[\mathcal{L}_t{}^4A_b - {}^4\mathcal{D}_b({}^4A\cdot t)] + \sigma^a\sigma^{b4}F_{ab} + i\sqrt{2}N^{-1}N^a\sigma^{b4}F_{ab}), \quad (19a)$$

where $({}^4A\cdot t)$ denotes $t^a({}^4A_{aA}{}^B)$; $\mathcal{L}_t{}^4A_b$ is the Lie derivative of ${}^4A_{aA}{}^B$ where internal indices are treated as if they were scalars; $\mathcal{D}_a := q_a{}^{b4}\mathcal{D}_b$, is the pull-of ${}^4\mathcal{D}$ to the 3-surface; and where we have used the identity $t^a({}^4F_{ab}) = \mathcal{L}_t{}^4A_b - {}^4\mathcal{D}_b({}^4A\cdot t)$. To further simplify this expression we define

$$\tilde{\sigma}^a{}_A{}^B := (\sigma)\sigma^a{}_A{}^B \qquad \text{and} \qquad \underset{\sim}{N} := (\sigma)^{-1}N \quad (20)$$

where (σ) denotes the inverse of $det(\sigma^a{}_A{}^B)$. Then, in terms of $g \equiv det(g_{ab})$ and $q \equiv det(q_{ab})$ we have $({}^4\sigma) = \sqrt{-g} = N\sqrt{q} = -N(\sigma)$. (Recall that, by convention, a tilde over a tensor denotes a density of weight one and a tilde below a tensor denotes a density of weight minus one.) Using these fields, we obtain

$$-({}^4\sigma)\sigma^a{}_A{}^{A'}\sigma^b{}_{BA'}{}^4F_{ab}{}^{AB} = \mathrm{tr}(i\sqrt{2}\tilde{\sigma}^b\mathcal{L}_t{}^4A_b - i\sqrt{2}\tilde{\sigma}^b\mathcal{D}_b({}^4A\cdot t) - i\sqrt{2}N^a\tilde{\sigma}^{b4}F_{ab}$$
$$-\underset{\sim}{N}\tilde{\sigma}^a\tilde{\sigma}^{b4}F_{ab}). \quad (19b)$$

Finally, we introduce pull-backs, $A_{aA}{}^B$ and $F_{abA}{}^B$, to Σ_t of ${}^4A_{aA}{}^B$ and ${}^4F_{abA}{}^B$- respectively; $A_{aA}{}^B := q_a{}^{b4}A_{bA}{}^B$ and $F_{abA}{}^B := q_a{}^c q_b{}^{d4}F_{cdA}{}^B$. Then, Eq.19 can be expressed in terms only of "3-dimensional" fields:

$$S_E = \int dt \int_{\Sigma_t} d^3x\ \mathrm{tr}(i\sqrt{2}\tilde{\sigma}^b\mathcal{L}_tA_b - i\sqrt{2}\tilde{\sigma}^b\mathcal{D}_b({}^4A\cdot t) - i\sqrt{2}N^a\tilde{\sigma}^bF_{ab}$$
$$-\underset{\sim}{N}\tilde{\sigma}^a\tilde{\sigma}^bF_{ab}), \quad (21)$$

where we have used the identity $\mathcal{L}_tq_a{}^b = 0$ to write $\mathrm{tr}(\tilde{\sigma}^b\mathcal{L}_t{}^4A_b)$ as $\mathrm{tr}(\tilde{\sigma}^b\mathcal{L}_tA_b)$ in the first term. This is the expression we were seeking. It is clear from its form that if we use $A_{aA}{}^B$ as the "configuration variable", apart from a numerical factor, $\tilde{\sigma}^a{}_A{}^B$ is the canonically conjugate momentum. As expected from chapters 3 and 4, time derivatives of ${}^4A\cdot t$, N^a and $\underset{\sim}{N}$ do not appear in the action. We shall see that this continues to be the case even when the matter contributions to the action are included. Therefore, they will continue to play the role of Lagrange multipliers in the theory. Their variations will lead to the constraint equations. From Eq.21

one can also read-off the gravitational contributions to the constraint equations. We can repeat the above procedure for matter fields. Let us begin with the Dirac action S_D. As in the gravitational case, we have to replace spinor fields ($\bar{\xi}^{A'}$ and $\eta^{A'}$) with primed indices by the corresponding Hermitian conjugate fields:

$$(\xi^\dagger)^A = -\bar{\xi}^{A'} G^A{}_{A'} \qquad \text{and} \qquad (\bar{\eta}^\dagger)^A = -\eta^{A'} G^A{}_{A'} \tag{22}$$

Using Eqs.17*b* and 22, and substituting $N^{-1}(t^a - N^a)$ for n^a we obtain:

$$\begin{aligned}({}^4\sigma)\sigma^a{}_{AA'}\bar{\xi}^{A'}{}^4\mathcal{D}_a\xi^A = \underset{\sim}{N}(\sigma)\tilde{\sigma}^a{}_A{}^B(\xi^\dagger)_B\mathcal{D}_a\xi^A + \frac{i}{\sqrt{2}}(\sigma)(\xi^\dagger)_A\mathcal{L}_t\xi^A \\ -\frac{i}{\sqrt{2}}(\sigma)({}^4A\cdot t)_B{}^A(\xi^\dagger)_A\xi^B - \frac{i}{\sqrt{2}}(\sigma)N^a(\xi^\dagger)_A\mathcal{D}_a\xi^A,\end{aligned} \tag{23}$$

where, as before, the Lie derivatives treat internal indices as scalars and $\mathcal{D}_a := q_a{}^{b\,4}\mathcal{D}_b$. Transcribing this result for $({}^4\sigma)\sigma^a{}_{AA'}\eta^{A'}{}^4\mathcal{D}_a\bar{\eta}^A$ and using the identity $\bar{\xi}^{A'}\eta_{A'} = (\xi^\dagger)^A(\bar{\eta}^\dagger)_A$ we obtain:

$$\begin{aligned}S_D = \int dt \int_{\Sigma_t} d^3x \Big\{ -\sqrt{2}\underset{\sim}{N}(\sigma)\tilde{\sigma}^a{}_A{}^B\Big[(\xi^\dagger)_B\mathcal{D}_a\xi^A - (\bar{\eta}^\dagger)_B\mathcal{D}_a\bar{\eta}^A\Big] \\ +i(\sigma)\Big[(\xi^\dagger)_A\mathcal{L}_t\xi^A - (\bar{\eta}^\dagger)_A\mathcal{L}_t\bar{\eta}^A\Big] - i(\sigma)({}^4A\cdot t)_B{}^A\Big[(\xi^\dagger)_A\xi^B - (\bar{\eta}^\dagger)_A\bar{\eta}^B\Big] \\ -i(\sigma)N^a\Big[(\xi^\dagger)_A\mathcal{D}_a\xi^A - (\bar{\eta}^\dagger)_A\mathcal{D}_a\bar{\eta}^A\Big] + \underset{\sim}{N}(\sigma)^2 im\Big[\xi^A\bar{\eta}_A - (\xi^\dagger)^A(\bar{\eta}^\dagger)_A\Big]\Big\}.\end{aligned} \tag{24}$$

This is the desired form of the action. Again, apart from numerical factors, we can pick out $(\xi^A, (\sigma)(\xi_A)^\dagger)$ and $(\bar{\eta}^A, (\sigma)(\bar{\eta}_A)^\dagger)$ as the canonically conjugate pairs and read off the contribution of Dirac fields to the constraint equations.

The treatment of the contributions due to the cosmological constant, the Klein-Gordon and the Yang-Mills fields is straightforward. The final (3 + 1)- forms of these terms are:

$$S_C = \int dt \int_{\Sigma_t} d^3x\, \underset{\sim}{N}(\sigma)^2\Lambda \tag{25}$$

$$\begin{aligned}S_{KG} = \int dt \int_{\Sigma_t} d^3x\,(4\pi)\Big\{ -\underset{\sim}{N}\,\mathrm{tr}(\tilde{\sigma}^a\tilde{\sigma}^b)\partial_a\phi\partial_b\phi \\ -\underset{\sim}{N}^{-1}(\mathcal{L}_t\phi - N^a\partial_a\phi)^2 + \underset{\sim}{N}(\sigma)^2\mu^2\phi^2\Big\}\end{aligned} \tag{26}$$

$$S_{YM} = \int dt \int_{\Sigma_t} d^3x\,\Big\{\underset{\sim}{N}^{-1}(\sigma)^{-2}\,\mathrm{tr}(\tilde{\sigma}^a\tilde{\sigma}^b)\,\mathrm{tr}(\mathcal{L}_t\mathbf{A}_a - \mathbf{D}_a({}^4\mathbf{A}\cdot t) - \frac{N^m}{2}\mathbf{B}_{ma})$$

$$\times(\mathcal{L}_t\mathbf{A}_b - \mathbf{D}_b({}^4\mathbf{A}\cdot t) - \frac{N^n}{2}\mathbf{B}_{nb}) + \frac{1}{8}\underset{\sim}{N}(\sigma)^{-2}\,\mathrm{tr}(\widetilde{\sigma}^a\widetilde{\sigma}^c)\,\mathrm{tr}(\widetilde{\sigma}^b\widetilde{\sigma}^d)\,\mathrm{tr}\,\mathbf{B}_{ab}\mathbf{B}_{cd}\Big\} \tag{27}$$

where $\mathbf{A}_a := q_a{}^{b4}\mathbf{A}_b$, $\mathbf{D}_a := q_a{}^{b4}\mathbf{D}_b$, and $\mathbf{B}_{ab} := 2q_a{}^c q_b{}^{d4}\mathbf{F}_{cd}$. ($\mathbf{B}_{ab}$ is the dual of the magnetic field of $\mathbf{A}_a$.) We can now collect the phase space variables. They are all represented by fields defined intrinsically on the space-like 3-manifolds Σ_t. We choose our "configuration variables" to be: $A_a{}^{AB}$, ϕ, ξ^A, $\bar{\eta}^A$ and $\mathbf{A}_a$. Then, from the expression of the Lagrangian, it follows that their canonically conjugate momenta are given, respectively, by: $-\sqrt{2}i\widetilde{\sigma}^a_{AB}$, $\widetilde{\pi} := -(8\pi)\underset{\sim}{N}^{-1}(\mathcal{L}_t\phi - \mathcal{L}_N\phi)$, $\widetilde{\pi}_A := i(\sigma)(\xi^\dagger)_A$, $\widetilde{\omega}_A := -i(\sigma)(\bar{\eta}^\dagger)_A$ and $\widetilde{\mathbf{E}}^a := 2\underset{\sim}{N}^{-1}(\sigma)^{-2}(\mathrm{tr}\,\widetilde{\sigma}^a\widetilde{\sigma}^b)(\mathcal{L}_t\mathbf{A}_b - \mathbf{D}_b({}^4\mathbf{A}\cdot t) - \frac{1}{2}N^m\mathbf{B}_{mb})$. (If Σ is non-compact, as in the source-free case, these canonically conjugate fields are subject to certain asymptotic fall-off conditions which we will specify in the next section.) Note that the Lagrangian also contains other variables: $\underset{\sim}{N}$, N^a, ${}^4A\cdot t$ and ${}^4\mathbf{A}\cdot t$. However, since the time derivatives of these variables never occur, they play the role of Lagrange multipliers. They do not obey any dynamical equations of motion and we can fix their values by a gauge choice. We now carry out variation of the action with respect to these variables. The variation with respect to $\underset{\sim}{N}$ yields:

$$\frac{\delta S_E}{\delta \underset{\sim}{N}} = -\,\mathrm{tr}(\widetilde{\sigma}^a\widetilde{\sigma}^b F_{ab}), \qquad \frac{\delta S_C}{\delta \underset{\sim}{N}} = (\sigma)^2\Lambda,$$
$$\frac{\delta S_{KG}}{\delta \underset{\sim}{N}} = -4\pi\,\mathrm{tr}(\widetilde{\sigma}^a\widetilde{\sigma}^b)\partial_a\phi\partial_b\phi + \frac{1}{16\pi}\widetilde{\pi}^2 + 4\pi(\sigma)^2\mu^2\phi^2,$$
$$\frac{\delta S_D}{\delta \underset{\sim}{N}} = i\sqrt{2}\widetilde{\sigma}^a{}_A{}^B\Big[\widetilde{\pi}_B\mathcal{D}_a\xi^A + \widetilde{\omega}_B\mathcal{D}_a\bar{\eta}^A\Big] + (\sigma)^2 im\xi^A\bar{\eta}_A - im\widetilde{\pi}^A\widetilde{\omega}_A,$$
$$\frac{\delta S_{YM}}{\delta \underset{\sim}{N}} = \frac{1}{8}(\sigma)^{-2}\,\mathrm{tr}(\widetilde{\sigma}^a\widetilde{\sigma}^c)\,\mathrm{tr}(\widetilde{\sigma}^b\widetilde{\sigma}^d)\,\mathrm{tr}(\mathbf{E}_{ab}\mathbf{E}_{cd} + \mathbf{B}_{ab}\mathbf{B}_{cd}),$$

where $\mathbf{E}_{ab}$ is the (metric independent) dual of the Yang-Mills electric field. The variation with respect to N^a yields:

$$\frac{\delta S_E}{\delta N^a} = -i\sqrt{2}\,\mathrm{tr}(\widetilde{\sigma}^b F_{ab}), \qquad \frac{\delta S_C}{\delta N^a} = 0$$
$$\frac{\delta S_{KG}}{\delta N^a} = -\widetilde{\pi}\partial_a\phi,$$
$$\frac{\delta S_D}{\delta N^a} = -(\widetilde{\pi}_A\mathcal{D}_a\xi^A + \widetilde{\omega}_A\mathcal{D}_a\bar{\eta}^A),$$
$$\frac{\delta S_{YM}}{\delta N^a} = -\frac{1}{2}\underset{\sim}{\eta}_{abc}\,\mathrm{tr}(\widetilde{\mathbf{E}}^b\widetilde{\mathbf{B}}^c) \quad \Big(= -\,\mathrm{tr}(\widetilde{\mathbf{E}}^b\mathbf{F}_{ab})\Big),$$

where $\underset{\sim}{\eta}_{abc}$ is the (c-number) Levi-Civita form-density. Next, we carry out the variation with respect to ${}^4A\cdot t$. The only non-zero terms are:

$$\frac{\delta S_E}{\delta({}^4A\cdot t)^{AB}} = -i\sqrt{2}\mathcal{D}_b\tilde{\sigma}^b{}_{AB}, \qquad \frac{\delta S_D}{\delta({}^4A\cdot t)^{AB}} = (\tilde{\pi}_{(B}\xi_{A)} + \tilde{\omega}_{(B}\bar{\eta}_{A)}).$$

Finally, the only non-vanishing variation with respect to ${}^4\mathbf{A}\cdot t$ comes from the Yang-Mills action:

$$\frac{\delta S_{YM}}{\delta({}^4\mathbf{A}\cdot t)} = \mathbf{D}_b\tilde{\mathbf{E}}^b.$$

Summing each of the above variations and setting the results equal to zero, we obtain four constraint equations:

$$\tilde{\tilde{C}}(\tilde{\sigma}, A; \tilde{\pi}, \phi; \tilde{\pi}, \tilde{\omega}, \xi, \bar{\eta}; \tilde{\mathbf{E}}, \mathbf{A}) := \frac{\delta S_T}{\delta \underset{\sim}{N}} = 0 \tag{28a}$$

$$\tilde{C}_a(\tilde{\sigma}, A; \tilde{\pi}, \phi; \tilde{\pi}, \tilde{\omega}, \xi, \bar{\eta}; \tilde{\mathbf{E}}, \mathbf{A}) := \frac{\delta S_T}{\delta N^a} = 0 \tag{28b}$$

$$\tilde{C}_A{}^B(\tilde{\sigma}, A; \tilde{\pi}, \phi; \tilde{\pi}, \tilde{\omega}, \xi, \bar{\eta}; \tilde{\mathbf{E}}, \mathbf{A}) := \frac{\delta S_T}{\delta({}^4A\cdot t)_B{}^A} = 0 \tag{28c}$$

$$\tilde{\mathbf{C}}(\tilde{\sigma}, A; \tilde{\pi}, \phi; \tilde{\pi}, \tilde{\omega}, \xi, \bar{\eta}; \tilde{\mathbf{E}}, \mathbf{A}) := \frac{\delta S_T}{\delta({}^4\mathbf{A}\cdot t)} = 0. \tag{28d}$$

$\tilde{\tilde{C}} = 0$ is called the *scalar* constraint, $\tilde{C}_a = 0$ is called the *vector* constraint, and $\tilde{C}_A{}^B = 0$ ($\tilde{\mathbf{C}} = 0$) is called the Einstein (Yang-Mills) *Gauss law* constraint. For the convenience of the reader, we have collected the contributions to these constraints from the various parts of the action in Table 1. The expressions of the Gauss law and vector constraints are manifestly polynomial in the basic canonical variables. In the expression of the scalar constraint, on the other hand, the Yang-Mills' contribution, as written, fails to be polynomial due to the appearance of the multiplicative factor $(\sigma)^{-2}$. Fortunately, however, since

$$(\sigma)^2 = q = -\frac{1}{3\sqrt{2}}\underset{\sim}{\eta}_{abc}\ tr(\tilde{\sigma}^a\tilde{\sigma}^b\tilde{\sigma}^c) \tag{29}$$

is polynomial in $\tilde{\sigma}^a$, we can just multiply the scalar constraint equation by $(\sigma)^2$, thereby restoring the polynomial character of *all* constraint equations. (Note, incidently, that, had there been a relative multiplicative factor of (σ) or $(\sigma)^{-1}$ between matter terms, this procedure would have failed.)

Constraint	Lagrange ×-ier	Gravity	Dirac	Klein-Gordon	Yang-Mills	C.C.
$\widetilde{C}_{AB}$	$t\cdot {}^4A^{AB}$	$-i\sqrt{2}\mathcal{D}_b\widetilde{\sigma}^b{}_{AB}$	$\widetilde{\pi}_{(A}\xi_{B)}+\widetilde{\omega}_{(A}\bar{\eta}_{B)}$	0	0	0
$\widetilde{C}_a$	N^a	$-i\sqrt{2}\,\mathrm{tr}(\widetilde{\sigma}^b F_{ab})$	$-(\widetilde{\pi}_A\mathcal{D}_a\xi^A+\widetilde{\omega}_A\mathcal{D}_a\bar{\eta}^A)$	$-\widetilde{\pi}\partial_a\phi$	$-\frac{1}{2}\underset{\sim}{\eta}_{abc}\,\mathrm{tr}(\widetilde{\mathbf{E}}^b\widetilde{\mathbf{B}}^c)$	0
$\widetilde{\widetilde{C}}$	$\underset{\sim}{N}$	$-\,\mathrm{tr}(\widetilde{\sigma}^a\widetilde{\sigma}^b F_{ab})$	$i\sqrt{2}\widetilde{\sigma}^a{}_A{}^B(\widetilde{\pi}_B\mathcal{D}_a\xi^A$ $+\widetilde{\omega}_B\mathcal{D}_a\bar{\eta}^A)$ $+im(\sigma^2\xi^A\bar{\eta}_A-\widetilde{\pi}^A\widetilde{\omega}_A)$	$-4\pi\,\mathrm{tr}(\widetilde{\sigma}^a\widetilde{\sigma}^b)\partial_a\phi\partial_b\phi$ $+\frac{(\widetilde{\pi})^2}{16\pi}+4\pi\sigma^2\mu^2\phi^2$	$\frac{1}{8(\sigma^2)}\,\mathrm{tr}(\widetilde{\sigma}^a\widetilde{\sigma}^c)\,\mathrm{tr}(\widetilde{\sigma}^b\widetilde{\sigma}^d)\times$ $\mathrm{tr}(\mathbf{E}_{ab}\mathbf{E}_{cd}+\mathbf{B}_{ab}\mathbf{B}_{cd})$	$\sigma^2\Lambda$
$\widetilde{\mathbf{C}}$	$t\cdot {}^4\mathbf{A}$	0	0	0	$\mathbf{D}_m\widetilde{\mathbf{E}}^m$	0

Table 1. Gravity and matter contributions to constraints.

Thus, the presence of Yang-Mills fields leads us to a scalar constraint with density weight *four*. In this case, a minor modification occurs in the discussion of dynamics: the lapse function must now be a density of weight minus three. In the absence of Yang-Mills fields, we can proceed in the same manner as in the source-free case and continue to use a lapse with density weight minus one.

5 Constraint algebra

In this section, we shall discuss the Poisson-bracket algebra generated by constraints. To compute these brackets in the case when Σ is non-compact, as in the source-free case, one needs to keep track of the precise boundary conditions satisfied by the fields since several integration-by-parts are involved. (See, e.g., chapter II.2 in [1] for a general discussion of this issue.) Therefore, let us begin by specifying these conditions. To handle the case when Σ is compact, one can just ignore these conditions and the subsequent discussion of surface integrals.

For simplicity, let us suppose that Σ has only one asymptotic region, i.e., that the complement of a compact subset of Σ is diffeomorphic to the complement of a closed ball in $\mathbb{R}^3$. Let $\mathring{\tilde{\sigma}}^a$ be a flat soldering form in this asymptotic region. Then, as in chapter 6, the gravitational variables $\tilde{\sigma}^a{}_A{}^B$ and $A_{aA}{}^B$ will be required to satisfy:

$$\tilde{\sigma}^a = (1 + \frac{M(\theta,\phi)}{r})^2 \mathring{\tilde{\sigma}}^a + O(1/r^2), \qquad \text{and,}$$
$$\mathring{\sigma}^a A_a = O(1/r^3), \qquad A_a + \frac{1}{3}(\mathrm{tr}\, A_m \mathring{\sigma}^m)\mathring{\sigma}_a = O(1/r^2).$$

where r is the radial coordinate defined by the flat metric $\mathring{q}_{ab} (= -\,\mathrm{tr}\,\mathring{\sigma}_a \mathring{\sigma}_b)$ and where the fall-off refers to the cartesian components of fields in the chart defined by this flat metric. For matter fields, we will simply require that all fields and their canonically conjugate momenta should fall-off as $1/r^2$. This will in particular ensure that the stress-energy will fall-off as $(1/r^4)$. These are not the weakest conditions necessary for our framework. However, these are the simplest to work with. (We can, in particular, weaken the conditions on the Yang-Mills potential to allow a non-zero magnetic charge by a more subtle choice of conditions.) Note that these boundary conditions are adapted to asymptotically Minkowskian space-times. That is, we are considering the sector of the theory for which Minkowski space can be thought of as the "classical vacuum". *Thus, from now on, we will set the cosmological constant to zero.* In the spatially non-compact situation now under consideration, we could also have used conditions which refer to asymptotically anti-De Sitter space-times [17].

We expect that the main results of this and the following section will go through in that case as well.

The phase-space Γ will now consist of fields satisfying these boundary conditions (in addition to the obvious algebraic conditions on their indices.) The symplectic structure on Γ is specified by the following fundamental (non-vanishing) Poisson brackets:

$$\begin{aligned}
&\left\{\tilde{\sigma}^a_{AB}(x), A_b{}^{CD}(y)\right\} = -\frac{i}{\sqrt{2}}\delta^3(x,y)\delta_b{}^a\delta_{(A}{}^C\delta_{B)}{}^D \\
&\left\{\tilde{\pi}_A(x), \xi^B(y)\right\} = -\delta^3(x,y)\delta_A{}^B \\
&\left\{\tilde{\omega}_A(x), \bar{\eta}^B(y)\right\} = -\delta^3(x,y)\delta_A{}^B \\
&\{\tilde{\pi}(x), \phi(y)\} = -\delta^3(x,y) \\
&\left\{\tilde{\mathbf{E}}^a_{\mathbf{AB}}(x), \mathbf{A}_b{}^{\mathbf{CD}}(y)\right\} = -\delta^3(x,y)\delta_b{}^a\delta_{\mathbf{AB}}{}^{\mathbf{CD}}
\end{aligned} \tag{30}$$

where boldface uppercase latin indices ($\mathbf{A}, \mathbf{B}$, etc.) are the internal indices of Yang-Mills fields.

Next, we define constraint functionals by smearing the constraints, Eqs.28 with suitable fields. Let us set:

$$C_{\underset{\sim}{N}} = \int_\Sigma d^3x\, \underset{\sim}{N}\tilde{\tilde{C}} \tag{31a}$$

$$\mathbb{C}_{\vec{N}} = \int_\Sigma d^3x\, N^a \tilde{C}_a \tag{31b}$$

$$C_{N,\mathbf{N}} = \int_\Sigma d^3x\, \mathrm{tr}(N\tilde{C} + \mathbf{N}\tilde{\mathbf{C}}) \tag{31c}$$

where $\underset{\sim}{N}$, $\vec{N} \equiv N^a$, $N_A{}^B$ and $\mathbf{N_A{}^B}$ are the smearing fields. Recall that, to qualify as generators of canonical transformations, functions on phase space must be differentiable. To ensure that the constraint functionals have this property, one must impose boundary conditions on the smearing fields as well. Conditions on the first three of these are the same as those in chapter 6 – $\underset{\sim}{N}$ and $\vec{N}$ should go to zero as $1/r$ and $N_A{}^B$ as $1/r^2$ – while the condition on the Yang-Mills smearing function $\mathbf{N_A{}^B}$ turns out to be that it should fall-off as $1/r$.

We now wish to find the canonical transformations that these constraint functionals generate. To do this we use the fundamental Poisson bracket relations

(Eqs.30) and the properties

$$\{A,B\} = -\{B,A\}, \tag{32a}$$
$$\{A,B+\lambda C\} = \{A,B\} + \lambda\{A,C\}, \tag{32b}$$
$$\{A,BC\} = B\{A,C\} + \{A,B\}C, \tag{32c}$$

obeyed by all Poisson brackets.

Let us begin with the Gauss law constraints. For conciseness, we will treat the gravitational and the Yang-Mills Gauss laws simultaneously. The gravitational part can be recovered by setting $\mathbf{N}$ equal to zero in Eq.31c and the Yang-Mills part by setting $N = 0$. An integration by parts of Eq.31c yields:

$$C_{N,\mathbf{N}} = \int_{\Sigma_t} d^3x \left\{ i\sqrt{2}\,\mathrm{tr}\,((\partial_b N + [A_b, N])\widetilde{\sigma}^b) + N_A{}^B(\widetilde{\pi}_B \xi^A + \widetilde{\omega}_B \bar{\eta}^A) \right.$$
$$\left. + \,\mathrm{tr}\,((\partial_b \mathbf{N} + [\mathbf{A}_b, \mathbf{N}])\widetilde{\mathbf{E}}^b) \right\}. \tag{33}$$

Therefore, Eqs.30 and 32 immediately imply:

$$\begin{aligned}
&\{C_{N,\mathbf{N}}, \widetilde{\sigma}^a{}_A{}^B\} = [N, \widetilde{\sigma}^a]_A{}^B, && \{C_{N,\mathbf{N}}, A_{aA}{}^B\} = -\mathcal{D}_a N_A{}^B,\\
&\{C_{N,\mathbf{N}}, \widetilde{\pi}_A\} = N_A{}^B \widetilde{\pi}_B, && \{C_{N,\mathbf{N}}, \widetilde{\omega}_A\} = N_A{}^B \widetilde{\omega}_B,\\
&\{C_{N,\mathbf{N}}, \xi^A\} = -\xi^B N_B{}^A, && \{C_{N,\mathbf{N}}, \bar{\eta}^A\} = -\bar{\eta}^B N_B{}^A,\\
&\{C_{N,\mathbf{N}}, \phi\} = 0, && \{C_{N,\mathbf{N}}, \widetilde{\pi}\} = 0,\\
&\{C_{N,\mathbf{N}}, \widetilde{\mathbf{E}}^a{}_{\mathbf{A}}{}^{\mathbf{B}}\} = [\mathbf{N}, \widetilde{\mathbf{E}}^a]_{\mathbf{A}}{}^{\mathbf{B}}, && \{C_{N,\mathbf{N}}, \mathbf{A}_{a\mathbf{A}}{}^{\mathbf{B}}\} = -\mathbf{D}_a \mathbf{N}_{\mathbf{A}}{}^{\mathbf{B}}.
\end{aligned} \tag{34}$$

Thus, the infinitesimal canonical transformations generated by $C_{N,\mathbf{N}}$ are precisely the infinitesimal rotations on $SU(2)$ spinor indices by $N_A{}^B$ and infinitesimal rotations of Yang-Mills indices by $\mathbf{N}_{\mathbf{A}}{}^{\mathbf{B}}$. As in the source-free case, this geometric interpretation gives us immediately all the Poisson brackets between the Gauss law and other constraints. In particular, we have:

$$\{C_{N,\mathbf{N}}, C_{M,\mathbf{M}}\} = -C_{[N,M],[\mathbf{N},\mathbf{M}]} \quad \text{and} \quad \{C_{N,\mathbf{N}}, C_{\underset{\sim}{M}}\} = 0. \tag{35}$$

(The bracket between the vector and the Gauss law constraints is also zero.) As in in chapter 7, while our vector constraint itself does not have direct geometrical interpretation a combination of this constraint with the Gauss law does. Therefore, following [18], let us define a new constraint functional $C_{\vec{N}}$ as follows:

$$C_{\vec{N}} := \mathbb{C}_{\vec{N}} - \int_{\Sigma} d^3x\ \mathrm{tr}(N^a(A_a\widetilde{C} + \mathbf{A}_a\widetilde{\mathbf{C}})) \tag{31d}$$

(i.e. we have effectively set $N_A{}^B = N^a A_{aA}{}^B$, $\mathbf{N}_{\mathbf{A}}{}^{\mathbf{B}} = N^a \mathbf{A}_{a\mathbf{A}}{}^{\mathbf{B}}$). To compute the infinitesimal canonical transformations generated by this new constraint, it is convenient to rewrite it. We have:

$$C_{\vec{N}} = \int_{\Sigma} d^3x \Big\{ -i\sqrt{2}\,\mathrm{tr}(N^a \tilde{\sigma}^b F_{ab}) - N^a(\tilde{\pi}_A \partial_a \xi^A + \tilde{\omega}_A \partial_a \bar{\eta}^A) - N^a \tilde{\pi} \partial_a \phi$$

$$- \mathrm{tr}(N^a \widetilde{\mathbf{E}}^b \mathbf{F}_{ab}) + i\sqrt{2}\,\mathrm{tr}(N^a A_a \mathcal{D}_b \tilde{\sigma}^b) + \mathrm{tr}(N^a \mathbf{A}_a \mathbf{D}_b \widetilde{\mathbf{E}}^b) \Big\}$$

$$= -\int_{\Sigma} d^3x \Big\{ i\sqrt{2}\,\mathrm{tr}(\tilde{\sigma}^b \mathcal{L}_{\vec{N}} A_b) + (\tilde{\pi}_A \mathcal{L}_{\vec{N}} \xi^A + \tilde{\omega}_A \mathcal{L}_{\vec{N}} \bar{\eta}^A)$$

$$+ \tilde{\pi} \mathcal{L}_{\vec{N}} \phi + \mathrm{tr}(\widetilde{\mathbf{E}}^b \mathcal{L}_{\vec{N}} \mathbf{A}_b) \Big\} \qquad (36)$$

where we used the result that

$$\begin{aligned} \mathrm{tr}(N^a \widetilde{\mathbf{E}}^b \mathbf{F}_{ab} - N^a \mathbf{A}_a \mathbf{D}_b \widetilde{\mathbf{E}}^b) &= \mathrm{tr}\{N^a \widetilde{\mathbf{E}}^b (2\partial_{[a} \mathbf{A}_{b]} + [\mathbf{A}_a, \mathbf{A}_b]) \\ &\quad - \mathbf{D}_b(N^a \mathbf{A}_a \widetilde{\mathbf{E}}^b) + \mathbf{D}_b(N^a \mathbf{A}_a)\widetilde{\mathbf{E}}^b\} \\ &= \mathrm{tr}\,(N^a \widetilde{\mathbf{E}}^b \partial_a \mathbf{A}_b + \widetilde{\mathbf{E}}^b \mathbf{A}_a \partial_b N^a - \partial_b(N^a \mathbf{A}_a \widetilde{\mathbf{E}}^b)) \\ &= \mathrm{tr}\,(\widetilde{\mathbf{E}}^b \mathcal{L}_{\vec{N}} \mathbf{A}_b) - \mathrm{tr}\,(\partial_b(N^a \mathbf{A}_a \widetilde{\mathbf{E}}^b)) \end{aligned} \qquad (37)$$

and also the fact that surface terms which arise in the simplification vanish because of our boundary conditions. It follows immediately from Eq.36 that canonical transformations generated by the new constraint correspond precisely to the diffeomorphisms generated by the smearing field N^a on the 3-manifold Σ (with spinor and Yang-Mills indices treated as scalars.) We have:

$$\{C_{\vec{N}}, \tilde{\sigma}^a{}_A{}^B\} = \mathcal{L}_{\vec{N}} \tilde{\sigma}^a{}_A{}^B, \qquad \{C_{\vec{N}}, A_{aA}{}^B\} = \mathcal{L}_{\vec{N}} A_{aA}{}^B, \qquad \text{etc.} \qquad (38)$$

for all the dynamical variables. Therefore, the new constraint (eq.31*d*) will be referred to as the *diffeomorphism* constraint. Now, we can once again use the geometrical interpretation of the canonical transformation to deduce the Poisson brackets between the diffeomorphism constraint and the other constraints. We have:

$$\{C_{\vec{N}}, C_{M,\mathbf{M}}\} = -C_{\mathcal{L}_{\vec{N}} M, \mathcal{L}_{\vec{N}} \mathbf{M}} \qquad (35c)$$

$$\{C_{\vec{N}}, C_{\vec{M}}\} = -C_{[\vec{N},\vec{M}]} \quad \text{where} \quad [\vec{N},\vec{M}]^a \equiv \mathcal{L}_{\vec{N}} M^a \qquad (35d)$$

$$\{C_{\vec{N}}, C_{\underset{\sim}{M}}\} = -C_{\mathcal{L}_{\vec{N}} \underset{\sim}{M}}. \qquad (35e)$$

Thus, by a judicious choice of constraints, one can compute all but one of the possible Poisson brackets between constraint functionals. The bracket yet to be evaluated is that between two scalar constraints. This computation is somewhat long. However, it is considerably simpler than the corresponding computation in geometrodynamics since now the functionals themselves are polynomial in the new variables. The final result is the expected one: as in the source free case, the Poisson bracket is a vector constraint. The evaluation of the bracket is straightforward. Again, as in the source-free case (see section 2 of chapter 7), one needs only to use the basic properties of Poisson brackets stated in Eqs.32 and 34. We have:

$$\{C_{\underset{\sim}{N}}, C_{\underset{\sim}{M}}\} = \mathbb{C}_{\vec{K}} \quad (= C_{\vec{K}} + C_{K\cdot A, K\cdot \mathbf{A}}) \tag{35f}$$

where

$$K^a := (\underset{\sim}{N}\partial_b \underset{\sim}{M} - \underset{\sim}{M}\partial_b \underset{\sim}{N})\, \mathrm{tr}(\tilde{\sigma}^a \tilde{\sigma}^b).$$

Note that in this calculation, we did *not* have to use non-degeneracy of $\tilde{\sigma}^a$. In fact, none of our evaluations of Poisson brackets are altered by the possible degeneracy. This is an important point to which we shall return in section 8.

The smearing field K^a in Eq.35f. depends on the dynamical variables $\tilde{\sigma}^a{}_A{}^B$. Thus, although the constraints are of *first class* in the Dirac-Bargmann terminology, they do not generate a proper Lie-group. In particular, as has been emphasized in the literature, e.g., by Bargmann and Komar [19], the algebra of constraints is *not* isomorphic to the Lie algebra of the obvious "gauge group" of tetrad-gravity, (i.e., the semi-direct product of the group of "internal" tetrad rotations with the four-dimensional diffeomorphism group of $\Sigma \times \mathbb{R}$.) In the BRST terminology, the algebra is *open*. Note, however, that the structure functions –and hence, also the BRST charge [18]– are polynomial in the canonical variables. Finally, although the constraint functionals now contain contributions from matter fields, the structure of the constraint algebra is identical to that in the source-free case [1,3]. As has been emphasized (especially by Hajman, Kuchăr and Teitelboim [20]) this comes about because the algebra has its roots in geometrodynamics.

6 Hamiltonians and reality conditions

We are now ready to discuss dynamics. The analysis here is quite similar to that carried out in chapter 7 for the source-free case. New issues arise only in the discussion of reality conditions, where it is the presence of spinor fields that adds new twists. Therefore, for brevity we shall focus on the Einstein-Dirac system and only comment at the end on the effect of other matter sources.

As mentioned in section 5, constraint functions $C_{\underset{\sim}{N}}$ and $C_{\vec{N}}$ are differentiable on the phase space only when the lapse and shift fields, $\underset{\sim}{N}$ and N^a tend to zero at infinity. The lapse-shift pairs generating space-time translations, on the other hand, tend to non-zero (constant) values at infinity. Therefore, constraints do *not* generate canonical transformations corresponding to space-time translations; they are not the Hamiltonians of the theory. However, the geometrical interpretation of the canonical transformations generated by the constraints suggests that the Hamiltonians should be closely related to the constraints. This is indeed the case. More precisely, we have the following. Let $\underset{\sim}{T}$ be a smooth scalar density of weight -1 which equals $(det\overset{\circ}{q}_{ab})^{-\frac{1}{2}}$ outside some compact subset of Σ and let T^a be a vector field which is a translational Killing field of $\overset{\circ}{q}_{ab}$ outside some compact set. The "time evolution" defined by the geometric lapse function $\underset{\sim}{T}\cdot(detq)^{\frac{1}{2}}$ and the diffeomorphism generated by T^a on Σ provide us with 1-parameter families of mappings of the phase-space onto itself. One can verify that these mappings preserve the symplectic structure. One can therefore compute the corresponding generating functionals. These are obtained by adding suitable boundary terms to the constraint functionals. We have:

$$H_{\underset{\sim}{T}}(A,\tilde{\sigma},\xi,\tilde{\pi},\bar{\eta},\tilde{\omega}) = \lim_{S\to\Sigma}\Big(-\int_S d^3x\underset{\sim}{T}\tilde{\tilde{C}} - 2\oint_{\partial S} dS_a\underset{\sim}{T}\tilde{\sigma}^{[a}\tilde{\sigma}^{b]}A_b\Big) \tag{36}$$

and,

$$H_{\vec{T}}(A,\tilde{\sigma},\xi,\tilde{\pi},\bar{\eta},\tilde{\omega}) = \lim_{S\to\Sigma}\Big(-\int_S d^3x T^a\tilde{C}_a - 2\sqrt{2}i\oint_{\partial S} dS_a T^{[a}\tilde{\sigma}^{b]}A_b\Big), \tag{37}$$

where, for reasons explained in chapter 7, the integral is first evaluated on a finite portion S of Σ and the limit of the result is then taken as S expands out to fill all of Σ. These are the Hamiltonians generating asymptotic translations, i.e., dynamics.

Note that, on the constraint surface, the numerical values of the Hamiltonians are given just by the surface integrals. Since these integrals explicitly involve $A_{aA}{}^B$, from one's experience in Yang-Mills theory, one might conclude that they are not gauge invariant. This is, however, *not* the case because our gravitational boundary conditions are different from those normally used in the Yang-Mills theory: since the connection $A_{aA}{}^{B}$ is now assumed to fall-off as $\frac{1}{r^2}$, the gauge invariance is in fact assured. Furthermore, due to this fall-off, we can replace the $\tilde{\sigma}^a{}_A{}^B$in the surface integrals by its asymptotic value $\overset{\circ}{\tilde{\sigma}}{}^a{}_A{}^B$. A simple calculation then shows that, when expressed in terms of the 3-metric and the extrinsic curvature of the 3-manifold Σ, these integrals reduce to the familiar ADM expressions. Thus, in particular, while the surface integral in Eq.36 is a *holomorphic* function of the connection $A_{aA}{}^{B}$, and therefore complex-valued at a generic point of the phase-space, its restriction to the constraint surface is in fact real (and, with our conventions, negative). Note also that, as is usual in theories without background structures, the energy and momentum integrals do not have an explicit dependence on matter fields. Matter fields make their presence felt in the expressions of the constraints and thus contribute indirectly to the asymptotic values of the gravitational field on which the energy and momentum integrals depend directly.

Let us examine in detail the infinitesimal canonical transformations generated by these Hamiltonians. As expected, those generated by $H_{\vec{T}}$ are the Lie derivatives of the dynamical variables along the asymptotic translations T^a:

$$\{\tilde{\sigma}^a, H_{\vec{T}}\} = \mathcal{L}_{\vec{T}}\tilde{\sigma}^a \qquad \text{and} \qquad \{A_a, H_{\vec{T}}\} = \mathcal{L}_{\vec{T}}A_a, \tag{38}$$

and similarly for the Dirac field. $H_{\underset{\sim}{T}}$ on the other hand generates evolution equations: denoting $\{f, H_{\underset{\sim}{T}}\}$ by $\dot{f}$, we have:

$$\begin{aligned}
\dot{\tilde{\sigma}}{}^m{}_{MN} &= -i\sqrt{2}\mathcal{D}_a(\underset{\sim}{T}\tilde{\sigma}^{[a}\tilde{\sigma}^{m]})_{MN} - \underset{\sim}{T}\tilde{\sigma}^m{}_{(M}{}^A\xi_{N)}\tilde{\pi}_A - \underset{\sim}{T}\tilde{\sigma}^m{}_{(M}{}^A\bar{\eta}_{N)}\tilde{\omega}_A, \\
\dot{A}_m{}^{MN} &= -\tfrac{i}{\sqrt{2}}\underset{\sim}{T}[\tilde{\sigma}^b, F_{mb}]^{MN} - \underset{\sim}{T}\left(\tilde{\pi}^{(M}\mathcal{D}_m\xi^{N)} + \tilde{\omega}^{(M}\mathcal{D}_m\bar{\eta}^{N)}\right) \\
&\quad + \tfrac{1}{2}m\underset{\sim}{T}(\xi^A\bar{\eta}_A)(\underset{\sim}{\eta}_{mbc}\tilde{\sigma}^b\tilde{\sigma}^c)^{MN}, \\
\dot{\xi}^M &= -i\sqrt{2}\underset{\sim}{T}\tilde{\sigma}^a{}_A{}^M\mathcal{D}_a\xi^A + im\underset{\sim}{T}\tilde{\omega}^M, \\
\dot{\tilde{\pi}}_M &= -i\sqrt{2}\mathcal{D}_a(\underset{\sim}{T}\tilde{\sigma}^a{}_M{}^B\tilde{\pi}_B) + im\sigma^2\underset{\sim}{T}\bar{\eta}_M,
\end{aligned} \tag{39}$$

where, as before, $\underset{\sim}{\eta}_{abc}$ is the metric independent Levi-Civita tensor density of weight minus one. We will need these equations in the analysis of the reality conditions that follows. To conclude the discussion of dynamics, let us consider the effect of inclusion of (real) Klein-Gordon and Yang-Mills sources. These fields contribute to the expressions of the constraint functions and hence also to the volume term in the Hamiltonian. Moreover, as remarked in section 4, the presence of Yang-Mills sources forces us to use for lapse fields densities of weight minus three. But these changes are minor and have no bearing on any of the conceptual issues.

The reality conditions in the source-free case have already been discussed in chapter 8. The inclusion of the cosmological constant, Klein-Gordon fields and Yang-Mills fields is completely straightforward. Let us therefore concentrate again on the Einstein-Dirac system. The reality conditions on $\widetilde{\sigma}^a{}_A{}^B$ and $A_{aA}{}^B$ are again that $\widetilde{\widetilde{q}}^{ab}$ and its time derivatives be real. However, since the time evolution equation of $\widetilde{\sigma}^a{}_A{}^B$ now involves the spinor fields as well, the explicit form of the condition is modified to:

$$\begin{gathered}\mathrm{tr}\,\widetilde{\sigma}^a\widetilde{\sigma}^b \ \text{ be real;} \qquad \text{and,}\\ \left(\mathrm{tr}\,\sqrt{2}i\mathcal{D}_m(\widetilde{\sigma}^{[m}\widetilde{\sigma}^{a]})\widetilde{\sigma}^b + \mathrm{tr}\,\sqrt{2}i\mathcal{D}_m(\widetilde{\sigma}^{[m}\widetilde{\sigma}^{b]})\widetilde{\sigma}^a - (\xi^A\widetilde{\pi}_A + \bar{\eta}^A\widetilde{\omega}_A)\widetilde{\widetilde{q}}^{ab}\right) \ \text{be real.}\end{gathered} \tag{40a}$$

There are, however, additional reality conditions involving spinor fields which arise simply from the definitions of the spinor field momenta, $\widetilde{\pi}_A = i\sigma\xi^\dagger_A$ and $\widetilde{\omega}_A = -i\sigma\bar{\eta}^\dagger_A$. Note, however, that these spinor reality conditions need to be imposed even in the traditional frameworks though they are often not stated explicitly. Without these conditions, the observable currents that arise as quadratic combinations of spinor fields will *not* be real in the phase-space description and the evolution of the initial data will *not* lead to real space-time geometry. Thus, the idea of first introducing a complex phase-space and then looking at a real section thereof is rather general; it is not peculiar to the use of new variables. Furthermore, in the specific case of spinor fields now under consideration, the conditions are precisely the ones that arise in the traditional framework. In terms of observable "currents", these conditions are:

$$\begin{aligned}\overline{(\widetilde{\sigma}^a{}_{AB}\xi^A\widetilde{\pi}^B)} &= (\widetilde{\sigma}^a{}_{AB}\xi^A\widetilde{\pi}^B), &\qquad \overline{(\widetilde{\sigma}^a{}_{AB}\bar{\eta}^A\widetilde{\omega}^B)} &= (\widetilde{\sigma}^a{}_{AB}\bar{\eta}^A\widetilde{\omega}^B),\\ \overline{(\widetilde{\sigma}^a{}_{AB}\widetilde{\pi}^A\widetilde{\pi}^B)} &= (\sigma)^2(\widetilde{\sigma}^a{}_{AB}\xi^A\xi^B), &\qquad \overline{(\widetilde{\sigma}^a{}_{AB}\widetilde{\omega}^A\widetilde{\omega}^B)} &= (\sigma)^2(\widetilde{\sigma}^a{}_{AB}\bar{\eta}^A\bar{\eta}^B).\end{aligned} \tag{40b}$$

The reality conditions for the Einstein-Dirac system are thus given by Eqs.40.

These conditions have been formulated in terms of physical, tensorial quantities, while the phase space is defined in terms of spinorial fields. Therefore, to single out the real section of the phase space explicitly, we have to eliminate some ambiguities by fixing conventions. This can be achieved as follows. First, we consider the complex phase-space spanned by the fields $(A_{aA}{}^B\,,\ \widetilde{\sigma}^a{}_A{}^B,\ \xi^A, \widetilde{\pi}_A, \bar{\eta}^A, \widetilde{\omega}_A)$ with the Poisson bracket relations given by Eq.30. Then, on objects with internal indices, we introduce a "dagger operation" satisfying the following relations: i) $(a\alpha_A + b\beta_A)^\dagger = \bar{a}\alpha^\dagger_A + \bar{b}\beta^\dagger_A$; ii) $(\alpha^\dagger_A)^\dagger = -\alpha_A$, iii) $(\alpha^A)^\dagger\alpha_A \geq 0$; iv) $(\epsilon_{AB})^\dagger = \epsilon_{AB}$; and, v) $(\alpha_A\beta_B)^\dagger = \alpha^\dagger_A\beta^\dagger_B$, for all fields α_A and β_A and complex functions a and b. The field $\alpha^{\dagger}_A$ will be called the adjoint of α_A. (There is considerable freedom in the initial choice of this operation. But once chosen, it is to be kept fixed.) Now, let us consider the reality condition on $\tilde{\sigma}^a$. Let $\tilde{\sigma}^a_0$ be such that $\tilde{\tilde{q}}^{ab}_0 := -\operatorname{tr}\tilde{\sigma}^a_0\tilde{\sigma}^b_0$ is real, and non-negative. Then, we can find an unique equivalence class of $\tilde{\sigma}^a$, each element of which is Hermitian and satisfies $-\operatorname{tr}\tilde{\sigma}^a\tilde{\sigma}^b = \tilde{\tilde{q}}^{ab}_0$, where two $\tilde{\sigma}^a$ are considered as equivalent if they are related by an $SU(2)$ group element. (Note that the initial $\tilde{\sigma}^a_0$ may not be in this equivalence class; there may be a $GL(2,\mathbb{C})$ transformation relating $\tilde{\sigma}^a_0$ and $\tilde{\sigma}^a$.) Each element of the equivalence class can lie on the real section of the phase space. Next, consider the spinor fields ξ^A and $\tilde{\pi}_A$. (The treatment of the pair $(\bar{\eta}^A, \tilde{\omega}_A)$ is completely analogous.) Let us suppose that this pair satisfies the reality condition (Eq.40*b*) on currents with respect to any one of the Hermitian $\tilde{\sigma}^a{}_A{}^B$ obtained above. Then, it is easy to show that we have $\tilde{\pi}_A = \pm i\sigma\xi^{\dagger}_A$. Of these two disjoint branches, we pick one by requiring that $i\tilde{\pi}^A\xi_A \geq 0$. We now have the appropriate conditions for $\tilde{\sigma}^a{}_A{}^B$ and the Dirac fields to lie on the real section of the phase space. It only remains to single out the compatible $A_{aA}{}^B$. For this, we use the second of Eqs.40*a*. Thus, a point $(\tilde{\sigma}^a{}_A{}^B\ ,\ A_{aA}{}^B\ ,\ \xi^A, \tilde{\pi}_A, \bar{\eta}^A, \tilde{\omega}_A)$ will lie on the real section if $\tilde{\sigma}^a{}_A{}^B$ and the Dirac fields satisfy the conditions given above *and* if $A_{aA}{}^B$ is such that the second of Eqs.40*a* ensuring the reality of $(\tilde{\tilde{q}}^{ab})^{\bullet}$ is satisfied. If a point of the phase space lies on a real section as well as the constraint hypersurface, its Hamiltonian evolution yields a real, Lorentzian space-time with Dirac spin-$\frac{1}{2}$ sources.

Let us conclude this section with a few remarks.

i) As in the source-free case, while the Hamiltonian (Eq.36) preserves the reality of tensorial quantities such as $\tilde{\tilde{q}}^{ab}$ and the Dirac currents, it does not preserve Hermiticity relations between the spinorial quantities such as $\tilde{\sigma}^a{}_A{}^B$or the Dirac fields themselves. The spinorial fields are in general "gauge rotated" by $SL(2,\mathbb{C})$ transformations. However, again as in the source-free case, this can be easily corrected by adding to this Hamiltonian a suitable multiple of the Gauss law constraint. The Hermiticity preserving Hamiltonian is given by:

$$H'_{\underset{\sim}{T}} = H_{\underset{\sim}{T}} + \lim_{S\to\Sigma}\left(\frac{i}{\sqrt{2}}\int_S (D_a\underset{\sim}{T})\tilde{\sigma}^a{}_{AB}\tilde{C}^{AB}\right), \tag{36$'$}$$

where $H_{\underset{\sim}{T}}$ is the Hamiltonian defined in Eq.36 and $\tilde{C}_A{}^B$ is the Gauss constraint (Eq.28*c*).

ii) Let us see how the reality conditions given above arise from the space-time geometry of solutions to the field equations. Let us suppose that we are given an anti-Hermitian soldering form $\sigma^a{}_A{}^{A'}$, a connection 1-form ${}^4A_{aA}{}^B$, and spin-$\frac{1}{2}$ fields $(\xi^A, \eta^{A'})$, satisfying the field equations of section 3 on a real 4-manifold $\Sigma \times \mathbb{R}$. Denote by ∇ the torsion-free derivative operator compatible with $\sigma^a{}_A{}^{A'}$; by ${}^4\mathcal{D}$ the derivative operator defined by ${}^4A_{aA}{}^B$; and by D the torsion-free derivative operator on a 3-dimensional submanifold Σ of M compatible with $\sigma^a{}_A{}^B$. Then, by equations of motion, 4 and 5, we know that $({}^4\mathcal{D}_a - \nabla_a)\alpha_A = {}^4C_{aA}{}^B\alpha_B$, and, by the definition of extrinsic curvature, K_{ab}, it follows that $(\underleftarrow{\nabla}_a - D_a)\alpha_A = -\frac{i}{\sqrt{2}}K_{aA}{}^B\alpha_B$ where $\underleftarrow{\nabla}_a$ is the pull-back of ∇_a to Σ and $K_{aA}{}^B = K_{ab}\sigma^b{}_A{}^B$. Therefore, using the fact that the pull-back of ${}^4\mathcal{D}$ to Σ is $\mathcal{D}$, we have: $(\mathcal{D}_a - D_a)\alpha_A \equiv (A_{aA}{}^B - \Gamma_{aA}{}^B)\alpha_B = (C_{aA}{}^B - \frac{i}{\sqrt{2}}K_{aA}{}^B)\alpha_B$, where, as before, $\Gamma_{aA}{}^B$ is the spin connection 1-form defined by D, and $C_{aA}{}^B = q_a{}^b$-${}^4C_{bA}{}^B$. If we now express $C_{aA}{}^B$ in terms of $\tilde{\sigma}^a{}_A{}^B$, the Dirac fields, and their momenta on Σ, we obtain an expression for the extrinsic curvature K_{ab} in terms of the phase-space variables:

$$\frac{i}{\sqrt{2}}K_a^{AB} = -A_a^{AB} + \Gamma_a^{AB} - \frac{i}{2\sqrt{2}(\sigma)^2}(\tilde{\pi}^M\xi^{(A} + \tilde{\omega}^M\bar{\eta}^{(A})\tilde{\sigma}_a{}^{B)}{}_M. \tag{41}$$

When $\tilde{\sigma}^a{}_A{}^B$ and the Dirac fields satisfy the reality conditions (i.e. first of Eqs.40a and Eqs.40b), the reality condition on $A_{aA}{}^B$ (i.e. the second of Eq.40a) is precisely the requirement that K_{ab} be real.

iii) Let us return to the difference in the Einstein-Cartan and the Einstein-Dirac systems discussed in section 3. How does this difference manifest itself in the Hamiltonian description? Since the two Lagrangians differ from each other by a term which is quartic in Dirac fields, the two Hamiltonians also differ by the same term. However, since the term does not contain any derivative couplings, the relations between the Hermitian adjoints of spinor fields and their momenta as well as the equation of motion for $\tilde{\sigma}^a{}_A{}^B$ remain unchanged. Therefore, the reality conditions are also unaffected. (One can also arrive at this conclusion from the space-time view point mentioned above. Since the equation of motion of $A_{aA}{}^B$ and the definitions of momenta conjugate to the Dirac fields are the same in the two theories, the expression for the extrinsic curvature in terms of the phase space variables remains unaffected by the quartic term. Hence, the condition that the extrinsic curvature be real is also unaffected.) Thus, the only difference in the two theories is that their scalar constraints and hence also the volume terms in their Hamiltonians differ by a quartic combination of Dirac fields.[4]

4 In Ref. [9], Jacobson raised the question of whether, in the Hamiltonian description, the Einstein-Cartan and the Einstein-Dirac theories correspond to just two different real sections of the complex phase-space. In the light of remark *iii*, the answer is in the negative at least in the present framework.

7 Polynomial matter actions

In the last three sections, we saw that, in the Hamiltonian description, the reality conditions, the constraints and the Hamiltonians are all polynomial in the basic canonical variables inspite of the fact that our Lagrangian, depends non-polynomially on the basic field $\sigma^a{}_A{}^{A'}$through the determinant, $({}^4\sigma)$ of the inverse of $\sigma^a{}_A{}^{A'}$. This is rather surprising. Could one not use an alternate but equivalent action which is polynomial in the basic fields? If so, at least some of the mystery surrounding the simplification of the Hamiltonian framework would be dispelled. Of course, as the discussion in the first part of chapter 4 shows, a polynomial action would not guarantee polynomial Hamiltonian equations: There may be second class constraints which, when solved, may force a complicated dependence of first class constraints and evolution equations on the basic canonical variables. Nonetheless, it *would* make the polynomial character of the Hamiltonian equations seem less miraculous. In the source-free case, it has been known for some time now (see Samuel [5], or pages 88-89 of Ref. [1]) that a polynomial action does exist if one uses ${}^4\sigma_{aA}{}^{A'}$, with a *covariant* vector index, as the basic field in place of $\sigma^a{}_A{}^{A'}$. Unfortunately, once the matter fields are brought in with their standard Lagrangians, Samuel's strategy does not work since the standard Lagrangians explicitly contain the *contravariant* space-time metric (see Eqs.2*c* and 2*d*). Are there alternate Lagrangians one can use? The answer is in the affirmative.[5] The key observation is that, since simplifications arise [5] in the source-free case when one uses a *first order* action for gravity, one should use first order actions also for matter fields. The reason I did not use this strategy in the main text is that, as we shall see, the use of a first order formalism requires the introduction of additional fields and equations and therefore makes the framework less transparent. Since the discussion of matter terms brings in many subtleties, I felt that it would be better to first make the main points using standard matter actions that most readers would be familiar with and then indicate how the whole discussion could be reformulated using first order actions.

In the first order formalism, we need the following fields to describe matter: A cosmological constant Λ as before; a scalar field ϕ *and* a vector field p^a to incorporate a (massive) Klein-Gordon field; a connection ${}^4\mathbf{A}_a$ *and* a skew tensor field $\mathbf{p}^{ab}$, both taking values in the Lie algebra of the gauge group, to incorporate a Yang-Mills field; and, as before, spinor fields ξ^A and $\eta_{A'}$, to incorporate a (massive) Dirac field. With these fields at hand, the first order actions which are to replace Eqs.2 can be

5 The material in this and the following three paragraphs is based on an unpublished comment [21] by Burnett *et al.* The basic point seems to be known to a number of experts but, to my knowledge, had not appeared in the literature prior to [21].

written as:

$$S^1_E[\sigma_a{}^{AA'}, {}^4A_a{}^{AB}] := \frac{1}{2i}\int_M \sigma_{aA}{}^{A'}\sigma_{bBA'}{}^4F_{cd}{}^{AB}\,\tilde{\eta}^{abcd}\,d^4x \qquad (2a')$$

$$S_C[\Lambda, \sigma_a{}^{AA'}] := -\int_M \Lambda\sqrt{-g}\,d^4x \qquad (2b')$$

$$S^1_{KG}[\phi, p^a, \sigma_a{}^{AA'}] := -8\pi\int_M (p^a\partial_a\phi - \frac{1}{2}g_{ab}p^ap^b + \frac{1}{2}\mu^2\phi^2)\sqrt{-g}\,d^4x \qquad (2c')$$

$$S^1_{YM}[{}^4\mathbf{A}_a, \mathbf{p}^{ab}, \sigma_a{}^{AA'}] := -\int_M \mathrm{tr}(\mathbf{p}^{ab\,4}\mathbf{F}_{ab} - \frac{1}{2}g_{ac}g_{bd}\mathbf{p}^{ab}\mathbf{p}^{cd})\sqrt{-g}\,d^4x \qquad (2d')$$

$$S_D[\xi^A, \eta^{A'}, \sigma^a{}_A{}^{A'}, {}^4A_{aA}{}^B] := -\sqrt{2}\int_M [\tilde{\sigma}^a{}_{AA'}(\overline{\xi}^{A'}\,{}^4\mathcal{D}_a\xi^A - ({}^4\mathcal{D}_a\overline{\eta}^A)\eta^{A'}) + \frac{im}{\sqrt{2}}(\overline{\eta}_A\xi^A - \overline{\xi}^{A'}\eta_{A'})\sqrt{-g}]\,d^4x, \qquad (2e')$$

where

$$\begin{aligned} g_{ab} &:= \sigma_a{}^{AA'}\sigma_{bAA'}, \\ \epsilon_{abcd} &= \frac{2}{i}\sigma_{[aA}{}^{A'}\sigma_{bA'}{}^B\sigma_{cB}{}^{B'}\sigma_{d]B'}{}^A, \\ \sqrt{-g} &:= \frac{1}{4!}\tilde{\eta}^{abcd}\epsilon_{abcd}, \\ \tilde{\sigma}^a{}_{AA'} &:= \frac{i}{3}\tilde{\eta}^{abcd}\sigma_{bA'}{}^B\sigma_{cB}{}^{B'}\sigma_{dB'A}, \\ {}^4F_{ab}{}^{AB} &= 2\partial_{[a}{}^4A_{b]}{}^{AB} + [{}^4A_a, {}^4A_b]^{AB}, \\ \text{and } {}^4\mathbf{F}_{ab} &= 2\partial_{[a}{}^4\mathbf{A}_{b]} + [{}^4\mathbf{A}_a, {}^4\mathbf{A}_b]. \end{aligned} \qquad (42)$$

Here 'tr' stands for the trace operation in some representation of the Lie algebra of the Yang-Mills gauge group, and $\tilde{\eta}^{abcd}$ is the natural metric independent Levi-Civita tensor density of weight 1 on $\mathcal{M}$. One minimally couples matter to gravity by defining a total action S^1_T via $S^1_T := S^1_E + S_C + S^1_{KG} + S^1_{YM} + S_D$, and extremizes S^1_T with respect to the gravitational and matter fields. Note that the actions defined above are *polynomial* in the basic fields, in contrast to the actions used in section 2. The inverse of $\sigma_a{}^{AA'}$ is *never* used.

The above description of the Klein-Gordon and Yang-Mills fields by *pairs*, (ϕ, p^a) and $({}^4\mathbf{A}_a, \mathbf{p}^{ab})$, differs from the standard '2nd order' treatment that was used in section 2. However, the two treatments are completely equivalent if the soldering form is invertible. We consider the Klein-Gordon field in detail– the Yang-Mills field can be treated in a similar manner. Extremizing $S_T^{\frac{1}{T}}$ with respect to both p^a and ϕ we find

$$8\pi\sqrt{-g}(\partial_a\phi - g_{ab}p^b) = 0 \quad \text{and} \quad -8\pi[\partial_a(\sqrt{-g}p^a) - \sqrt{-g}\mu^2\phi] = 0. \tag{43}$$

When $\sigma_a{}^{AA'}$ is invertible, the inverse metric g^{ab} exists, allowing us to solve the first equation for p^a. We obtain $p^a = g^{ab}\partial_b\phi$, which upon substitution into the second equation yields the standard Klein-Gordon equation $(\Box - \mu^2)\phi = 0$. By pulling back the total action $S_T^{\frac{1}{T}}$ to the solution space defined by $p^a = g^{ab}\partial_b\phi$, we recover the action (Eq.3) initially used in section 2. The general argument about the equivalence of the action and its 'pulled-back' version given in section 2 tells us that the first order actions considered here are indeed equivalent to the standard actions considered in section 2. In particular, the (3+1)-decomposition of this pulled-back action yields the same constraint and evolution equations found in sections 4 and 5. If we chose, instead, to (3+1)-decompose $S_T^{\frac{1}{T}}$ *before* solving for p^a, we would have found that the covariant (spatial) component $\overline{p}_a$ of p^a acts as a Lagrange multiplier of the theory. The constraint equations obtained by varying $\overline{p}_a$ could be solved by the 'gauge' choice $\overline{p}_a = D_a\phi$; whence we would again obtain the standard results of sections 4 and 5.

8 Discussion

A primary motivation behind the "new variables" framework as a whole comes from the possibility that the micro-physics of the gravitational interaction may be simpler to formulate in terms of the "Yang-Mills type of variables" than the standard geometrodynamical ones. Thus, for example, in the source-free case, using these new variables Rovelli and Smolin [22] were able to obtain a large class of physical states, i.e., solutions to all quantum constraints. By contrast, as far as the full theory is concerned, not a single physical state is known in the more thoroughly investigated metric representation of quantum geometrodynamics. There are two reasons underlying this success of the new framework. First, a significant technical simplification occurs because constraints are polynomial in terms of $A_{aA}{}^B$ and $\tilde{\sigma}^a{}_A{}^B$. The second and perhaps more important reason is that the use of new variables opens up fresh directions in the canonical quantum gravity program. In particular, the "Yang-Mills formulation" of general relativity enables one to introduce two new representations of quantum states –the connection representation (chapter 11) in which states are functionals of the connection 1-form $A_{aA}{}^B$ and the loop representation (chapter 13) in which they arise as functions on the loop space of the 3-manifold Σ– which in turn facilitates the problem of solving the quantum constraints. The viewpoint underlying the present program is that this shift of emphasis from geometrodynamics to gauge theory has a deep significance. Observables associated with, e.g., the parallel transport of "spinors" around closed loops are to be regarded as the fundamental quantities in the Planck regime while the space-time geometry is to be regarded as a secondary concept which comes on its own only in the semi-classical and classical approximations. This viewpoint is supported by the recent results in $(2+1)$-gravity [23, 24].

An important check on the viability of these ideas is whether or not the attractive features of the framework survive the introduction of matter. For, it is often the case that elegant features of source-free gravity are destroyed when sources are brought in. Let us therefore examine this question in some detail in the light of the results presented in this chapter. In the spirit of the program, it is natural to regard $A_{aA}{}^{B}$ as the configuration variable and $\tilde{\sigma}^a{}_A{}^B$, its conjugate momentum, as being analogous to the Yang-Mills electric field. Therefore, in quantum theory, we wish to represent $\tilde{\sigma}^a{}_A{}^B$ by an operator of the type $\delta/\delta A_{aA}{}^B$. But such a representation is permissible only if the classical momentum variable takes values in a vector space, i.e., in the present case, only if $\tilde{\sigma}^a{}_A{}^B$ is allowed to become degenerate. Therefore, to import gauge theory ideas into canonical quantum gravity, we must make sure that the classical Hamiltonian theory itself is meaningful if $\tilde{\sigma}^a{}_A{}^B$ is degenerate. Now, in section 2, we began by assuming that $\sigma^a{}_A{}^{A'}$ –and hence the 4-metric g_{ab}– are non-degenerate. In our presentation, the non-degeneracy was essential to perform the Legendre transform and to pass to the Hamiltonian description. However, in the final Hamiltonian formulation –which is complete in itself– the requirement can be dropped entirely. The symplectic structure, the constraint and evolution equations, the Hamiltonian and the reality conditions, all continue to be meaningful even when we allow $\tilde{\sigma}^a{}_A{}^B$ to become degenerate. A key question is whether the proof that the constraints are first class depends on non-degeneracy. For, even if the constraints themselves are polynomial in the basic canonical pair, the structure functions in the Poisson algebra may involve the inverse of $\tilde{\sigma}^a{}_A{}^B$. That is, while the expressions of the Poisson brackets may be polynomial, dependence on the inverse of $\tilde{\sigma}^a{}_A{}^B$ may creep in while re-expressing them as a linear combination of constraints.[6] Fortunately, this does *not* happen. Thus, the key features which

6 That this is not just an abstract possibility is illustrated by the following example. In the source-free case, assuming that $\tilde{\sigma}^a{}_A{}^B$ is non-degenerate, the vector and the scalar constraint can be combined into a single expression: $\tilde{\sigma}^a\tilde{\sigma}^b F_{ab} = 0$. The trace of this equation gives the scalar constraint and the trace-free part, the vector constraint. This form of the constraints is again polynomial and has a further attractive feature that, as in $(2+1)$-gravity, the only free indices are the internal ones. Although the resulting algebra is again of first class - as it must be - the structure functions involve the inverses of $\tilde{\sigma}^a{}_A{}^B$!

enabled one to introduce the connection and the loop representations in the source-free case are robust and survive the coupling to matter sources. Furthermore, since every equation in the final canonical description is polynomial in the basic variables, one hopes that the Jacobson-Rovelli-Smolin solutions to quantum constraints can be extended to include sources. Recently, Rovelli has extended the loop representation to pure Yang-Mills theory (private communication) and Brügmann has investigated lattice QCD in the loop representation ([25]). Therefore, there is now an attractive possibility on the horizon that the use of loop representation may enable one to construct, in an unified way, the non-perturbative quantum theory of all four basic interactions.

The above discussion brings out the fact that the Hamiltonian framework presented in sections 5 and 6 is in fact a slight generalization of Einstein's theory with matter sources, reducing to it in the case when $\tilde{\sigma}^a{}_A{}^B$ is non-degenerate. The new equations do admit a wider class of solutions in which $\tilde{\sigma}^a{}_A{}^B$ is degenerate. In particular, there is no *a priori* reason why $\tilde{\sigma}^a{}_A{}^B$ couldn't vanish identically even though $A_{aA}{}^B$ and its curvature do not vanish: these would be solutions in which there is curvature but no metric! While such solutions have no obvious significance in the classical theory –at least when $\tilde{\sigma}^a{}_A{}^B$ vanishes on a set of non-zero measure– they may be of considerable interest in quantum theory. Indeed, in 2+1 gravity, the most natural description of the quantum vacuum involves precisely the configuration in which $\tilde{\sigma}^a{}_A{}^B$ vanishes identically [23].

We will conclude by pointing out a curious feature of the Einstein Yang-Mills system in the present framework. Now, the Einstein fields are represented by pairs $(A_a, \tilde{\sigma}^a)$ and the Yang-Mills, by analogous pairs $(\mathbf{A}_a, \tilde{\mathbf{E}}^a)$. Both are subject to Gauss constraints: $\mathcal{D}_a\tilde{\sigma}^a = 0$ and $\mathbf{D}_a\tilde{\mathbf{E}}^a = 0$. The symmetry extends also to the vector constraint which has the form: $\mathrm{tr}\,\tilde{\sigma}^a F_{ab} = (constant)\,\mathrm{tr}\,\tilde{\mathbf{E}}^a\mathbf{F}_{ab}$. It is tempting to conjecture that this symmetry may be a reflection of a new type of possible unification. Perhaps there is a way to modify the present framework to larger internal symmetry groups so that a part of the new A_a captures the Einstein self dual connection and the remainder, the Yang-Mills connection 1-form, such that the source-free Einstein constraints on the new A_a and its conjugate momentum are equivalent to the Einstein Yang-Mills constraints of section 4. This would be complementary to the Kaluza-Klein type of unification. Now, the space-time dimension would continue to be $3+1$ but the internal symmetry group would be enlarged to encompass both the Einstein and the Yang-Mills fields and it is the gravitational field that would emerge as a part of an enlarged Yang-Mills type connection rather than the Yang-Mills field emerging as a part of a metric in a higher dimensional space-time.

References

[1] A.Ashtekar (with invited contributions), *New Perspectives in Canonical Gravity* (Bibliopolis, Naples, 1988).

[2] A.Sen, J. Math. Phys. **22**, 1718 (1981); Phys. Lett. **119B**, 89 (1982).

[3] A.Ashtekar, in *Quantum Concepts in Space and Time*, C.J.Isham and R.Penrose (eds.), 302 (Oxford University Press, Oxford, 1986);

[4] A.Ashtekar, Phys. Rev. Lett. **57**, 2244 (1986); Phys. Rev. **D36**, 1587 (1987).

[5] J.Samuel, Pramana J. Phys. **28**, L429 (1987).

[6] T.Jacobson and L.Smolin, Phys. Lett. **196B**, 39 (1987); Class. & Quant. Grav.**5**, 583 (1988).

[7] A.Ashtekar, J.D.Romano, R.S.Tate, Phys. Rev. **D40**, 2572 (1989).

[8] S.Koshti, *private communication.*

[9] T.Jacobson, Class. & Quant. Grav. **5**, L143 (1988).

[10] A.Ashtekar, A.P.Balachandran and S.Jo, Intl. J. Mod. Phys. **4**, 1493 (1989).

[11] P.Renteln, Ph.D. Thesis (Harvard University, 1988); P.Renteln and L.Smolin, Class. & Quant. Grav. **6**, 275 (1989).

[12] H.Weyl, Phys. Rev. **77**, 699 (1950);

[13] R.Penrose and W.Rindler, *Spinors and Space-time*, vol.1(Cambridge University Press, Cambridge, 1984).

[14] A.Ashtekar, G.T.Horowitz and A.Magnon, Gen. Rel. & Grav. **14**, 411 (1982).

[15] R.Wald, *General Relativity* (The University of Chicago Press, Chicago, 1984), see appendix E.

[16] T.Jacobson, Class. & Quant. Grav. **5**, 923 (1988).

[17] A.Ashtekar and A.Magnon, Class. & Quant. Grav. **1**, L39 (1984).

[18] A.Ashtekar, P.Mazur and C.T.Torre, Phys. Rev. **D36**, 2955 (1987).

[19] P.Bergmann and A.Komar, Intl J. Theo. Phys. **5**, 15 (1972).

[20] S.J. Hajman, C.Teitelboim and K.Kuchař, Nature **245**, 97 (1973); Ann. Phys. (N.Y.) **96**, 88 (1974).

[21] G. Burnett, J. D. Romano and R. S. Tate, *Polynomial coupling of matter to gravity using Ashtekar variables*, Syracuse University pre-print.

[22] C.Rovelli and L.Smolin, Nucl. Phys. **B331**, 80 (1990).

[23] E.Witten, Nucl. Phys. **B311**, 41 (1988).

[24] S.Carlip, Nucl. Phys. **B324**, 106 (1989); S.Martin, Nucl. Phys. **B327**, 78 (1989); A.Ashtekar, V.Husain, C.Rovelli, J.Samuel and L.Smolin, Class. & Quant. Grav. **6**, L185 (1989).

[25] B.Brügmann, Phys. Rev. **D**, in print (1990).

III QUANTUM THEORY

I think it must be the case that the all-pervading use of the continuum in physics stems from its mathematical utility *rather than from any essential physical reality that it may possess. However, it is not even quite clear that such use of the continuum is not, to some extent,* a historical accident. ... *My own view is that ultimately physical laws should find their most natural expression in terms of essentially* combinatorial *principles, that is to say, in terms of finite processes such as* counting or *other basically simple manipulative procedures.* Thus in accordance with such a *view, should emerge some form of discrete or combinatorial space-time.*

- Roger Penrose, *On the nature of quantum geometry*

In this part, we shall use the Hamiltonian framework of Part II as the point of departure for canonical quantization of general relativity.

In chapter 11, we present the general quantization program. The overall procedure is the same as the one proposed by Dirac for quantization of constrained systems. However, the program goes beyond the Dirac theory in that it provides a strategy for finding the inner product on the space of physical states: Choose that inner product with respect to which the operators corresponding to any two classical variables which are complex conjugates of one another become Hermitian adjoints. Thus, far from being a nuisance, the classical "reality condition" are to be thought of as the vehicles that are lead us to the correct inner product. At a certain level, the strategy is an obvious one. However, it is not the one that is generally employed. In Minkowskian field theories, for example, the inner product is chosen by making an appeal to Poincaré invariance. We implement the quantization program in a number of examples with and without constraints and show that it does indeed succeed. Although these examples have only a finite number of degrees of freedom, they all mimick different features of general relativity and are also of considerable interest in their own right. In particular, we show how one can unambiguously quantize a harmonic oscillator in the z, q variables which are analogous to the pair $A_a^i, \widetilde{E}_i^a$ of general relativity and how the program enables one to quantize parametrized particle systems *without* having to first single out time.

The next three chapters are devoted to the development [1] and applications of the connection representation in which quantum states arise as holomorphic functions of the self dual connection A_a^i. In chapter 11, we show how to recover the Fock quantization of linearized gravity in this representation. We implement the general program step by step and find that the reality conditions are indeed strong enough to pick out the correct inner product without having to explicitly demand Poincaré invariance at any stage. In chapter 12, the connection representation is introduced for the full theory. We argue that this representation is well-suited to address the issue of time in canonical quantum gravity. In a suitably truncated version of the theory, one can implement these ideas in a concrete fashion to isolate an internal time among the components of the connection A_a^i in such a way that the quantum scalar constraint reduces to the familiar Schrödinger equation for gravity plus matter [2].[1] In chapter 13, we present another application of the connection representation: the analysis of the gravitational CP problem [4]. There has been some confusion on this issue in the geometrodynamical framework. The connection representation is well-suited to clarify these points since it is closely related to the representation one normally uses in QCD where the strong CP problem is well understood. In broad terms, the connection representation seems to be useful in analysing issues which have direct analogs in non-gravitational interactions and thus to make a contact with everyday physics. It has not been very useful in extracting the exact solutions to all quantum constraints, i.e., to probe in detail the Planck regime itself.

In the next three chapters, we introduce the loop representation which is better adapted for this task. Now, the quantum states arise as functionals of closed loops on the ("spatial") 3-manifold. This is quite unusual because quantum states are normally represented by functionals of some dynamical variables and closed loops have no dynamical content what so ever in classical general relativity. Recall, however, that, already in non-relativistic quantum mechanics, one does not *have to* use dynamical variables as arguments of wave functions. Indeed, it is sometimes extremely useful to make a different choice. As was pointed out in chapter 2, in the quantum theory of hydrogen atom, it is particularly convenient to use a representation in which states $\psi(n,l,m) =< n,l,m|\psi >$ are functions of three *integers*, the principal quantum numbers, eventhough the triplet n,l,m has no significance at all to the classical dynamics of the Coulomb problem. It turns out that this analogy is quite good. Just as the problem of finding the energy eigenvalues is simpler in this representation of the hydrogen atom, the problem of finding solutions to the constraints of general relativity is simpler in the loop representation; this representation is useful in analysing the micro-structure of space-time at Planck scale. To answer semi-classical questions, such as the ones pertaining to the position of the electron, the x-representation is more useful in the hydrogen atom problem. Similarly, the A_a^i-representation is better adapted to the analysis of the semi-classical regime in quantum gravity.

In chapter 14, we introduce the key ideas behind the loop representation through an example: the quantum Maxwell field in Minkowski space-time. In chapter 15, we discuss the classical theory behind the loop representation for general relativity and in chapter 16, we summarize the current status of the quantum theory. There

1 Kodama [3] has recently proposed a different approach to the problem of time in the context of quantum cosmology. His approach differs from the one proposed here in that his wave functions are not *entire* in the component of the connection that corresponds to time. The relation between the two approaches is not well understood.

are three related but different approaches to quantization [5-7]. We present the first and the most developed approach, due to Rovelli and Smolin. The end result is that there is a large class of exact solutions to quantum constraints. The solutions suggest that the quantum geometry in the Planck regime is best described in terms of *discrete structures*, associated with knot and link classes of closed loops. Approximation methods are now being developed to compute physical quantities using these structures. The hope is that these techniques will enable one to pinpoint where and why the standard perturbation theory fails. Using the ideas from quantum gravity, the loop representation has now been developed also for Yang-Mills theory both on lattice and continuum [8]. It appears to be especially well suited for certain numerical lattice calculations [9]. Thus, there is now a synergetic exchange of ideas and techniques between quantum gravity and QCD.

While considerable progress has thus been made, the quantization program is still incomplete. There are of course many important and fascinating issues of interpretation that are yet to be addressed satisfactorily and the quantum measurement problems remain almost entirely unexplored. Furthermore, even as a program in mathematical physics, several steps remain to be completed: construction of a complete set of physical states, the introduction of an inner product and isolation of physically interesting operators on the physical sub-space. To gain insight into these issues, in chapter 17, the program is carried out in a simpler model: 2+1 dimensional quantum gravity. This system has all the conceptual problems that we encounter in the 3+1 theory but is technically more manageable because it has only a finite number of degrees of freedom. We will find that the program *can* be carried out to completion and the solution offers several interesting lessons for the 3+1 theory.

References

[1] A. Ashtekar, Phys. Rev. Lett.**57**, 2224 (1986).

[2] A. Ashtekar, in *Conceptual Problems of Quantum Gravity*, edited by A. Ashtekar and J. Stachel (Birkhäuser, Boston 1991).

[3] H. Kodama, Phys. Rev.**D42**, 2548 (1990).

[4] A. Ashtekar, A.P. Balachandran and S. Jo, Int. J. Theo. Phys.**A4**, 1493 (1990).

[5] C. Rovelli and L. Smolin, Phys. Rev. Lett.**61**, 1155 (1988); Nucl. Phys.**B331**, 80 (1990).

[6] M. Blencowe Nucl. Phys.**341**, 213 (1990).

[7] R. Gambini, Loop space representation of quantum general relativity and the group of loops (Montevideo pre-print, 1990).

[8] C. Rovelli and L. Smolin, Loop representation for lattice gauge theory, (Syracuse pre-print, 1990).

[9] B. Brügmann, Phys. Rev.**D** (in press).

10 THE QUANTIZATION PROGRAM

1 Introduction

In this chapter we will spell out the steps for passage to quantum theory. We will begin with simple quantum mechanical systems to illustrate the main ideas and then arrive at the desired quantization program for general relativity. In chapters 11–16 we will report on the progress that has been made so far in implementing this program. We will see that while several important issues have been resolved by now, a number of key problems are yet to be faced. The entire program has, however, been completed in a number of simpler models that mimic the structure of full general relativity. One such model will be discussed in the last chapter of this Part.

The program is based on an algebraic approach to quantum theory. We shall assume that the phase space of the system is a cotangent bundle over some configuration space, an assumption that is satisfied in the case of general relativity. In section 2, we begin with quantum mechanical systems without constraints and recover the Schrödinger representation –discussed in appendix C– from an algebraic approach. In section 3, we focus on the harmonic oscillator and show how the algebraic approach can be completed using the "hybrid" (q, z) variables which are analogous to the new variables $(\widetilde{E}_i^a, A_a^i)$ for general relativity. Section 4 discusses the modifications needed in the program to accommodate systems with first class constraints. It must be emphasized that the program constitutes only a set of guidelines; it is not a "crank" that can be turned to convert a classical theory to a quantum one. In particular, the procedure requires two key inputs on which the final answer depends crucially. The choices involved have to be made using physical intuition derived from one's experience with simple systems. To illustrate this point, we apply the program to three "toy" models and work out in detail the resulting quantum theories. These are: *i*) a parametrized non-relativistic particle; *ii*) a free particle in Minkowski space; and, *iii*) two harmonic oscillators "coupled" by the constraint that the difference between their energies is fixed. These systems are rather simple. However, they mimic various aspects of the constraints of Einstein's theory. All these ideas are finally distilled in section 5 to arrive at a non-perturbative, canonical quantization program for general relativity. We conclude with a discussion of how the program bypasses the traditional difficulty of "isolating time" by adopting a "covariant" approach to the problem of finding the inner product on the space of physical states.

2 Simple systems

Fix a physical system whose configuration space C is a finite dimensional, real manifold. The phase space Γ is then the cotangent bundle over C. In this section, we assume that there are no constraints. The classical mechanics of the system is then governed by the symplectic geometry of Γ (see appendix B) while the Schrödinger quantum mechanics is dictated essentially by the differential geometry of C (see appendix C).

In the algebraic approach to quantum mechanics, one begins by choosing certain *elementary classical variables* which are to have unambiguous quantum analogs. For reasons that will become clear as we proceed, the choice of the set S of elementary variables is subject to two conditions:

i) S should be a complex vector space which is closed under the operation of taking Poisson brackets.

ii) S should be "large enough" so that any (sufficiently regular) function on Γ can be represented as (possibly a limit of) a sum of products of elements of S.

Before going on to construct the algebra of quantum operators based on S, let us examine some natural candidates for the space S, in order to develop a feeling for the type of structures that will arise. From text-book quantum mechanics we expect that, in order to be physically acceptable, the final quantum theory should have configuration and momentum operators. In the classical theory, configuration variables are the functions on Γ which are independent of momenta –i.e., which are the pull-backs to Γ of smooth functions on C– while the momentum variables are the functions on Γ which are linear in momenta. Let us therefore begin by considering the vector space S_0 spanned by all complex-valued functions on Γ which are either independent of or linear in momenta. Since any observable which is polynomial in momenta and has arbitrary dependence on the configuration variables can be expressed as a finite sum of products of elements of S, this space is clearly sufficiently large; requirement *ii)* above is met by S_0. Requirement *i)* is also satisfied because the Poisson bracket between any two configuration variables vanishes, that between any configuration and any momentum variable is a configuration variable while that between any two momentum variables is a momentum variable:

$$\begin{aligned} \{f(q), g(q)\} = 0; \qquad \{V^a(q)p_a, f(q)\} = -(\mathcal{L}_V f)(q); \\ \{V^a(q)p_a, W^b(q)p_b\} = -(\mathcal{L}_V W^a)(q)p_a \,. \end{aligned} \tag{1}$$

What is the situation with respect to variables which are of higher order in momenta? The Poisson bracket between two functions which are quadratic in momenta is *cubic* in momenta. More generally, the bracket between any two functions which are respectively of order m, n, in momenta is of order $m + n - 1$ in momenta. Therefore, had we also allowed as elementary variables those functions which are quadratic in momenta, the requirement *i)* of the closure under the Poisson algebra would have driven us to include among the elementary observables arbitrary polynomials in momenta. The resulting vector space of elementary observables would then have been essentially as large as the space of all functions on the phase space. From text-book quantum mechanics we know that this is inadmissible; due to factor ordering problems, we cannot associate a quantum operator with an arbitrary function on the phase space in a natural way. These considerations suggest that the "largest space" we can choose for S is S_0. This is of course not a watertight conclusion; the argument has a number of caveats. In particular, we could have used a totally different scheme based on a few, judiciously chosen variables which are *non-polynomial* in momenta. However, as a general rule of thumb, the conclusion is quite useful.

Note, however, that the vector space S_0 is really "overcomplete" in the sense that elements of S_0 are constrained by algebraic relations. Thus, given any configuration variable $f(q)$ and $g(q)$, their linear combination $af(q) + bg(q)$ with arbitrary complex coefficients a and b, their product, $h(q) := f(q)g(q)$ and their complex conjugates $\overline{f(q)}$ and $\overline{g(q)}$ are all configuration variables and appear as distinct elements of the vector space S_0. Similarly, given any two momentum variables, their linear combinations, complex conjugates and their products with arbitrary configuration variables are again momentum variables. Therefore, if we associate a quantum operator with each element of S_0, we must make sure that the resulting set of operators faithfully captures the algebraic relations between their classical counterparts. We shall see that this can indeed be arranged.

In certain cases, however, we can avoid much of this over-completeness by choosing for the set S of elementary variables, a small subspace of S_0[1] For example, if the phase space Γ itself has the structure of a vector space, one may choose for S

1 An example is provided by general relativity. We shall see that both the connection and the loop representations of quantum general relativity use for S spaces which are smaller than S_0; however, the two spaces are different from one another.

the space of *linear* (and constant) functions on Γ. This space is closed under the Poisson-bracket operation and also "sufficiently large" since every (regular) function on the phase space can be expressed as a sum of products of its elements. However, now, products of elements of $\mathcal{S}$ are not again in $\mathcal{S}$ whence the anti-commutation relations become redundant. Consequently, we have only to accommodate the linear and the complex-conjugate relations in the quantum theory. This is the choice one generally makes in non-relativistic quantum mechanics, where $\mathcal{S}$ is taken to be the seven dimensional vector space generated by complex linear combinations of the (coordinate) functions $(1, x, y, z, p_x, p_y, p_z)$ on Γ. Even when the configuration space $\mathcal{C}$ is a genuine manifold, by introducing a preferred chart on $\mathcal{C}$, one can generally reduce the over-completeness. For example, if $\mathcal{C}$ happens to be a circle, S^1, one can let $\mathcal{S}$ be the four dimensional vector space generated by the functions $(1, \cos\theta, \sin\theta, p)$ on Γ, where θ is the obvious angular parameter along S^1. In this case, in addition to the linear and the complex-conjugate relations, one has only one algebraic relation to take into account in quantum theory: $\cos^2\theta + \sin^2\theta = 1$. On the other hand, the space $\mathcal{S}_0$ constructed above is infinite dimensional even in this case. Consequently, if one uses $\mathcal{S}_0$ as the space of elementary variables, there are infinitely many (anti-commutation) relations –discussed above– that one must incorporate in the construction of the quantum algebra.

Let us now construct the quantum operator algebra. Associate with every element F of $\mathcal{S}$, an abstract operator $\hat{F}$ such that the linear relations between elements of $\mathcal{S}$ are faithfully reflected among the operators:

$$\sum_i a_i \hat{F}_i = 0 \quad \Leftrightarrow \quad \sum_i a_i F_i = 0 \quad \text{in } \mathcal{S}\,. \tag{2}$$

The $\hat{F}$ are the ***elementary quantum operators***. All other quantum operators are to be obtained by taking sums of products of these elementary ones. However, since the quantum algebra is going to be non-commutative, its product structure cannot be the same as that of the classical algebra. Rather, the quantum algebra $\mathcal{A}$ is constructed as follows. Consider first the free associative algebra over complex numbers generated by the vector space $\mathcal{S}$. This algebra is generated by elements of $\mathcal{S}$: each element of $\mathcal{A}$ is just a finite sum of products of the elements of $\mathcal{S}$ where two such sums are set equal to each other *only* if one can be obtained from the other by manipulations permitted just by the axioms defining an associative algebra. (For a precise definition of a free associative algebra generated by a vector space, see, e.g. [1].) On this free algebra, we have to impose two sets of relations. The first is the canonical commutation relations:

$$[\hat{F}, \hat{G}] - i\hbar\widehat{\{F, G\}} = 0, \tag{3}$$

for all F, G in $\mathcal{S}$. These are well-defined because the vector space $\mathcal{S}$ of elementary classical variables is closed under Poisson brackets.

The second set consists of certain *anti-commutation* relations which arise when the set S is overcomplete; they capture the algebraic relations that exist among the elementary classical variables. Let us begin with the simplest case. The idea is that, *if all three functions F, G, FG on Γ are in the space S*, then we should require:

$$\hat{F} \cdot \hat{G} + \hat{G} \cdot \hat{F} - 2\widehat{FG} = 0\,, \tag{4a}$$

More generally, if $F_1, F_2, \cdots, F_n$, as well as their product, $F_1 F_2 \cdots F_n$, belong to the space S, we require that

$$\hat{F}_{(1}\hat{F}_2 \cdots \hat{F}_{n)} - (\widehat{F_1 \cdots F_n}) = 0. \tag{4b}$$

Technically, the imposition of these relations simply amounts to taking the quotient of the free algebra by the Lie ideal generated by the left sides of Eqs.3 and 4. The result is the required algebra $\mathcal{A}$ of quantum operators. Note that eventhough we are imposing anti-commutation relations, there is nothing fermionic about the system. It must also be emphasized that the purpose of Eq.4 is *not* to resolve any factor ordering ambiguity in the definition of the operator $\widehat{FG}$: Since FG is, by assumption, an elementary variable, $\widehat{FG}$ is already a well-defined, –in fact *elementary*– quantum operator. Rather, Eq.4 is imposed to remove unwanted sectors in the final quantum description. To see this, suppose that we had not imposed Eq.4. Then in the resulting algebra of quantum operators, $\hat{\mathcal{Q}} := \hat{F}_{(1}\hat{F}_2 ... \hat{F}_{n)} - (\widehat{F_1 ... F_n})$ would commute with all operators because of Eq.3. Hence, $\hat{Q}$ would super-selected and we would obtain the correct classical limit *only* in the sector on which $\hat{Q}$ takes the value zero. By imposing Eq.4, we avoid the spurious sectors right from the beginning.

Again, to get a feel for this second step in the algebraic approach, let us consider some examples. If one chooses the space S_0 for S, then Eqs.3 are the canonical commutation relations between the configuration and the momentum operators:

$$\begin{aligned} &[\widehat{f(q)}, \widehat{g(q)}] = 0; \qquad [\widehat{V^a p_a}, \widehat{f(q)}] = -i\hbar(\widehat{\mathcal{L}_V f})(q); \\ &[\widehat{V^a p_a}, \widehat{W^b p_b}] = -\, i\hbar(\widehat{\mathcal{L}_V W})^a p_a\,. \end{aligned} \tag{3.a}$$

while Eqs.4 are the anti-commutation relations:

$$\begin{aligned} \{\widehat{f(q)},\, \widehat{g(q)}\}_+ &= 2(\widehat{f(q)g(q)}) \\ \text{and} \quad \{\widehat{f(q)},\, \widehat{V^a p_a}\}_+ &= 2(\widehat{fV^a p_a})\,. \end{aligned} \tag{4.a}$$

between these operators. While the commutation relations may be familiar from the text book treatment of quantum mechanics, the anti-commutation relations may seem surprising at first. It is important to note, however, that these relations *must* be imposed if one is to obtain the correct quantum description at the end. (For details, see appendix C.) As we remarked earlier, the text book treatments use for $\mathcal{S}$ the seven dimensional vector space generated by $(1, x, y, z, p_x, p_y, p_z)$ whence Eq.4 becomes redundant while Eqs.3 provide the familiar commutation relations between the three position and the three momentum operators. In the example with the configuration space S^1, on the other hand, in addition to the commutation relations, there *is* a non-trivial anti-commutation relation $(\widehat{\cos\theta})^2 + (\widehat{\sin\theta})^2 = 1$, in the quantum algebra $\mathcal{A}$. If this is left out, the quantization procedure acquires ambiguities and one can get inequivalent quantum theories. We shall see in chapters 14 and 15 that such non-trivial anti-commutation relations arise also in the loop representation in quantum general relativity.

The third step in the construction is to introduce $\star$-relations on the algebra $\mathcal{A}$. These are to reflect the relations between the classical variables and their complex conjugates. More precisely, we introduce on $\mathcal{A}$ an involution operation, denoted by $\star$, as follows. Set

$$(\hat{F})^\star = \hat{G} \quad \text{iff} \quad \overline{F} = G \tag{5a}$$

for all F and G in $\mathcal{S}$, where as usual $\overline{F}$ is the complex conjugate of the function F on Γ, and extend the $\star$-operation to the full algebra $\mathcal{A}$ via the axioms of the involution operation:

$$(\alpha A + \beta B)^\star = \overline{\alpha} A^\star + \overline{\beta} B^\star, \quad (A \cdot B)^\star = B^\star \cdot A^\star, \quad (A^\star)^\star = A, \tag{5b}$$

where A, B are arbitrary elements of $\mathcal{A}$ and α, β are any complex numbers. Note that in this procedure, we have *not* added new elements to $\mathcal{A}$. We have merely equipped $\mathcal{A}$ with a new operation. Nonetheless, it is useful to distinguish between the algebra with and without the $\star$-operation. When $\mathcal{A}$ *is* equipped with this involution, we will denote it by $\mathcal{A}^{(\star)}$. Note that, once the initial choice of the space $\mathcal{S}$ of elementary classical variables is made, the $\star$-algebra $\mathcal{A}^{(\star)}$ is uniquely determined.

The fourth step in the program is to find a representation of the algebra $\mathcal{A}$ by linear operators on a complex vector space V. This is the second and the last choice one must make in the quantization program. At this stage, the $\star$-relations are ignored and we focus just on the commutation and anti-commutation relations, Eqs.3 and 4, between elements of $\mathcal{A}$. For example, had we initially chosen $\mathcal{S}$ to be $\mathcal{S}_0$, we can choose for V the space of smooth complex-valued densities $\Psi(q)$ of weight $\frac{1}{2}$ on the configuration space $\mathcal{C}$ and represent the abstract operators $\widehat{f(q)} \equiv \hat{Q}(f)$ and $\widehat{V^a p_a} \equiv \hat{P}(V)$ by the concrete expressions:

$$\hat{Q}(f) \cdot \Psi(q) = f(q)\Psi(q) \quad \text{and} \quad \hat{P}(V) \cdot \Psi(q) = -i\hbar\, \mathcal{L}_V \Psi(q)\,, \tag{6}$$

which automatically satisfy Eqs.3 and 4.[2] However, at this stage, we could have chosen other representations of $\mathcal{A}$ as well. For example, we could have let V be the complex vector space of smooth functions –rather than densities– of compact support on C but continued to represent the configuration operator $\hat{Q}(f)$ by a multiplication operator and the momentum operator $\hat{P}(V)$ by just the Lie derivative term. Again, Eqs.3 and 4 would have been satisfied. However, as we shall see below, this choice would not have led to a satisfactory quantum theory.

The last step in the mathematical construction of the quantum theory is the introduction of a suitable Hermitian inner product on the space V of states. It is here that we need the $\star$-relations which have been ignored so far. The idea is to use these relations to select an inner product; we seek a Hermitian inner product on the linear space V such that the abstract $\star$-relations in $\mathcal{A}^{(\star)}$ become the Hermitian-adjoint relations with respect to that inner product. There is no guarantee that such an inner product will exist. For it to exist, the linear representation has to be chosen properly. However, if the product does exist then on each irreducible sub-space of V, the resulting $\star$-representation of $\mathcal{A}^{(\star)}$ is unique (upto of course unitary equivalence).

In the example considered above, the unique inner product that we thus obtain is precisely the one used in appendix C:

$$\langle \Psi \mid \Phi \rangle = \int_C d^n q \, \overline{\Psi(q)} \, \Phi(q) \, , \tag{7}$$

where the integral is well-defined because the integrand is a density of weight 1. On the other hand, if we choose V to be the space of functions rather than densities but continue (unlike in footnote 1) to represent the momentum operators via just the Lie derivatives, the situation is very different. Now, we need a volume element on C to define the inner product and a given momentum operator would be its own Hermitian adjoint *only* if it is divergence-free with respect to that volume element, i.e., only if the Lie derivative of the volume element with respect to that vector field vanishes.[3] Since, in the case under consideration, our initial set S includes momentum operators associated with *any* (sufficiently regular) vector field on C, for all these operators to be self-adjoint, we would need a volume element which is

2 Alternatively, as in appendix C, we can fix a fiducial volume element ϵ on C, choose for V the space of smooth functions $\psi(q)$ of compact support and replace Eq. 6 by: $\hat{Q}(f)\cdot\psi(q) = f(q)\psi(q)$ and $\hat{P}(V)\cdot\psi(q) = -i\hbar(\mathcal{L}_V\psi(q) + \frac{1}{2}(\mathrm{Div}_\epsilon V)\psi(q)$, where $\mathrm{Div}_\epsilon V$ is the divergence of V^a with respect to the volume element ϵ. This would be completely equivalent to the use of densities and Eqs. 6.

3 For simplicity, throughout these notes, we will ignore the distinction between self-adjoint and symmetric operators.

Lie-dragged by *all* vector fields. Clearly, no such volume element exists. Therefore, there is *no* Hermitian inner-product which can incorporate all the $\star$-relations with this choice of the linear representation. Note, finally, that if one chooses a smaller set S initially, the requirements on the volume element are less severe and then it is often possible to satisfy them. For example, consider the case when Γ has the structure of a vector space and S is simply the space of constant and linear functions on Γ. In this case, one can indeed choose for V the space of smooth functions of compact support on C and represent the momentum operators simply by Lie derivatives since now the vector fields defining momenta in S are the constant ones and the standard translation invariant volume element $d^n q$ on C is Lie-dragged by all constant vector fields.

Thus, for the program to ultimately succeed, once the set S is chosen, the linear representation of the resulting quantum algebra $\mathcal{A}$ has to be selected judiciously.

3 (q, z) variables for the oscillator

The Hamiltonian description of general relativity obtained in Part II is of-course substantially more complicated than that of the simple systems considered above. At the conceptual level, the major differences are the following: *i*) In order to simplify the constraints and use techniques from QCD, we are led to consider "hybrid" variables $(\widetilde{E}_i^a, A_a^i)$ on the gravitational phase space, where $\widetilde{E}_i^a$ is real but A_a^i is complex; *ii*) the system has non-trivial first class constraints; and, *iii*) the system has an infinite number of degrees of freedom. In this section, we shall see how the above program can be carried out with "hybrid" variables. In the next section, we shall present the modifications required due to the presence of constraints. The last difference will confront us with difficult regularization issues. While, as we shall see, some work has been done on these issues, it will be some time before one has a conclusive resolution.

In Part II, we saw that there are two ways of looking at the canonical formulation of general relativity in terms of new variables. In the first, presented in chapter 3, one obtains a clean canonical formulation of complex general relativity and then recovers the real theory by imposing the reality conditions by hand. In the second, presented in chapters 6-8, one works throughout with the real theory but for convenience uses complex variables on its real phase space. Throughout this Part, I will adopt the second point of view. In this section, therefore, we will carry out the quantization program for a harmonic oscillator from this viewpoint.

Let us then focus on a harmonic oscillator with unit mass and spring constant. The phase space Γ is a 2 real dimensional vector space, coordinatized by the real variables q and p. The symplectic structure is given by $\Omega = dp \wedge dq$. We can now make a complex canonical transformation on Γ which is analogous to the one used in general relativity (see section 6.4): $q \to q$ and $p \to (dF(q)/dq) - ip$ with $F(q) = \frac{1}{2}q^2$. The new variables are then $(q, z = q - ip)$. The symplectic structure on Γ can be re-expressed in terms of this pair as $\Omega = idz \wedge dq$. Because of this, as in Part II, we will misuse terminology somewhat and refer to (q, z) as canonically conjugate variables even though q is real and z is complex. If one is simply given a 3-real dimensional space spanned by a real variable q and a complex variable z, one can recover from it the 2-real dimensional phase space Γ by imposing the reality conditions: $\bar{q} = q$ and $\bar{z} = 2q - z$. As explained in chapter 8, one can obtain both the classical and the quantum description of the system working entirely with the pair (q, z) and imposing the reality restriction at the end of any calculation. Our goal now is to go through the algebraic quantization program of section 2 using the (q, z) pair.

Consider the complex vector space spanned by the functions $(1, q, z)$ on Γ. It is closed under Poisson brackets and every complex-valued polynomial function on Γ can be expressed as a sum of products of elements of this vector space. We can therefore choose this vector space to be the space S in the algebraic program. To generate the algebra $\mathcal{A}$, one needs only to impose the canonical commutation relations (Eqs.3) (the anti-commutation relations (Eqs.4) are redundant with this choice of S). The $\star$-relations are also straightforward to impose. We simply define $(\hat{q})^\star = \hat{q}$ and $(\hat{z})^\star = 2\hat{q} - \hat{z}$ and extend the definition of the $\star$-operation using the properties (Eq.5b) of the involution operation. It is straightforward to check that the resulting $\star$-algebra $\mathcal{A}^{(\star)}$ is isomorphic to the standard $\star$-algebra constructed from the operators $\hat{q}$ and $\hat{p}$ one finds in text books.

The next step is to find a representation of this algebra. It is here that the presence of the hybrid variables suggests a new avenue. Let V now be the vector space of entire holomorphic functions of z and let the concrete operators representing $\hat{q}$ and $\hat{z}$ be:

$$\hat{q} \cdot \Psi(z) = \hbar \frac{d\Psi(z)}{dz}, \quad \text{and} \quad \hat{z} \cdot \Psi(z) = z\Psi(z)\,, \tag{8a}$$

so that the canonical commutation relations $[\hat{q}, \hat{z}] = \hbar$ are satisfied. Now, the last question is whether there is an inner product on V which realizes the $\star$-relations: $(\hat{q})^\star = \hat{q}$ and $(\hat{z})^\star = 2\hat{q} - \hat{z}$. Let us begin by introducing a general measure $\mu(z, \bar{z})$ on the complex z-plane on which the wave functions are defined and set the inner-product to be:

$$\langle \Psi |(z), \Phi(z)| \rangle = i \int \mu(z, \bar{z})\, dz \wedge d\bar{z}\, \overline{\Psi(z)}\, \Phi(z)\,. \tag{9}$$

(The factor of i arises because $dp \wedge dq \equiv \Omega = 2i dz \wedge d\bar{z}$.) Positivity of norms requires that $\mu(z,\bar{z})$ must be real. This condition ensures that the requirement that $\hat{q}$ is its own Hermitian adjoint is satisfied by $\mu = \mu(z+\bar{z})$. The second $\star$-relation now determines the form of μ completely: $\mu(z,\bar{z}) =$ $\exp(-\frac{1}{4\hbar}(z+\bar{z})^2)$. Thus, the Hilbert space of quantum states consists of entire holomorphic functions of z which are normalizable with respect to this measure. Note that there is freedom to add to the expression of the operator $\hat{q}$ any holomorphic function of z; this addition will not alter the commutation relations. It is easy to work out the change in the measure caused by this addition and to show that the resulting quantum theory is unitarily equivalent to the one obtained above.

Given a $\Psi(z)$ in the Hilbert space, set $f(z) = \Psi(z)\exp(-\frac{z^2}{4\hbar})$. Then, the finiteness of the norm of $\Psi(z)$ implies simply that the integral $i\int dz \wedge d\bar{z}\ \exp -\frac{(|z|^2}{2\hbar)}$ $|f(z)|^2 < \infty$. Our experience with the Bargmann representation [2] of the harmonic oscillator tells us that such holomorphic functions $f(z)$ are precisely the polynomials. Let us translate the action of the operators defined in Eq.8*a* to the space of Bargmann states $f(z)$, using the unitary transform $\Psi(z) \mapsto \exp(-\frac{z^2}{4\hbar})\Psi(z)$. We find:

$$\hat{q}\cdot f(z) = \hbar\,\frac{df(z)}{dz} + \frac{z}{2}f(z) \quad \text{and} \quad \hat{z}f(z) = zf(z)\,. \tag{8b}$$

It is easy to check that Eqs.8*b* give precisely the operators $\hat{q}$ and $\hat{z}$ in the Bargmann representation. Thus, the representation we constructed using the hybrid (q,z)-variables in the algebraic quantization program is unitarily equivalent to the Bargmann representation.

The above choice of V and the representation (Eq.8*a*) was motivated by the fact that z is complex and $\hat{q}$ and $\hat{z}$ satisfy the canonical commutation relations. However, we can also arrive at this choice systematically [3] using the fact that the passage from (q,p) to (q,z) was carried out via the canonical transformation $(q,p) \to (q, dF/dq - ip)$ with generating function $F(q) = \frac{1}{2}q^2$. Had we used the pair (q,p) as our basic variables, the algebraic quantization program would have led us, as in section 2, to the Schrödinger representation of the harmonic oscillator. In this picture, the states are represented by square-integrable functions $\psi(q)$ of the real variable q and the basic operators are given by $\hat{q}\cdot\psi(q) = q\psi(q)$ and $\hat{p}\cdot\psi(q) = -i\hbar\,(d\psi/dq)$. The canonical transformation now lets us pass to the new "momentum" or z-representation via the usual transform between the configuration and the momentum representations:

$$\Psi(z) := \int dq\ \exp\left(\frac{zq}{\hbar} - \frac{q^2}{2\hbar}\right)\ \psi(q)\,. \tag{10}$$

The function $\Psi(z)$ is clearly holomorphic and, given a square-integrable $\psi(q)$, the integral converges for all (complex values of) z. Thus, the result of the transform of any Schrödinger wave function is an entire holomorphic function in the z-representation. Using the expressions of the operators $\hat{q}$ and $\hat{p}$ in the Schrödinger representation, and the definition $\hat{z} = \hat{q} - i\hat{p}$ of $\hat{z}$ we can now transform the operators $\hat{q}$ and $\hat{z}$ from the q to the z-representation. The result is precisely Eq.8a.

4 Constraints

Let us now suppose that the given physical system is subject to certain first class constraints, $C_I(q,p) = 0$, and discuss the modifications needed in the quantization program to accommodate these in the quantum theory.

For the first four steps in the program, we can merely ignore the presence of the constraints. Thus, as before, we construct a $\star$-algebra $\mathcal{A}^{(\star)}$, choose a representation space V and, ignoring the $\star$-relations, represent the elements of $\mathcal{A}$ as linear operators on V. It is at this stage that we bring in the constraints. Following Dirac, we will use the quantum constraints to select the physical states. Thus, our first task is to single out the operator analogs $\hat{C}_I$ of the classical constraint functions $C_I(q,p)$ from the elements of $\mathcal{A}$. If the constraints are quadratic or higher order in momenta, this step will require a resolution of the factor ordering problem for the constraint functions. Once we have the operators $\hat{C}_I$, we can single out the physical states $|\Psi\rangle_{phy}$ by solving the equations $\hat{C}_I|\Psi\rangle_{phy} = 0$. Thus, the subspace V_{phy} of physical states consists of the kernel of all the constraint operators $\hat{C}_I$. We need to introduce the Hermitian inner-product *only* on this subspace. Indeed, as simple examples show, for states which lie outside V_{phy}, their norms with respect to the physical inner product are typically infinite. Hence, for constrained systems, it seems inappropriate to attempt to introduce a preferred inner product on all of V (except perhaps as a technical device, e.g., for regularization of the constraint operators in field theories).

It is at this point that one has to be careful. For unconstrained systems, one can use the $\star$-relations to single out the inner-product. For constrained systems, on the other hand, the algebra $\mathcal{A}$ does not even have a well-defined action on the space of physical states. A typical operator in $\mathcal{A}$ will map elements of V_{phy} to vectors in V which do not belong to V_{phy}. Hence one must first single out physical operators, i.e., operators which map V_{phy} to itself and then select an inner product that will carry over the $\star$-relations on *these* operators to Hermiticity conditions. Thus, now the procedure has an extra step in it.

Let us explore the space of physical operators. An operator $\hat{A}$ in $\mathcal{A}$ will leave V_{phy} invariant if and only if $[\hat{A}, \hat{C}_I] = \sum_J \hat{f}_I^J \hat{C}_J$. The collection of all such operators forms a sub-algebra of $\mathcal{A}$. Denote it by $\mathcal{A}'_{phy}$. Since constraints annihilate all physical states, the algebra $\mathcal{A}_{phy}$ of physical observables can be obtained "by setting the constraint operators to zero" in $\mathcal{A}'_{phy}$. More precisely, one constructs the two sided ideal $\mathcal{I}_C$ of $\mathcal{A}'_{phy}$ generated by the constraints and takes the quotient of $\mathcal{A}'_{phy}$ by this ideal: $\mathcal{A}_{phy} := \mathcal{A}'_{phy}/\mathcal{I}_C$. Now, if $\hat{A}$ in $\mathcal{A}$ belongs to $\mathcal{A}'_{phy}$, its $\star$-adjoint, $\hat{A}^\star$ may not belong to $\mathcal{A}'_{phy}$. Hence, in general the $\star$-relation on $\mathcal{A}$ does not induce an involution on $\mathcal{A}_{phy}$. If this occurs, no obvious prescription is available to select the Hermitian inner product on the space V_{phy} of physical states. If, on the other hand,

$$\hat{A}^\star \in \mathcal{A}_{phy} \text{ for a sufficient number of } \hat{A} \in \mathcal{A}_{phy}, \tag{11}$$

then we do obtain an involution on $\mathcal{A}_{phy}$ (denoted again by $\star$) and hence a physical $\star$-algebra $\mathcal{A}^{(\star)}_{phy}$. In this case the $\star$-relations on $\mathcal{A}^{(\star)}_{phy}$ can again be used to select an inner product. Are there physical systems for which the condition 11 is guaranteed to be satisfied? The answer is in the affirmative. Consider, in particular, the case when the constraint operators satisfy: $\hat{C}_I{}^\star = \hat{C}_I$, $\forall I$. Now, if a physical operator $\hat{A}$ commutes strongly with the constraints, i.e. if $[\hat{A}, \hat{C}_I] = 0$, then $\hat{A}^\star$ is also a physical operator. We will see that this situation occurs in a number of model systems. In these cases, the quantization program can be completed successfully. Note however, that it is *not essential* that the operators to commute strongly with the constraints for the $\star$-relations to be well-defined on the physical algebra. For example, if there exists a sufficient number of physical operators which commute only weakly with the constraints but which *are* their own $\star$-adjoints, the $\star$-relations on $\mathcal{A}$ filter down unambiguously to $\star$-relations on $\mathcal{A}_{phy}$. This occurs, in particular, in 2+1-dimensional general relativity (chapter 17).

5 Constrained systems: Examples

To illustrate the above ideas and to refine them, in this section we will consider three examples. These are all simple: there are no non-trivial anti-commutation relations; the constraint operators are their own $\star$-adjoints, and the elementary physical operators commute strongly with the constraints. However, each mimics an important aspect of the constraints of general relativity. The first two examples have been discussed exhaustively in the literature and we will, of course, recover the standard answers. Indeed, if we had not, the program would not have been viable! In these cases, what interests us is the method adopted: the examples test the general program in interesting ways.

5.1 Non-relativistic parametrized particle

One often thinks of general relativity as an "already parametrized theory" [4]. The idea is that "time" resides in one of the canonical variables and has to be singled out to see the unfolding of dynamics of the "true" degrees of freedom. (This idea will be discussed in some detail in chapter 12.) It is often argued that such a "deparametrization" of the theory is *essential* for introducing a viable inner-product in quantum theory. Our quantization program, on the other hand, adopts a different strategy: the inner-product is to be selected by the reality conditions. Can this strategy then avoid the difficult problem of singling out "time" prior to quantization? To gain insight into this issue, we shall now consider the problem of quantization of a non-relativistic, parametrized particle, which is generally used to illustrate this particular aspect of the "problem of time" in quantum gravity.

Consider a non-relativistic particle moving in a potential $V(q_i)$ in Euclidean space. The motion is captured by the Hamiltonian $H(q_i, p_i) = \frac{1}{2m}\Sigma p_i p_i + V(q_i)$. This simple system can be "parametrized" by adding to the 3-dimensional configuration space the time variable. Thus, the (enlarged) configuration space, C, is now 4-dimensional, coordinatized by (q_0, q_i); and the phase space is 8-dimensional. There is one (first class) constraint: $C(q,p) := p_0 + H(q_i, p_j) = 0$, where q and p stand for (q_0, q_i) and (p_0, p_j) respectively. The constraint reduces the the fictitious 4 degrees of freedom to the original 3 "true degrees": classically, the constrained system is equivalent to the original system evolving in the 6 dimensional phase space spanned by (q_i, p_i) via the Hamiltonian $H(q_i, p_i)$.

Let now carry out the quantization program. Let the space S of elementary observables be the complex vector space spanned by the 9 functions $(1, q, p)$ on the phase space Γ. The $\star$-algebra $\mathcal{A}^{(\star)}$ is then the familiar operator algebra. Choose for the representation space V the space of smooth functions on the *4-dimensional* configuration space C. The quantum constraint is now given by:

$$\hat{C} \cdot \Psi(q) \equiv -i\hbar \frac{\partial \Psi(q)}{\partial q_0} + \hat{H} \cdot \Psi(q) = 0\,. \tag{13}$$

The space of physical states, V_{phy}, now consists of solutions of this equation. Note that the solutions are *not* functions of q_i alone; they necessarily depend on q_0 as well. Furthermore, as was expected in the general discussion above, none of the elementary quantum operators, $\hat{q}_0, \hat{q}_i, \hat{p}_0$ or $\hat{p}_i$, from which the algebra $\mathcal{A}$ was constructed, is a physical operator. Therefore, the $\star$-relations between the elementary operators cannot be directly used to single out an inner-product on V_{phy}. Our first task then is to find a sufficient number of physical operators. Fortunately, this is not difficult to accomplish. Set

$$\hat{Q}_i := U(q_0)\hat{q}_i U(-q_0), \quad \text{and} \quad \hat{P}_i := U(q_0)\hat{p}_i U(-q_0)\,, \tag{14}$$

with $U(q_0) = \exp(-\frac{i}{\hbar}q_0\hat{H})$. A simple algebraic calculation shows that these six operators, $\hat{Q}_i$ and $\hat{P}_i$, commute with the constraint, and furthermore, are their own *s. Hence we can now look for an inner product on V_{phy} with respect to which these operators are Hermitian. For this, let us begin by introducing a measure $\mu(q_0, q_i)$ on the configuration space and set:

$$\langle\, \Psi(q_0,q_i) \,|\, \Phi(q_0,q_i)\,\rangle = \int_C d^4q\; \mu(q_0,q_i)\; \overline{\Psi}(q_0,q_i)\; \Phi(q_0,q_i), \tag{15a}$$

for all physical states $\Psi(q_0, q_i)$ and $\Phi(q_0, q_i)$. To determine the measure, we impose the Hermiticity requirements. The condition that $\hat{Q}_i$ be Hermitian does not constrain the inner product in any way. The condition that $\hat{P}_i$ be Hermitian requires that the measure be independent of q_i. Thus, the inner-product now reduces to:

$$\begin{aligned}\langle\, \Psi(q_0,q_i) \,|\, \Phi(q_0,q_i)\,\rangle &= \int_C d^4q\; \mu(q_0)\; \overline{\Psi}(q_0,q_i)\; \Phi(q_0,q_i)\\ &= \int dq_0\; \mu(q_0) \int d^3q_i\; \overline{\Psi}(q_0,q_i)\; \Phi(q_0,q_i)\\ &= K \int d^3q_i \overline{\Psi}(q_0,q_i)\; \Phi(q_0,q_i),\end{aligned} \tag{15b}$$

where the constant K is given by $K = \int dq_0\mu(q_0)$. Here, in the last step, we have used the fact that, because $\Psi(q_0, q_i)$ and $\Phi(q_0, q_i)$ are physical states, i.e., because they satisfy Eq.13, the second integral in the second step is independent of q_0. Thus, the reality conditions do indeed select a unique inner product on V_{phy} (upto the usual overall constant) and the resulting quantum description is completely equivalent to the quantum theory of the original unconstrained particle moving in a potential V in the Euclidean space. However, nowhere in the procedure did we have to "deparametrize" the theory; the reality conditions on V_{phy} suffice to give us the inner product.

Note that the Hermitian inner product is defined on V_{phy}, i.e., on the space of solutions $\Psi(q_0, q_i)$ to Eq.13. That in the final step we can perform the integral on a constant q_0 surface is "accidental"; it is only a calculational device. The situation is rather similar to that encountered in the covariant symplectic description of fields on Minkowski space where the expression of the symplectic structure involves an integration over a spatial slice although the structure itself is defined on the space of solutions to the field equations on the entire space-time (see, e.g. [5]). In this sense, the above quantum description of the parametrized particle is also "covariant". Finally, we note that it is trivial to extend this discussion to allow for a q_0-dependence in the expression of the Hamiltonian or to replace the Euclidean space by a 3-manifold.

5.2 Relativistic free particle

In the previous example, the constraint function was linear in one of the momenta, p_0, while in general relativity, the scalar constraint is quadratic in momenta – both in geometrodynamics and in connection dynamics. Let us therefore consider a second system, a free particle of mass m moving in Minkowski space. Now, the configuration space is Minkowski space, with coordinates x^μ, with μ = 0-3; the system is already "parametrized". The phase space is therefore 8-dimensional and coordinatized by (x^μ, p_μ). There is again a first class constraint, which is, however, quadratic in momenta:

$$C \equiv p^\mu p_\mu + m^2 \approx 0, \tag{16}$$

(as well as a non-holonomic constraint that requires p^μ to be future pointing). Due to the form of the constraint it is natural –although by no means essential– to work in the p-representation, in which the constraint is independent of "momenta" x_μ. The configuration space $\mathcal{C} = \{p_\mu\}$ has the flat metric $\eta_{\mu\nu} = diag(-1,1,1,1)$. Since the constraint is independent of momenta, it defines the reduced configuration space $\hat{\mathcal{C}} = \{p_\mu \, | p^\mu p_\mu + m^2 = 0\}$, which is the future mass shell in $\mathcal{C}$.

It will be convenient to choose for the space $\mathcal{S}$ of elementary classical variables $\mathcal{S}_0$, the vector space spanned by all complex functions on Γ linear in or independent of momenta x^μ. Functions on Γ independent of momenta are the pull-backs to Γ of functions $f(p)$ on the configuration space $\mathcal{C}$. Functions linear in the momenta, x^μ, are of the form $V_\mu x^\mu$ where V_μ is some vector field on $\mathcal{C}$. These variables satisfy the Poisson bracket relations of Eq.1. The construction of the quantum operator algebra $\mathcal{A}$ is straightforward, following the steps detailed in section 2. The $\star$-relations on $\mathcal{A}$ reflecting the reality conditions on the classical variables are:

$$(\widehat{f(p)})^\star = \widehat{f(p)} \quad \text{and} \quad (\widehat{V_\mu x^\mu})^\star = \widehat{V_\mu x^\mu} \tag{17}$$

for all real functions $f(p)$ and real vector fields V_μ on $\mathcal{C}$.

The next step in the program is to find a linear representation of the algebra $\mathcal{A}$. As in an example in section 2, we choose the space of functions $\psi(p)$ on $\mathcal{C}$ as the carrier space for the representation, and represent the abstract operators $\widehat{f(p)}$ and $\widehat{V_\mu x^\mu}$ by

$$\widehat{f(p)} \cdot \psi(p) = f(p)\psi(p) \quad \text{and} \quad \widehat{V_\mu x^\mu} \cdot \psi(p) = i\hbar \mathcal{L}_V \psi(p). \tag{18}$$

Since the constraint is independent of momenta, Eq.18 above implies $\hat{C} \cdot \psi(p) = (p^\mu p_\mu + m^2)\psi(p)$. Thus the physical space V_{phy} consists of those functions $\psi(p)$ which have support only on the mass shell $\hat{\mathcal{C}}$.

Having solved the constraint, we now need to find the physical operators which leave V_{phy} invariant, i.e., we need to find the physical sub-algebra $\mathcal{A}_{phy}$. Obviously, all operators corresponding to variables independent of momenta commute with the constraint and leave the physical subspace invariant. A simple calculation shows that the space of physical operators linear in momenta is spanned by the operators corresponding to the three boosts and the three rotations:

$$B_{(i)} = p_i(\partial/\partial p_0) - p_0(\partial/\partial p_i) \quad \text{and} \quad R_{(i)} = \epsilon_{ijk} p_j (\partial/\partial p_k), \tag{19}$$

where $i, j, k = 1, 2, 3$, and ϵ_{ijk} is the alternating symbol.

Since physical states have support only on the mass shell, the inner product we seek is of the form

$$\langle \phi \,|\, \psi \rangle \;=\; \int_{\hat{\mathcal{C}}} \mu(p) d^3p \, \overline{\phi(p)} \psi(p) \,, \tag{20a}$$

where the measure $\mu(p)$ is to be determined by the "reality conditions". Now, the boost vector fields $B_{(i)}$ span the tangent space to $\hat{\mathcal{C}}$ at all points, thus to fix the inner product it is sufficient to impose the $\star$-relations on the momentum operators corresponding to the boosts only; the rotation operators can be generated by commuting the boosts. The requirement that $B_{(i)}$ be self-adjoint does indeed determine the measure uniquely (up to the usual overall constant). We have: $\mu = (\overline{p}{\cdot}\overline{p}{+}m^2)^{-\frac{1}{2}}$, where $\overline{p} = \{p_1, p_2, p_3\}$. Thus we have obtained an inner product on the physical states:

$$\langle \phi\psi \,|\, \rangle = \int_{\hat{\mathcal{C}}} \frac{d^3\overline{p}}{\sqrt{\overline{p} \cdot \overline{p} + m^2}} \, \overline{\phi} \, \psi. \tag{20b}$$

This is precisely the inner product used in standard textbook treatments of massive Klein-Gordon field. We see again that the reality conditions do indeed suffice to select the inner product without having to first deparametrize the theory.

Finally, note that, in this example, we used the momentum representation only for simplicity. We could equally well have worked in the position representation. The calculation leading to the determination of the inner-product would have been longer. However, it *is* possible to carry out the quantization program step by step and arrive at the correct quantum description in the position representation as well.

5.3 Two oscillators constrained to have a fixed energy difference

The scalar constraint in geometrodynamics is not only quadratic in momenta but also has a potential term (of indefinite sign). To mimic this situation we shall now consider a system of two harmonic oscillators (each with unit mass and spring constant) which are constrained to have a fixed energy difference $\delta > 0$. Thus, the unconstrained system now has two degrees of freedom and the full phase space can be coordinatized by (x_I, p_I) with $I = 1, 2$. The constraint is given by:

$$\tfrac{1}{4}(p_1^2 + x_1^2 - p_2^2 - x_2^2) = \delta. \tag{21}$$

For definiteness, let us consider the generic (and more interesting) case when δ is *not* an integer. Note that, as in general relativity, the "metric" which features in the kinetic term of the constraint has an indefinite signature.

Let us choose as our elementary classical variables the standard "creation" and "annihilation" functions on the phase space:

$$z_I = \tfrac{1}{\sqrt{2}}(x_I - ip_I) \quad \text{and} \quad \bar{z}_I = \tfrac{1}{\sqrt{2}}(x_I + ip_I). \tag{22}$$

The quantum $\star$-algebra is straightforward to construct. To make the notation transparent, let us denote the elementary quantum operators $\hat{z}_I$ by $\hat{c}_I$ and $\hat{\bar{z}}$ by $\hat{a}_I$. The quantum algebra $\mathcal{A}$ is then generated by the set of elementary quantum operators $(1, \hat{a}_1, \hat{c}_1, \hat{a}_2, \hat{c}_2)$ satisfying the canonical commutation relations:

$$[\hat{a}_I, \hat{a}_J] = 0 = [\hat{c}_I, \hat{c}_J] \quad \text{and} \quad [\hat{a}_I, \hat{c}_J] = \delta_{IJ}, \quad I, J = 1, 2, \tag{23}$$

and subject to the $\star$-relation $\hat{a}_I^\star = \hat{c}_I$. In terms of these operators, the constraint we wish to impose is

$$\hat{C}|\,\psi\,\rangle_{phy} := \Big[\tfrac{1}{2}(\hat{c}_1\hat{a}_1 - \hat{c}_2\hat{a}_2) - \delta\Big]|\,\psi\,\rangle_{phy} = 0. \tag{24}$$

The next step in the quantization program is to represent the algebra $\mathcal{A}$ by means of concrete operators on a vector space V. Recall that the $\star$-relations are ignored at this stage. Indeed, had we imposed them, i.e., had we introduced a $\star$-representation of $\mathcal{A}^{(\star)}$ by operators on a Hilbert space, we would have run into the following difficulty. The number operators $\hat{c}_1\hat{a}_1$ and $\hat{c}_2\hat{a}_2$ would have taken on only integral values, and since δ is *not* an integer, the only state in the kernel of the constraint operator would have been the zero state. Thus, we would have been led to the incorrect conclusion that V_{phy} is zero dimensional![4] Thus, our strategy of holding off the imposition of the $\star$-relations until after the physical states are isolated is *essential* in this example to obtain an acceptable quantum theory.

Let us choose the vector space representation of $\mathcal{A}$ as follows. Since any complete set of commuting operators consists of only two of the elementary operators, let us choose as V the complex vector space spanned by states of the form $|\, j, m\,\rangle$, where (to begin with) j and m are any *complex* numbers, and represent the elementary quantum operators as follows:

$$\begin{aligned} \hat{a}_1|\, j,m\,\rangle &= \alpha_1(j+m)|\, j-\tfrac{1}{2}, m-\tfrac{1}{2}\,\rangle, \\ \hat{c}_1|\, j,m\,\rangle &= \gamma_1(j+m+1)|\, j+\tfrac{1}{2}, m+\tfrac{1}{2}\,\rangle, \\ \hat{a}_2|\, j,m\,\rangle &= \alpha_2(j-m)|\, j-\tfrac{1}{2}, m+\tfrac{1}{2}\,\rangle \\ \text{and} \quad \hat{c}_2|\, j,m\,\rangle &= \gamma_2(j-m+1)|\, j+\tfrac{1}{2}, m-\tfrac{1}{2}\,\rangle; \end{aligned} \tag{25}$$

where the coefficients, $\alpha_I(k)$ and $\gamma_I(k)$, functions only of their argument k, are subject to the conditions

$$\alpha_1(k)\gamma_1(k) = k \quad \text{and} \quad \alpha_2(k)\gamma_2(k) = k \tag{26}$$

It is straightforward to check that the commutation relations, Eqs.23, are satisfied by this choice of representation. The notation $|\, j,m\,\rangle$ to represent the kets may seem strange at first. Note, however, that each $|\, j,m\,\rangle$ is an eigenket of the total number operator $\hat{N} = \hat{c}_1\hat{a}_1 + \hat{c}_2\hat{a}_2$ with eigenvalue $2j$ as well as of the constraint operator C with eigenvalue $m - \delta$. Thus, had we represented states as wave functions $\psi(z_1, z_2)$, the ket $|\, j,m\,\rangle$ would have corresponded to the function $z_1^{j+m} z_2^{j-m}$. These angular momentum like states arise naturally because the the constraint surface $\overline{\Gamma}$ is the group manifold of $SO(2,1)$ and the commutator algebra of physical observables provides us with a representation of $SO(2,1)$.

The next step is to find the physical states. These belong to the kernel of the constraint operator $\hat{C}$. Hence, a basis for the *physical* space V_{phy} is given simply by the kets $\{|\, j, \delta\,\rangle\}$. Physical operators are the elements of $\mathcal{A}$ that map V_{phy} to itself. It is straightforward to verify that a complete (in the sense specified below) set of physical operators is generated by $\{1, \hat{N}_I, \hat{J}_+, \hat{J}_-\}$, where $\hat{N}_I := \hat{c}_I\hat{a}_I$ are the two number operators, and $\hat{J}_+ := \hat{c}_1\hat{c}_2$, $\hat{J}_- := \hat{a}_1\hat{a}_2$ are the raising and the lowering

4 This is an incorrect conclusion because in this case the reduced phase space is a 2-dimensional (non-compact) manifold; the system has one "true" degree of freedom.

operators. They satisfy the following $\star$-relations:

$$(\hat{J}_+)^\star = \hat{J}_- \quad \text{and} \quad (\hat{N}_I)^\star = \hat{N}_I. \tag{27}$$

The last step in the program is to select an inner-product by requiring that the $\star$-relations of Eq.27 become Hermitian adjointness relations on the resulting Hilbert space. To carry out this step, let us first understand the sense in which the set $\{1, \hat{N}_I, \hat{J}_\pm\}$ is complete. The classical analogs of these operators are the functions $\{1, z_1\bar{z}_1, z_2\bar{z}_2, z_1 z_2$ and $\bar{z}_1\bar{z}_2\}$ on phase space. These functions form an (over)complete set in the sense that their gradients span the cotangent space of the reduced phase space $\hat{\Gamma}$ almost everywhere (the exception being a single point.) In this sense, then, the set of physical operators $\{1, \hat{N}_I, \hat{J}_\pm\}$ is also complete in $\mathcal{A}_{phy}$. Note, however, that the above set of functions do not form a *global* chart on $\hat{\Gamma}$: they are all invariant under the (discrete) parity map $(x_I, p_I) \longmapsto (-x_I, -p_I)$, or, $(z_I) \longmapsto (-z_I)$. In quantum theory, this symmetry corresponds to the operation $\psi(z_I) \longmapsto \psi(-z_I)$, or, equivalently,

$$|\, j, m\,\rangle \longmapsto \mathbf{P}|\, j, m\,\rangle := (-1)^{2j}|\, j, m\,\rangle, \tag{28}$$

where, to evaluate the right hand side we will take the principal value, namely, $(-1)^{2j} = \exp(i2\pi\epsilon)$, where $\epsilon = \mathrm{frac}(j)$ is the fractional part of j. The operator $\mathbf{P}$ is *super-selected*: not only is it a physical operator but it commutes with *all* operators in the physical algebra. Consequently, each eigenspace V^ϵ_{phy} of this operator will provide a representation of the algebra $\mathcal{A}_{phy}$. Thus, we have a 1-parameter family of ambiguities in quantization of the system, labelled by the parameter $\epsilon \in [0, 1)$. We will see in chapter 16 that the situation is quite analogous in the connection representation of 2+1 gravity.

For definiteness, let us just consider the representation V^0_{phy} with eigenvalue $+1$ of $\mathbf{P}$. Then, j is restricted to be an integer. It now remains to introduce on V^0_{phy} an inner product. It is straightforward to check that the Hermiticity conditions corresponding to Eq.27 is satisfied if and only if:

$$\langle\, j', \delta\,|\, j, \delta\,\rangle = \delta_{j',j} \tag{29}$$

Thus, the physical Hilbert space is spanned by the kets of the type $|\, j, \delta\,\rangle$ where j is an integer and these kets form an orthonormal basis. This representation of $\mathcal{A}^{(\star)}_{phy}$ is irreducible. Note, however, that this representation has a peculiar feature. In the classical theory, the total Hamiltonian of the system:

$$H(q_I, p_I) := \tfrac{1}{2}(q_1^2 + p_1^2 + q_2^2 + p_2^2), \tag{30}$$

is non-negative. Quantum mechanically, however, the Hamiltonian operator $H = \hbar(\hat{c}_1\hat{a}_1 + \hat{c}_2\hat{a}_2 + 1)$ is *unbounded from below*! This comes about because the physical

Hilbert space allows kets $|\, j, \delta \,\rangle$ with arbitrarily negative values j and the eigenvalue of the Hamiltonian on this ket is simply $\hbar(2j+1)$. Note that, one cannot restrict oneself just to the subspace corresponding to positive energy because this subspace fails to be invariant under the action of the physical algebra $\mathcal{A}_{phy}$. It thus appears that in the presence of non-trivial constraints, extreme quantum tunneling can occur in which "half" the physical Hilbert space corresponds to states that are classically forbidden. That such a phenomenon can arise was pointed out [6] in the context of quantum geometrodynamics already in the early eighties. However, the analysis of the example considered then was not as complete. It is extremely important to find out if the scalar constraint of non-perturbative canonical gravity allows such a phenomenon to occur.

In the above discussion, we restricted ourselves to states with eigenvalue $+1$ of the parity operator $\mathbf{P}$. The situation with eigenvalue -1 is quite similar. However there seems *no* compelling reason to restrict ourselves to states with eigenvalues ± 1 of $\mathbf{P}$. To allow other eigenvalues seems to violate classical intuition. Recall however that a similar situation occurs in systems of identical particles in 2-spatial dimensions where the use of eigenvalues other than ± 1 for the parity or the permutation operators leads to the interesting quantum phenomena of fractional statistics. In the present case, one can even make a strong argument for such states with fractional eigenvalues of $\mathbf{P}$: in some of these representations –with $\epsilon =$ fractional part of δ– the Hamiltonian *is* bounded from below. A detailed discussion of these representations (as well as the special case when δ is an integer) will be given elsewhere [7].

To summarize, the quantization program can be implemented in this example and does indeed lead to viable quantum theories. However, due to global issues related to the completeness of the set of physical operators $\{\hat{N}_I, \hat{J}_\pm\}$, there exists a 1-parameter family of inequivalent quantum theories, parametrized by $\epsilon \in [0,1)$. Additional physical input is needed to remove this quantization ambiguity.

6 General relativity

Using the ideas presented above we can now formulate a program for non-perturbative, canonical quantization of general relativity. The steps in the program are as follows.

1. Find a suitable subspace $\mathcal{S}$ of the space of complex-valued functions on the real phase space of general relativity which is closed under Poisson brackets and which is large enough so that any sufficiently regular function on the phase space can be obtained as (possibly a suitable limit of) a sum of products of

elements of S; each function in S represents an elementary classical variable which is to have an unambiguous quantum analog.

2. Associate with each element F in S an abstract operator $\hat{F}$. Construct the free algebra generated by these elementary quantum operators. Impose on it the canonical commutation relations, and if appropriate, also the anti-commutation relations which capture the algebraic relations between the classical elementary observables. Denote the resulting algebra by $\mathcal{A}$.
3. On this algebra, introduce an involution operation, denoted by $\star$, by requiring that if two elementary classical variables F and G are related by $\bar{F} = G$, then $\hat{F}^{\star} \equiv \hat{G}$ in $\mathcal{A}$. Denote the resulting $\star$-algebra by $\mathcal{A}^{(\star)}$.
4. Introduce a linear representation of $\mathcal{A}$ on a vector space V; for now the $\star$-relations are ignored.
5. Obtain explicit operators on V representing the quantum constraints. At this stage, one would have to invent suitable regularization schemes to make the quantum constraints well-defined. Find the kernel V_{phy} of these operators. This is the space of physical states.
6. Extract the physical $\star$-algebra of operators that leave V_{phy} invariant. Introduce on V_{phy} a Hermitian inner product so that the abstract $\star$-relations on $\mathcal{A}^{(\star)}_{phy}$ –which have been ignored so far– are represented as Hermitian adjoint relations on the resulting Hilbert space.

The new ingredients of this program can be summarized as follows.

i) The introduction of anti-commutation relations: We saw in section 2 that for generic choices of the space S of elementary classical variables, anti-commutation relations *must be* introduced in the quantum algebra $\mathcal{A}$ to correctly incorporate the algebraic relations that exist between the elementary variables. We will see that in the loop representation in 2+1 as well as 3+1 gravity, the natural choice of S is indeed of this type.

ii) The procedure for selecting the inner-product: The idea of using the "reality conditions" to select the inner product is, at a certain level, rather trivial. After all, in elementary quantum mechanics, the physical observables are always Hermitian on the Hilbert space of quantum states. However, the idea of using this as a general principle to *select* the inner product appears to be new. The procedure usually adopted is quite different. In Minkowskian quantum field theories, for example, the inner-product –or, equivalently, the vacuum state– is selected by making an appeal to Poincaré invariance: the vacuum is the unique Poincaré invariant state and the vacuum expectation values of all operators provide us (e.g., through the Gelfand-Naimark-Segal [8] construction) the Hilbert space, including the inner-product. In quantum gravity, on the other hand, Poincaré invariance cannot play a fundamental role. In the canonical program, one might imagine using the spatial diffeomorphism group in the place of the Poincaré group. However, we cannot select the vacuum expectation value in this manner: as we will see, now the constraints require that *every* physical state be diffeomorphism invariant! ***Thus, a brand new guiding principle is needed.*** Unfortunately, the Dirac theory of constrained systems does not furnish any guidelines. We have attempted to supplement this theory by suggesting a strategy: select the inner product through the simple requirement that the classical reality conditions become Hermiticity relations in quantum theory. In sections 2, 3 and 5 we saw that in a number of (unconstrained and constrained) examples with a finite number of degrees of freedom, the strategy does indeed succeed. Later in this Part we will see that it also works in 2+1 gravity and linear theories in 3+1-dimensional Minkowski space including Maxwell theory and linearized gravity.

iii) Issue of time: As remarked in section 4, it is generally believed that "time" is hidden among the canonical variables of general relativity and must be isolated prior to the imposition of the inner-product. (See, e.g. [4].) As we saw in the three examples in section 5, the imposition of "reality conditions" enable one to select the inner-product on the physical states in a "covariant manner"; the inner-product is defined directly on the space of solutions to the constraints, without having first to decompose the arguments of these wave functions into dynamical variables and time. Thus, as far as the issue of constructing the Hilbert space is concerned, it appears that one *need not* first isolate time. Of course, for the physical interpretation of the framework and particularly for the quantum measurement theory, it would be very useful –perhaps essential– to isolate time in some fashion. However, for such interpretational purposes, it may suffice to have only an approximate notion of time. (See, e.g. [9].) Had it been essential to single out time for obtaining the inner product, on the other hand, an approximate notion would not have sufficed.

If the general program can be completed, one would have a consistent mathematical framework for quantum gravity. One would then use it to analyze physical problems and to make predictions. In the next four chapters we will present examples of the issues that can be discussed within such a framework. One is still be left with the important and difficult problems related to the quantum measurement theory. The ideas on this subject are still in infancy.

References

[1] A. Ashtekar, Comm. Math. Phys. **71**, 59 (1980); R.P. Geroch, *Mathematical Physics*, (University of Chicago Press, Chicago, 1985), pages 98-101.

[2] V. Bargmann, Proc. Natl. Acad. Sci.(U.S.A.) **48**, 199 (1962); J.M. Jauch, *Foundations of Quantum Mechanics*, (Addison Wesley, Reading, 1968), pages 215-219.

[3] H. Kodama, Phys. Rev.**D42**, 2548 (1990).

[4] K. Kuchař, In:*Quantum gravity 2: An Oxford Symposium*, eds. C.J. Isham, R. Penrose and D.W. Sciama (Oxford University Press, Oxford 1981).

[5] A. Ashtekar, L. Bombelli and O. Reula, In: *Analysis, Geometry and Mechanics: 200 Years after Lagrange*, eds. M. Francaviglia and D. Holm (North Holland, Amsterdam, in press).

[6] A. Ashtekar, G.T. Horowitz, Phys. Rev. **D26**, 3342 (1982).

[7] A. Ashtekar and R.S. Tate (in preparation).

[8] I.M. Gelfand and M.A. Naimark, Mat. Sobernik **12**, 197 (1943); I.E. Segal, Bull. Amer. Math. Soc. **53**, 73 (1947).

[9] C. Rovelli, Phys. Rev. **D**, (in press).

11 CONNECTION REPRESENTATION: LINEARIZED GRAVITY

1 Introduction

Recall that in the new Hamiltonian formulation of general relativity, the basic fields are a connection 1-form $A_{aA}{}^{B}$ and its conjugate momentum $\tilde{\sigma}^{a}{}_{A}{}^{B}$. As was emphasized in Part II, these variables enable us to take over several techniques from Yang-Mills theory to general relativity. Recall, however, that in Yang-Mills theory, upon quantization, the connection yields a family of massless *spin-one* particles. On the other hand, one expects the quanta of the gravitational field to be massless *spin-two* particles. The question therefore arises: How does the multiplet of spin-one quanta reduce to a single spin-two quantum? The natural arena for analyzing these issues is the weak field limit of the theory. Hence we are led to ask whether, by applying the quantization program of chapter 10 to linearized gravity, we can manage to extract the correct excitations of the gravitational field inspite of the apparent problem. This limiting case thus provides a powerful viability criterion for the general program. In this chapter, therefore, we will first study the weak field limit of general relativity in terms of $(\tilde{\sigma}^{a}{}_{A}{}^{B}, A_{aA}{}^{B})$ and then apply our quantization program to this limit. We will find that it is the presence of the four additional constraints, absent in Yang-Mills theory, together with the geometrical interpretation of the momentum $\tilde{\sigma}^{a}{}_{A}{}^{B}$, that are responsible for the reduction of the spin-one multiplet to a single spin-two quantum.

In section 2, we introduce the Hamiltonian description of linearized gravity starting from the framework developed in chapters 6 and 7 for the full theory.[1] Thus, our basic variables will be the linearized soldering form and the linearized self-dual connection. We will see that, in the linear approximation, it is easy to isolate the true degrees of freedom by explicitly solving the constraints and fixing the gauge appropriately. Therefore, in this simplified theory, we can carry out our

1 This material in this section based on joint work with Joohan Lee [1].

quantization program in two ways.[2] The first method uses, as its starting point, the phase space of the true degrees of freedom; constraints are eliminated at the classical level. Quantization can then proceed along the lines of section 10.2. Alternatively, one can regard linearized gravity as a constrained system and using the strategy of section 10.4, impose constraints *a la* Dirac as conditions on physically admissible quantum states. The first method is closer to the one normally used in the Fock quantization of linearized gravity. To show that the general program does indeed produce the familiar answer, in section 3, we use it to carry out quantization of the true degrees of freedom isolated in section 2. The result is the Bargmann representation for spin-2 gravity in which states arise as holomorphic functions of positive frequency connections; the positive frequency condition is achieved by decomposing the self dual connection into the two helicity states. This method of quantization does not, however, carry over to the full theory: In the non-linear theory, neither do we know how to isolate the true degrees of freedom nor can we decompose the connection into its helicity states in a natural way. . Therefore, to gain insight into the viability of the program for full general relativity, we consider next the issue of constructing the quantum theory in the linear case in a way that can extend to the full, non-linear context. The idea now is to adopt the second of the two strategies outlined above. Thus, we wish to retain the constraints in the classical theory and impose them *a la* Dirac to select the physical quantum states, and we wish to work in a representation in which the self dual –rather than the positive frequency– connection is diagonal. The use of the self dual representation raises new issues which, to my knowledge, have not been discussed in the literature. Therefore, in section 4, we make a detour and address these issues in the context of a harmonic oscillator. In section 5, we return to linear gravity and construct the required self dual representation.

In chapters 12 and 13, we will introduce the self dual representation for full general relativity, discuss two of the main results obtained in this framework and summarize the present status of the resulting quantization program. As mentioned in the introduction to this Part, this representation is better suited to analyze semi-classical issues, and more generally, to answer questions that require the presence of at least an approximate space-time geometry. In chapters 14-16, we will introduce another framework –the loop representation– which appears to be better suited to analyze issues related to the extreme Planck regime.[3]

---

2 For a general discussion of the two methods, see, e.g., appendix D. Since the constraints in linearized gravity are rather simple, the two methods yield the same quantum theory in this case.

3 Linearized gravity can also be treated in the loop representation [2]. In chapter 14, we will illustrate the ideas involved in a simpler case, that of the Maxwell field.

2 Canonical framework

In the weak field limit, one ignores the self-interaction of the field. One is therefore led to linearize the field equations off a "classical vacuum". In general relativity (without cosmological constant), the required background geometry is provided by Minkowski space. Therefore, in this chapter, we will consider linearization off the point in the phase space corresponding to the trivial initial data: flat $\tilde{\sigma}^a{}_A{}^B$ and vanishing $A_{aA}{}^B$.

Geometrically, linearized fields are represented by tangent vectors at the "background" point about which one studies linearization. Thus, we now have to analyze the structure of the tangent space $T_p\Gamma =: \Gamma_{lin}$ to the phase space at the point $p := (\overset{\circ}{\tilde{\sigma}}{}^a, \overset{\circ}{A}_a = 0)$. Consider a 1-parameter curve $p(\lambda)$ in Γ passing through this point, with $p(\lambda = 0) = p$. Consider the linearized gravitational field defined by the tangent vector to this curve at p, i.e., by fields $\tilde{h}^a{}_A{}^B$ and $C_{aA}{}^B$ on Σ defined by

$$\begin{aligned} \tilde{h}^a{}_A{}^B &:= \frac{d}{d\lambda}\tilde{\sigma}^a{}_A{}^B(\lambda)|_{\lambda=0} \\ C_{aA}{}^B &:= \frac{d}{d\lambda}A_{aA}{}^B(\lambda)|_{\lambda=0}. \end{aligned} \tag{1}$$

The fact that we have a reference soldering form $\overset{\circ}{\tilde{\sigma}}{}^a{}_A{}^B$ allows us to also consider $\tilde{h}$ and C as tensors. Let us set

$$\tilde{h}^{ab} := -\operatorname{tr}(\tilde{h}^a\overset{\circ}{\sigma}{}^b) \quad \text{and} \quad C_{ab} := -\operatorname{tr}(C_a\overset{\circ}{\sigma}_b), \tag{2}$$

where $\overset{\circ}{\sigma}$ is the undensitized reference soldering form. These are our basic linearized fields.

We now want to obtain the constraint equations induced on these perturbations by the constraints of the full non-linear theory. Let us restrict the curve $p(\lambda)$ to lie wholly in the constraint surface $\bar{\Gamma}$ of Γ. Then $(\tilde{\sigma}^a{}_A{}^B(\lambda), A_{aA}{}^B(\lambda))$ satisfies Eqs.6.10′, 6.14′ and 6.15′ for each value of λ. Taking the derivative of these equations with respect to λ and evaluating at p one obtains:

$$\partial_a\tilde{h}^a + [C_a, \overset{\circ}{\tilde{\sigma}}{}^a] = 0, \quad \operatorname{tr}(\overset{\circ}{\tilde{\sigma}}{}^b\partial_{[a}C_{b]}) = 0, \quad \text{and} \quad \operatorname{tr}(\overset{\circ}{\tilde{\sigma}}{}^a\overset{\circ}{\tilde{\sigma}}{}^b\partial_{[a}C_{b]}) = 0, \tag{3}$$

where we have used $\overset{\circ}{A} = 0$. In terms of space-time tensors, these equations can be rewritten as:

$$\partial_a\tilde{h}^{ap} - \sqrt{2}\,\tilde{\eta}^{amp}C_{am} = 0, \quad \partial_a C - \partial_b C_a{}^b = 0, \quad \text{and} \quad \partial_{[a}C_{bc]} = 0, \tag{4}$$

where $C := C_{ab}e^{ab}$ is the trace of C_{ab}, and all indices are raised and lowered with the background (Euclidean) metric $e_{ab} = -\operatorname{tr}(\overset{\circ}{\sigma}_a\overset{\circ}{\sigma}_b)$.

In order to compute the canonical transformations generated by these constraints, we need to spell out the structure of the linearized phase space. As remarked earlier, as a vector space, Γ_{lin} is the tangent space to Γ at $p = (\mathring{\tilde{\sigma}}, \mathring{A})$:

$$\Gamma_{lin} := T_{(\mathring{\tilde{\sigma}},\mathring{A})}\Gamma = \{(\tilde{h}^a{}_A{}^B, C_{aA}{}^B)\} \equiv \{(\tilde{h}^{ab}, C_{ab})\}, \tag{5a}$$

The boundary conditions of Eqs.6.5 and 6.6 satisfied by points $(\tilde{\sigma}^a{}_A{}^B,\ A_{aA}{}^B)$ of Γ imply that $(\tilde{h}^{ab}, C_{ab})$ must have the following fall-off:

$$\begin{aligned} \tilde{h}^{ab} - \tfrac{1}{3}\tilde{h}\, e^{ab} &= O(1/r^2), \qquad & \tilde{h} &= O(1/r), \\ C_{ab} - \tfrac{1}{3}C\, e_{ab} &= O(1/r^2), \qquad & C &= O(1/r^3), \end{aligned} \tag{5b}$$

where $\tilde{h} = \tilde{h}^{ab}e_{ab}$ is the trace of $\tilde{h}^{ab}$. Finally, the symplectic structure Ω_{lin} on Γ_{lin} is the restriction to $T_p\Gamma$ of the full symplectic structure Ω. Since $\tilde{\sigma}^a{}_A{}^B$ and $A_{aA}{}^B$ have canonical Poisson bracket relations (Eq.9.30), it follows that the non-vanishing Poisson brackets between the linearized variables are given by:

$$\{h^{ab}(\vec{x}), C_{cd}(\vec{x}')\} = -\frac{i}{\sqrt{2}}\delta^a_c\delta^b_d\delta^3(\vec{x},\vec{x}'), \tag{6}$$

Let us smear the left sides of the constraints, Eqs.4, by suitable test fields to obtain the constraint functions:

$$f_\omega(\tilde{h}, C) := i\sqrt{2}\int_\Sigma d^3x\, \omega_a(\partial_b\tilde{h}^{ba} - \sqrt{2}\,\tilde{\eta}^{bma}C_{bm}) = 0 \tag{7a}$$

$$f_V(\tilde{h}, C) := i\sqrt{2}\int_\Sigma d^3x\, \tilde{V}^a(\partial_a C - \partial_b C_a{}^b) = 0 \tag{7b}$$

$$f_\Lambda(\tilde{h}, C) := i\sqrt{2}\int_\Sigma d^3x\, \Lambda\tilde{\eta}^{abc}\partial_{[a}C_{bc]} = 0, \tag{7c}$$

where the test fields ω_a, $\tilde{V}^a$ and $\tilde{\Lambda}$ vanish at infinity. The changes in the basic canonical variables generated by these constraint functions are given by:

$$\Delta_\omega\tilde{h}^{ab} = -\frac{i}{\sqrt{2}}\frac{\delta f_\omega}{\delta C_{ab}} = -\sqrt{2}\,\tilde{\eta}^{abc}\omega_c; \tag{8a}$$

$$\Delta_V\tilde{h}^{ab} = -\frac{i}{\sqrt{2}}\frac{\delta f_V}{\delta C_{ab}} = \partial^b\tilde{V}^a - (\partial_c\tilde{V}^c)\, e^{ab}; \tag{8b}$$

$$\Delta_\Lambda\tilde{h}^{ab} = -\frac{i}{\sqrt{2}}\frac{\delta f_\Lambda}{\delta C_{ab}} = -\tilde{\eta}^{abc}\partial_c\Lambda \tag{8c}$$

and

$$\Delta_\omega C_{ab} = \frac{i}{\sqrt{2}} \frac{\delta f_\omega}{\delta \tilde{h}^{ab}} = \partial_a \omega_b; \tag{8d}$$

$$\Delta_V C_{ab} = \frac{i}{\sqrt{2}} \frac{\delta f_V}{\delta \tilde{h}^{ab}} = 0; \tag{8e}$$

$$\Delta_\Lambda C_{ab} = \frac{i}{\sqrt{2}} \frac{\delta f_\Lambda}{\delta \tilde{h}^{ab}} = 0. \tag{8f}$$

Thus, two points of the linearized phase space which satisfy the constraints (Eqs.4) are gauge-related if and only if they differ by some combination of Eqs.8. The reduced phase space $\hat{\Gamma}_{lin}$ is obtained by quotienting the constraint surface by the canonical transformations generated by the constraints. Points of $\hat{\Gamma}_{lin}$ represent the "true degrees of freedom" of linearized gravity. To coordinatize this space in a convenient manner, let us further investigate the effect of the various gauge transformations generated by Eqs.8. First, note that since all the infinitesimal transformations are constant vector fields on Γ_{lin}, the finite gauge transformations are of the same form. Hence we can use Eqs.8 directly as finite transformations.

Since Σ is topologically $\mathbb{R}^3$, the general solution to Eq.7c is given by $C_{[ab]} = \partial_{[a}\mathbb{C}_{b]}$, for some covector $\mathbb{C}_a$. Now, by Eq.8d, we can transform

$$C_{ab} \mapsto C_{ab} + \partial_a \omega_b. \tag{9}$$

Thus $C_{[ab]} \mapsto C_{[ab]} + \partial_{[a}\omega_{b]}$, which suggests that we choose

$$\omega_b = -\mathbb{C}_b + \partial_b \omega, \tag{10}$$

to make C_{ab} symmetric, and use the remaining freedom, the choice of ω, to make C_{ab} trace-free, by requiring

$$e^{ab}(C_{ab} + \partial_a \partial_b \omega) = 0, \qquad \text{or} \qquad \nabla^2 \omega = -e^{ab} C_{ab}. \tag{11}$$

Thus, by solving the scalar constraint, Eq.7c and using Eq.8d to gauge fix ω_a, we can make C_{ab} symmetric and trace-free.

Let us now use the remaining gauge transformations, Eqs.8b and 8c, to simplify the form of $\tilde{h}^{ab}$ in a similar way. Set ${}^*h_a := \frac{1}{2} \underset{\sim}{\eta}_{abc} \tilde{h}^{bc}$. Under Eq.8$c$ we have

$$\tilde{h}^{[ab]} \mapsto \tilde{h}^{[ab]} - \tilde{\eta}^{abc}\, \partial_c \Lambda \tag{12}$$

and therefore,

$$^*h_a \mapsto {}^*h_a - \partial_a \Lambda, \tag{13}$$

which allows us to fix Λ by requiring that the longitudinal part of $^*h_a$ vanish. The transformation induced by $\widetilde{V}^a$ (Eq.8*b*) now implies

$$^*h_a \mapsto {}^*h_a + \tfrac{1}{2}\underset{\sim}{\eta}_{abc}\,\partial^c \widetilde{V}^b, \tag{14}$$

and we can choose $\widetilde{V}^c$ to kill the transverse part of $^*\widetilde{h}_a$, which determines $\widetilde{V}^c$ up to a gradient. Thus, we have made the antisymmetric part of $\widetilde{h}^{ab}$ vanish. Let us now use the remaining freedom in $\widetilde{V}^c$ to make $h_{ab} = h_{(ab)}$ traceless: if $\widetilde{V}^c = \partial^c \widetilde{V}$,

$$\widetilde{h}^{ab} \mapsto \widetilde{h}^{ab} + \partial^b \partial^a \widetilde{V} - (\partial_c \partial^c \widetilde{V})\, e^{ab}, \tag{15}$$

and, if we choose $\widetilde{V}$ to satisfy

$$\partial_c \partial^c \widetilde{V} = \tfrac{1}{2}\widetilde{h}, \tag{16}$$

the transformation of Eq.15 leads to $\widetilde{h} := \widetilde{h}_a{}^a = 0$. Thus, using Eqs.8*b* and 8*c*, we have gauge fixed Λ and $\widetilde{V}$ to make $\widetilde{h}^{ab}$ symmetric and trace-free.

To summarize, we can always fix the gauge, using Eqs.8, and solve the scalar constraint (Eq.7*c*) so that both $\widetilde{h}^{ab}$ and C_{ab} are symmetric traceless. The remaining (i.e., first two) linearized constraints then tell us that the basic fields have to be transverse:

$$\partial_a \widetilde{h}^{ab} = 0 \quad \text{and} \quad \partial_a C^{ab} = 0. \tag{17}$$

Thus, the true degrees of freedom of the linearized gravitational field are represented by two *symmetric traceless transverse* (STT) tensor fields $\widetilde{h}^{ab}$ and C_{ab}. Put differently, the symmetric traceless transverse parts, $(\widetilde{h}^{STT\,ab}, C^{STT}_{ab})$ of elements of Γ_{lin} serve to coordinatize the reduced linearized phase space. This shows that we are indeed dealing with a spin-2 field rather than a triplet of spin-1 fields. If, as in Yang-Mills theory, we had only the Gauss law constraint to impose, we would have obtained the triplet. It is the restrictions imposed and the gauge transformations generated by the additional four constraints that led us to a single spin-2 quantum.

We can now ask for the relation between these fields and the ones normally used, the linearized 3-metric δq_{ab} and the linearized extrinsic curvature δK_{ab}. It is straightforward to show that one has

$$\begin{aligned} \widetilde{h}^{STT\,ab} &= \tfrac{1}{2}\,(e)^{1/2}\delta q^{TT\,ab} \\ \sqrt{2}C^{TT}_{ab} &= -\epsilon_b{}^{mn}\partial_m\, h^{TT}_{na} - i\,(\delta K)^{TT}_{ab}, \end{aligned} \tag{18}$$

where the superscript TT stands for "transverse-traceless".

This concludes the discussion of kinematics. Let us now turn to linearized dynamics. For this, we use the following general result from symplectic geometry. (See, e.g.,[1]). Let (Γ, Ω) be a symplectic manifold with a Hamiltonian H. Let p be a point of Γ at which the Hamiltonian vector field X_H of H vanishes. Then, the dynamical flow on Γ leads to a linear mapping on the tangent space T_p of Γ at the point p. This is the linearized dynamics of the theory at p. On the linearized phase space $(T_p, \Omega|_p)$, this dynamical flow is generated by the Hamiltonian $h(v) := \frac{1}{2}(\partial_\alpha \partial_\beta H)_p \, v^\alpha v^\beta$ for all $v \in T_p$. Furthermore, if the non-linear system (Γ, Ω) admits a set of first class constraints which are preserved by the Hamiltonian flow X_H, the linearized dynamical flow as well as the linearized Hamiltonian $h(v)$ project down to the reduced, linearized phase space. To apply this result to general relativity, we must choose a lapse $\underset{\sim}{T}$ and a shift T^a on Σ such that the resulting Hamiltonian vector field on the full non-linear phase space vanishes at the point $p := (\overset{\circ}{\widetilde{\sigma}}{}^a, \overset{\circ}{A}_a = 0)$. This is easily achieved by setting $\underset{\sim}{T} = (\det q)^{-1/2}$ (so that the geometric lapse T equals 1) and $T^a = 0$; with this choice the fiducial point p is left invariant, implying the vanishing of X_H. Recall that for this lapse-shift pair, the full Hamiltonian can be recast as (see Eq.7.14):

$$H(\widetilde{\sigma}, A) = -\int_\Sigma d^3x \, q^{1/2} (\overline{A}_{ab} A^{ba} - \overline{A} A), \tag{19}$$

where $A_{ab} = -\operatorname{tr}(A_a \sigma_b)$. This is a particularly convenient form for linearization. Using the general result quoted above and noting that in our case v stands for the linearized data $(\widetilde{h}^{ab}, C_{ab})$, we have

$$\begin{aligned} h(\widetilde{h}^{ab}, C_{ab}) &= -\tfrac{1}{2} \int_\Sigma d^3x \, (\det e)^{1/2} \, (\overline{C}_{ab} C^{ba} - \overline{C} {*} C) \\ &\approx -\tfrac{1}{2} \int_\Sigma d^3x \, (\det e)^{1/2} \, \overline{C}_{ab} \, C^{ab}, \end{aligned} \tag{20}$$

The first of these two integrals gives us the Hamiltonian on the entire linearized phase space, while the second integral gives us the "reduced Hamiltonian" on the gauge-fixed subspace, which, in our case, is preserved by the Hamiltonian flow.

This form of the Hamiltonian bears a striking resemblance to the expression $H(q,p) = \bar{z} z$ of the Hamiltonian of a simple harmonic oscillator. Indeed, we shall see in the next section that the analogy goes deeper: the variables C_{ab} will naturally lead us to the Bargmann representation for linearized gravity, with one oscillator (per point in the momentum space) for each helicity of linearized gravitons.

3 Bargmann representation

We can now carry out the quantization program of section 10.2 using the reduced phase space of the linearized theory as the point of departure.

It is convenient –although by no means essential– to work in the momentum space. Then, the symmetric-transverse-traceless conditions on the fields $h^{TT}_{ab}(\vec{x})$, $(\delta K)^{TT}_{ab}$ and $C^{TT}_{ab}(\vec{x})$ imply that their Fourier transforms, $h_{ab}(\vec{k})$, $(\delta K)_{ab}(\vec{k})$ and $C_{ab}(\vec{k})$, have the following algebraic form:

$$\begin{aligned} h_{ab}(\vec{k}) &= q_1(\vec{k})m_a(\vec{k})m_b(\vec{k}) + q_2(\vec{k})\bar{m}_a(\vec{k})\bar{m}_b(\vec{k}),\\ (\delta K)_{ab}(\vec{k}) &= p_1(\vec{k})m_a(\vec{k})m_b(\vec{k}) + p_2(\vec{k})\bar{m}_a(\vec{k})\bar{m}_b(\vec{k})\\ \text{and}\quad C_{ab}(\vec{k}) &= z_1(\vec{k})m_a(\vec{k})m_b(\vec{k}) + z_2(\vec{k})\bar{m}_a(\vec{k})\bar{m}_b(\vec{k}), \end{aligned} \tag{21}$$

where $\sqrt{2}m_a \equiv (\partial_a\theta + i\sin\theta\,\partial_a\varphi)$ is the complex vector in the momentum space transverse to the radial vector $\vec{k}$, with normalization $m^a\bar{m}_a = 1$. (For notational simplicity, we have "de-densitized" $\tilde{h}^{TT}_{ab}$ by dividing it by the square-root of the determinant of the background Euclidean metric e_{ab}; $h^{TT}_{ab} = \tilde{h}^{TT}_{ab}/\sqrt{e}$.) The algebraic relations in Eq.18, the canonical Poisson-bracket relations of Eq.6, and the reality conditions on h^{TT}_{ab} and $(\delta K)^{TT}_{ab}$ now provide us with the following relations between the dynamical variables $q_j(\vec{k}), p_j(\vec{k})$ and $z_j(\vec{k})$:

$$z_1(\vec{k}) = -\tfrac{1}{\sqrt{2}}(|\vec{k}|q_1(\vec{k}) + ip_1(\vec{k})), \qquad z_2(\vec{k}) = \tfrac{1}{\sqrt{2}}(|\vec{k}|q_2(\vec{k}) - ip_2(\vec{k})), \tag{22a}$$

$$\{q_j(\vec{k})\,,\, p_i(\vec{k'})\} = \delta_{ij}\delta^3(\vec{k}+\vec{k'}), \tag{22b}$$

$$\{z_1(\vec{k}),\,\bar{z}_1(\vec{k'})\} = -i|\vec{k}|\delta^3(\vec{k}-\vec{k'}), \qquad \{z_2(\vec{k}),\,\bar{z}_2(\vec{k'})\} = i|\vec{k}|\delta^3(\vec{k}-\vec{k'}), \tag{22c}$$

$$\bar{q}_j(\vec{k}) = q_j(-\vec{k}), \qquad \bar{p}_j(\vec{k}) = p_j(-\vec{k}), \tag{22d}$$

where $j = 1, 2$. Thus, the linearized gravitational field can indeed be considered as an assembly of harmonic oscillators, there being two oscillators, corresponding to the two helicity states, for each momentum $\vec{k}$. In our conventions, the oscillators with subscript 1 are of negative helicity and those with subscript 2 are of positive helicity.

Note however that there is an *asymmetry* between the two families due to certain sign differences. (See Eqs.22*a* and 22*b*.) It arises because we are dealing here with *self dual* rather than positive frequency connections. More precisely, the reason is the following: As is well-known (see, e.g.,[3]), the positive helicity corresponds to positive frequency self dual fields and negative helicity to negative frequency self dual ones. The modes described by $z_2(\vec{k})$ thus correspond to positive frequency connections while those described by $z_1(\vec{k})$, to negative frequency ones. In the standard treatment of spin-2 fields, one considers just the positive frequency fields. In the above framework, they are represented by $(\bar{z}_1(\vec{k}),\ z_2(\vec{k}))$. It is in terms of these modes that the description is manifestly symmetric. Since the aim of this section is to show that we can indeed recover the standard description of spin-2 fields through our quantization program, let us define new dynamical variables $\varsigma_j(\vec{k})$ as follows:

$$\varsigma_1(\vec{k}) = \bar{z}_1(\vec{k}) \quad \text{and} \quad \varsigma_2(\vec{k}) = z_2(\vec{k}) \tag{23}$$

In terms of these *positive frequency* fields, the basic (non-vanishing) Poisson brackets are given by:

$$\{\varsigma_i(\vec{k}),\ \bar{\varsigma}_j(\vec{k'})\} = i|\vec{k}|\delta_{ij}\delta^3(\vec{k}-\vec{k'}), \tag{24}$$

where $\bar{\varsigma}_j(\vec{k})$ is the complex conjugate of $\varsigma_j(\vec{k})$. Thus, in terms of positive frequency connections, the description is indeed entirely symmetric between the two helicity modes.

Let us now apply the quantization program of section 10.2 to this system. The space S of elementary variables will be the one spanned by $\varsigma_j(\vec{k}),\ \bar{\varsigma}_j(\vec{k})$. (More precisely, elements of S are the functions on the reduced phase space obtained by integrating $\varsigma_j(\vec{k})$ and $\bar{\varsigma}_j(\vec{k})$ with smooth test fields of compact support on the 3-dimensional k-space.) In the construction of the algebra $\mathcal{A}$ of quantum operators, we have to impose only the canonical commutation relations; anti-commutation relations are not needed since there are no algebraic relations between the elementary variables (other than the complex conjugation relation). The commutation relations are given by:

$$[\hat{\varsigma}_i(\vec{k}),\ \hat{\bar{\varsigma}}_j(\vec{k'})] = -\hbar|\vec{k}|\delta_{ij}\delta^3(\vec{k}-\vec{k'}). \tag{25a}$$

Next, let us introduce the $\star$-relations on the elementary operators. These are the obvious ones:

$$(\hat{\varsigma}_j(\vec{k}))^\star = \hat{\bar{\varsigma}}_j(\vec{k}). \tag{25b}$$

The properties of the involution operator now endow the entire algebra $\mathcal{A}$ with a $\star$-relation. The remaining problem then is to find a $\star$-representation of the resulting $\star$-algebra $\mathcal{A}^{(\star)}$.

Following the general program, in the first step, we ignore the $\star$-relations and seek only a linear representation of $\mathcal{A}$ by operators on a vector space V. Let us choose for V the space of entire holomorphic functionals $\Psi(\varsigma_1, \varsigma_2)$ of the complex variables $\varsigma_j(\vec{k})$ and represent the operators $\hat{\varsigma}_j(\vec{k})$ as follows:

$$\hat{\varsigma}_j(\vec{k}) \cdot \Psi(\varsigma_1, \varsigma_2) = \varsigma_j(\vec{k})\Psi(\varsigma_1, \varsigma_2), \qquad \hat{\bar{\varsigma}}_j(\vec{k}) \cdot \Psi(\varsigma_1, \varsigma_2) = \hbar|\vec{k}|\frac{\delta}{\delta\varsigma_j(\vec{k})} \cdot \Psi(\varsigma_1, \varsigma_2) \quad (26)$$

It is straightforward to check that the commutation relations of Eq.25*a* are satisfied. To introduce the inner product on V, we use the reality conditions of Eq.25*b*. Thus, we seek an inner-product that makes the operator $\hat{\bar{\varsigma}}_j(\vec{k})$ the adjoint of the operator $\hat{\varsigma}_j(\vec{k})$. For this, as in the description of the harmonic oscillator in section 10.3, we begin with a general measure $\mu(\varsigma_j, \bar{\varsigma}_j)$ on the reduced phase space, set

$$\langle\, \Psi(\varsigma_1, \varsigma_2) \mid \Psi'(\varsigma_1, \varsigma_2)\,\rangle = \int \Pi_j\; d\!\!I\varsigma_j(\vec{k}) \wedge d\!\!I\bar{\varsigma}_j(\vec{k})\; \mu(\varsigma_j, \bar{\varsigma}_j)\; \overline{\Psi(\varsigma_1, \varsigma_2)}\; \Psi'(\varsigma_1, \varsigma_2), \quad (27a)$$

and attempt to determine the measure by requiring that $\hat{\bar{\varsigma}}_j$ be the Hermitian adjoint of $\hat{\varsigma}_j$. It turns out that this requirement determines the measure uniquely (up to the usual overall constant):

$$\mu(\varsigma_j, \bar{\varsigma}_j) = \exp\left[-\int \frac{d^3k}{2\hbar|k|}\left(|\varsigma_1(\vec{k})|^2 + |\varsigma_2(\vec{k})|^2\right)\right]. \quad (27b)$$

Thus, the reality conditions have provided us with the standard Poincaré invariant inner product on the Hilbert space of linearized gravity; Once the $\varsigma_j(\vec{k})$, $\bar{\varsigma}_j(\vec{k})$ are chosen as the elementary classical variables, the quantization program leads us directly to the standard Bargmann representation of linearized gravity.

To conclude, we note that in this representation, $\hat{\varsigma}_j(\vec{k})$ are the creation operators; $\hat{\bar{\varsigma}}_j(\vec{k})$, the annihilation operators; the vacuum state is given by $\Psi_0(\varsigma_1, \varsigma_2) = 1$; and the normal ordered Hamiltonian operator $\hat{h}$ (corresponding to h of Eq.20) is given by:

$$\hat{h} \cdot \Psi(\varsigma_1, \varsigma_2) = -\hbar \int d^3k\, |k| \sum_j \varsigma_j(\vec{k}) \frac{\delta}{\delta\varsigma_j} \Psi(\varsigma_1, \varsigma_2). \quad (28)$$

4 Mathematical digression

As explained in the Introduction, although the Bargmann representation constructed in the last section is quite satisfactory in the linear approximation, it does not extend to the full theory. Therefore, to gain insight into what is feasible in the full theory, in the next section, we will work with self dual rather than positive frequency connections. The elementary classical variables will then be $q_j(\vec{k}), z_j(\vec{k})$, whose analogs do exist in the full, non-linear theory. Thus, the description will be quite similar to the description of the harmonic oscillator in the "hybrid" variables (q, z) that we studied in detail in section 10.3.

In fact, for helicity $+1$, the description is completely analogous: the commutation relations, $[\hat{q}_2(-\vec{k}),\ \hat{z}_2(\vec{k'})] = (\hbar/\sqrt{2})\delta^3(\vec{k} - \vec{k'})$, and the reality conditions, $\hat{q}_2^\star(\vec{k}) = \hat{q}_2(-\vec{k}),\ \hat{z}_2^\star(\vec{k}) = -\hat{z}_2(-\vec{k}) + \sqrt{2}|k|\hat{q}_2(-\vec{k})$, that now arise are the analogs of the relations, $[\hat{q}, \hat{z}] = \hbar,\ \hat{q}^\star = \hat{q}, \hat{z}^\star = -\hat{z} + 2\hat{q}$ we encountered in the case of the harmonic oscillator. Therefore, we can complete the quantization program, step by step, using the procedure used in section 10.3. For the helicity -1 modes, however, there is a crucial difference: the commutation relations are the same as before, $[\hat{q}_1(-\vec{k}),\ \hat{z}_1(\vec{k'})] = (\hbar/\sqrt{2})\delta^3(\vec{k} - \vec{k'})$, but there is a change of sign in the second reality condition: $\hat{z}_1^\star(\vec{k}) = -\hat{z}_1(-\vec{k}) - \sqrt{2}|k|\hat{q}_1(-\vec{k})$. Thus, the analogy now is with a harmonic oscillator with the same commutation relations,

$$[\hat{q}, \hat{z}] = \hbar, \tag{29a}$$

as before, but with modified reality conditions on $\hat{z}$:[4]

$$\hat{q}^\star = \hat{q} \quad \text{as before but} \quad \hat{z}^\star = -\hat{z} - 2\hat{q} \tag{29b}$$

In this section, we will make a small detour to show how a system satisfying Eqs.29 can be handled within our program.

There are two ways of carrying out the required quantization which correspond to making different choices in the two inputs that the program of chapter 10 requires. The first input needed is a choice of the space S of elementary classical variables. We may choose, in place of (q, z), the pair $(q, \overline{z})$ as our elementary variables. If this choice is made, it turns out that one can proceed exactly as in section 10.2 and complete the program without any difficulty. However, in linearized gravity, this choice would lead us precisely to the use of the variable $\varsigma_2(\vec{k})$ in place of $z_z(\vec{k})$ and is therefore uninteresting from the present standpoint. Let us therefore continue to use (q, z) as the elementary variables. Then using the commutation relation and the reality conditions, we are led to construct, as in section 10.3, a $\star$-algebra $\mathcal{A}^{(\star)}$ of quantum operators. The second input needed in the program is the choice of a linear space representation. It turns out that we can indeed pick a different representation to obtain a consistent quantum theory.

---

4 Thus, we are effectively setting $z = -q - ip$ rather than $z = q - ip$. This is actually equivalent to keeping the reality conditions the same but changing the sign of the symplectic structure and hence of the CCR.

To appreciate the change needed in the choice of V, let us first keep V as before and see what goes wrong. Thus, let V consist of entire holomorphic functions of z and let the operators be represented via:

$$\hat{q}\cdot\Psi(z) = \hbar\frac{d\Psi(z)}{dz}, \quad \text{and} \quad \hat{z}\cdot\psi(z) = z\Psi(z). \tag{30}$$

Then, the canonical commutation relations of Eq.29a are satisfied and we have a representation of the quantum algebra $\mathcal{A}$ without the $\star$-relations. Our final task is to introduce on V an inner product so that the $\star$-relations become concrete Hermitian-adjointness relations on the resulting Hilbert space. As before, let us set:

$$\langle\, \Psi(z) \mid \Psi'(z)\,\rangle = \int idz\wedge d\overline{z}\,\mu(z,\overline{z})\;\overline{\Psi(z)}\Psi'(z), \tag{31a}$$

and attempt to determine the measure $\mu(z,\overline{z})$ by imposing the reality conditions. As before, the measure is uniquely determined. However, since there is a change in sign in the reality condition, the sign in the exponent of the measure is now the opposite of what it was in section 10.3. We obtain:

$$\mu(z,\overline{z}) = \exp\left(+\tfrac{1}{4\hbar}(z+\overline{z})^2\right). \tag{31b}$$

Consequently, the arguments that led us in section 10.3 to the conclusion that the Hilbert space of normalizable states is infinite dimensional (and naturally isomorphic to the Bargmann Hilbert space), now implies that there are no (non-zero) entire holomorphic functions which are normalizable with respect to the inner product of Eq.31. Thus, a new strategy is needed in the choice of the linear representation.[5]

Let us begin by introducing some additional structure. Let us define a holomorphic generalized function (or a distribution) $\delta(z)$ as follows: it is a complex linear mapping from the space of functions of the type $\sum f_i(z)g_i(\overline{z})$, where $f_i(z)$ are entire

---

5 The new strategy is motivated by the transform from the q to the z-representation discussed at the end of section 10.3 (See the discussion surrounding Eq. 10 of chapter 10.) Note also that the choice of the vector space V we are about to introduce would be necessary also in Bargmann quantization had the symplectic structure been of opposite sign. More precisely, if the symplectic structure had been $\Omega = dp\wedge dq$ and we had continued to represent the wave functions by holomorphic functions - or, alternatively, if we keep the symplectic structure as it is, $\Omega = dq\wedge dp$, but use anti-holomorphic wave functions - we would have found that the measure needed to ensure the correct reality conditions would have been $\exp(+\frac{z\overline{z}}{2\hbar})$, whence no entire holomorphic function would have been normalizable. In this case, we need to use for states the holomorphic distributions introduced below.

holomorphic functions and $g_i(\bar{z})$ are entire anti-holomorphic functions, to the space of entire anti-holomorphic functions:

$$\delta(z) \circ \sum_i f_i(z) g_i(\bar{z}) = \sum_i f_i(0) g_i(\bar{z}). \tag{32a}$$

We can also define the anti-holomorphic distribution $\delta(\bar{z})$ simply by taking the complex conjugate of $\delta(z)$ and this new distribution has the action:

$$\delta(\bar{z}) \circ \sum_i f_i(z) g_i(\bar{z}) = \sum_i f_i(z) g_i(0) \tag{32b}$$

We will also need the product of a polynomial $a(z, \bar{z})$ with a distribution $\mathcal{F}(z)$. This is a distribution defined by:

$$[a(z,\bar{z})\mathcal{F}(z)] \circ \sum_i f_i(z) g_i(\bar{z}) := \mathcal{F}(z) \circ \, a(z,\bar{z}) \sum_i f_i(z) g_i(\bar{z}) \tag{33a}$$

Using, as usual, the Leibnitz rule as a motivation, one can define the derivative of a distribution $\mathcal{F}(z)$, as

$$\left[\frac{d}{dz}\mathcal{F}(z)\right] \circ \sum_i f_i(z) g_i(\bar{z}) := \frac{d}{dz}\left(\mathcal{F}(z) \circ \sum_i f_i(z) g_i(\bar{z})\right) - \mathcal{F}(z) \circ \frac{d}{dz} \sum_i f_i(z) g_i(\bar{z}), \tag{33b}$$

and similarly for the derivative w.r.t. $\bar{z}$. Applying this to Eq.32a, for example, we find

$$\frac{d}{d\bar{z}}\delta(z) = 0, \quad \text{and} \quad [\frac{d}{dz}\delta(z)] \circ \sum_i f_i(z) g_i(\bar{z}) = -\sum_i \frac{df_i(z)}{dz}\Big|_{z=0} g_i(\bar{z}). \tag{34}$$

Thus, *the δ-distribution is "holomorphic" and its derivative with respect to z is a distribution with the expected property.* Finally, we notice that the product of the two distributions Eqs.32a and 32b, is well-defined; it is just the two dimensional δ-distribution and therefore admits the standard integral representation:

$$\begin{aligned} [\delta(z)\delta(\bar{z})] \circ \sum_i f_i(z) g_i(\bar{z}) &= \sum_i f_i(0) g_i(0) \\ &\equiv \int dq \wedge dp \, \delta^2(q,p;0,0) \sum_i f_i(z) g_i(\bar{z}), \end{aligned} \tag{35}$$

where, we have used $z = q - ip$. Thus, one can regard $\delta(z)$ as the "holomorphic square-root" of the standard 2-dimensional δ-distribution on the 2-plane, (picked out by the complex structure).

Now, let us use as the carrier space for the representation the space V spanned by the holomorphic distributions of the type $\Psi = \sum(a_n(z))\ (d^n\delta(z)/dz^n)$ where each $a_n(z)$ is a polynomial in z. Then, we can continue to represent the operators $\hat{q}$ and $\hat{z}$ by Eq.30. We define an inner product on V by: $\langle \Psi \mid \Phi \rangle := \overline{\Psi}\Phi \circ \mu$, where $\mu = \mu(z, \bar{z})$ is a measure. From Eq.35 we see that this inner product is formally the same as the ansatz Eq.31a, used earlier. Thus, we again conclude that the measure must be given by Eq.31b. (The previous calculations go through step by step because they only need the states to be holomorphic, i.e., to be annihilated by the operator $d/d\bar{z}$.) But now, because of the presence of the δ distributions, all integrals converge and the inner product is well-defined on all of V.[6]

In this representation, it is the *annihilation* operator that is represented by $\hat{z}$. The vacuum state is simply the normalized state $\Psi_0(z) = \delta(z)$. An orthogonal basis in the Hilbert space is provided by the states $d^n\delta(z)/dz^n$. (Thus, we could also have let the representation space V to be the linear span of states of the type $\sum a_n\ (d^n\delta(z)/dz^n)$, where a_n are constants.) The Hamiltonian is given by

$$\hat{H} = \tfrac{1}{2}(\hat{z}^\star\hat{z} + 1) \equiv -\tfrac{1}{2}\left((\hat{z} + 2\hbar\frac{d}{dz})\hat{z} - 1\right). \tag{36}$$

We can now return to linearized gravity. If one begins with $q_j(\vec{k}), z_j(\vec{k})$ as the elementary variables, care must be exercised while choosing vector space V underlying the representation of the quantum algebra. In the sector $j = 1$, we can continue to use holomorphic functionals. However, in the sector $j = 2$, we will have to use the analogs of the holomorphic distributions encountered here.

5 Self dual representation

Let us now turn to the self dual representation of linearized gravity.

In exact general relativity, we do not know how to isolate the true degrees of freedom in a convenient way. Let us therefore ignore the reduced phase space of the linearized theory and use instead the full phase space. The natural choice for the space S of elementary classical variables –a choice which extends to the full theory– is now the complex vector space generated by the canonical pair $h^{ab}(\vec{x}), C_{ab}(\vec{x})$.

6 Strictly speaking, *all* calculations can be performed using just the definitions Eqs.32 and 33 and the form $\langle \Psi \mid \Phi \rangle := \overline{\Psi}\Phi\circ\mu$ for the inner product. However, as usual, the integral representation of the action of distributions (Eqs. 35 and 31) is convenient to use because it naturally codes the permissible algebraic manipulations.

As pointed out earlier, this is a "hybrid" choice, similar to the choice (q, z) of elementary variables we made in section 10.3 for the harmonic oscillator. Recall from Eq.6 that the only non-vanishing Poisson brackets between these elementary variables are:

$$\{h^{ab}(\vec{x}), C_{cd}(\vec{x}')\} = -\frac{i}{\sqrt{2}}\delta_c^a \delta_d^b \delta^3(\vec{x}, \vec{x}'). \tag{37a}$$

From Eq.7.21, the linearized reality conditions become:

$$\overline{h}_{ab}(\vec{x}) = h_{ab}(\vec{x}), \quad \overline{C}_{ab}(\vec{x}) = -C_{ab}(\vec{x}) + 2(\delta\Gamma)_{ab}(\vec{x}), \tag{37b}$$

where $(\delta\Gamma)_{ab} = -\operatorname{tr}(\delta\Gamma)_a \mathring{\sigma}^a$ is the linearized spin-connection of h_a^{AB}. It is straightforward to construct the quantum $\star$-algebra $\mathcal{A}^{(\star)}$ based on this $\mathcal{S}$: there are no anti-commutation relations, the commutation relations make the operators $\hat{h}_{ab}(\vec{x})$ and $\hat{C}_{ab}(\vec{x})$ canonically conjugate and the $\star$-relations are dictated by Eq.37b. For our representation space V, we now choose the space of entire holomorphic (possibly generalized) functionals of C_{ab} and choose the obvious action of the basic operators:

$$\hat{h}_{ab}(\vec{x}) \cdot \Psi(C) = \frac{\hbar}{\sqrt{2}} \frac{\delta}{\delta C_{ab}(\vec{x})} \cdot \Psi(C), \qquad \hat{C}_{ab} \cdot \Psi(C) = C_{ab} \cdot \Psi(C) \tag{38}$$

The next step is to impose the constraints and select the quantum states. This task is also straightforward: Since all equations are linear in the dynamical variables, there are neither factor ordering ambiguities nor regularization problems. The general solution is given by:

$$\Psi_{phy}(C) = \Psi(C^{STT}) e^{K(C^A)}, \tag{39}$$

where $\Psi(C^{STT})$ is any holomorphic (generalized) function of the symmetric, transverse, traceless part of C_{ab}, and where $K(C^A)$ is a *fixed* function of the anti-symmetric part of C_{ab}.[7] Thus, as one might have expected, in essence the physical states are again (generalized) functionals only of the true degrees of freedom. It is therefore convenient to work in the momentum space. Then, the physical subspace V_{phy} consists of functionals $\Psi(z_1(\vec{k}), z_2(\vec{k}))$ of the two modes $z_j(\vec{k})$. A complete set

---

7 The explicit form of the function is: $K(C^A) = \frac{G}{\sqrt{2}} \int d^3x \mathbf{C}_a C_{[bc]} \epsilon^{abc}$. Here, $\mathbf{C}_a$ is a vector potential for $C_{[ab]}$ which, by the scalar constraint, is guaranteed to be curl-free. We will not need this explicit form in what follows. In fact, had we not only linearized the field equations but also taken the limit $G \rightarrow 0$, which is appropriate for weak fields, we would have: $K = 0$.

of physical operators is easy to find. The basic physical operators are simply: $\hat{z}_j(\vec{k})$ and $\hat{q}_j(\vec{k})$. They satisfy the commutation relations:

$$[\hat{q}_i(-\vec{k}),\ \hat{z}_j(\vec{k'})] = \frac{\hbar}{\sqrt{2}}\delta_{ij}\delta^3(\vec{k}-\vec{k'}), \tag{40a}$$

and the reality conditions:

$$(\hat{z}_1(\vec{k}))^\star = -\hat{z}_1(-\vec{k}) - \sqrt{2}|k|\hat{q}_1(-\vec{k}), \qquad (\hat{z}_2(\vec{k}))^\star = -\hat{z}_2(-\vec{k}) + \sqrt{2}|k|\hat{q}_2(-\vec{k}) \tag{40b}$$

Thus, the mode labelled 1 is analogous to the harmonic oscillator treated in section 10.3 while the mode labelled 2 is analogous to the oscillator treated in section 4 above. Finally, on the physical sub-space V_{phy}, the representation Eq.38 yields:

$$\begin{aligned} \hat{q}_j(-\vec{k})\cdot\Psi(z_1(\vec{k}), z_2(\vec{k})) &= \frac{\hbar}{\sqrt{2}}\frac{\delta}{\delta z_j(\vec{k})}\Psi(z_1(\vec{k}), z_2(\vec{k})), \\ \text{and}\quad \hat{z}_j(\vec{k})\cdot\Psi(z_1(\vec{k}), z_2(\vec{k})) &= z_j(\vec{k})\Psi(z_1(\vec{k}), z_2(\vec{k})). \end{aligned} \tag{41}$$

Our job now is to select the inner product using the reality conditions. The similarity of the two modes to the two treatments of the harmonic oscillator enables us to follow the two procedures step by step. The calculations are straightforward and final results are as follows. The inner product is again expressible as:

$$\langle\,\Psi(z_1,z_2)\,|\,\Psi'(z_1,z_2)\,\rangle = \int \left[\Pi_j\ \mathbb{d}z_j(\vec{k})\wedge \mathbb{d}\bar{z}_j(\vec{k})\right]\ \mu(z_j(\vec{k}),\bar{z}_j(\vec{k}))\,\bar{\Psi}\,\Psi', \tag{42a}$$

where, as usual, $\mathbb{d}$ is the infinite dimensional exterior derivative on the space of radiative modes $(z_j(\vec{k}),\bar{z}_j(\vec{k}))$. The reality conditions tell us simply that the measure is given by:

$$\mu(z_j(\vec{k}),\bar{z}_j(\vec{k})) = \exp\left[\sum_j \tfrac{(-1)^j}{4\hbar}\int\frac{d^3k}{|k|}\,(z_j(\vec{k})+\bar{z}_j(-\vec{k}))(z_j(-\vec{k})+\bar{z}_j(\vec{k}))\right]. \tag{42b}$$

The normalized states are therefore of the form:

$$\begin{aligned} \Psi(z_1,z_2) = \int\frac{d^3k_1}{|k|_1}\cdots\int\frac{d^3k_n}{|k|_n}\,f(k_1,...k_n)\,\frac{\delta}{\delta z_1(\vec{k}_1)}\cdots\frac{\delta}{\delta z_1(\vec{k}_n)}\,\delta(z_2(\vec{k})) \\ \times\ P(z_2(\vec{k}))\,\exp\left[\tfrac{1}{4\hbar}\int d^3k|\vec{k}|^{-1}(z_2(\vec{k})z_2(-\vec{k}))\right] \end{aligned} \tag{43a}$$

where $P(z_2(\vec{k})$ is a polynomial in $z_2(\vec{k})$. Their norm is norm is given by:

$$\|\Psi(z_1,z_2)\|^2 = \{\int \frac{d^3k_1}{|k|_1} \dots \int \frac{d^3k_n}{|k_n|} |f(k_1,\dots k_n)|^2\} \times$$
$$\{\int d\mathrm{I}z_2(\vec{k}) \wedge d\mathrm{I}\bar{z}_2(\vec{k})\, exp\Big[-\frac{1}{2\hbar}\int \frac{d^3k}{|\vec{k}|}|z_2(\vec{k})|^2\Big]\, |P(z_2(\vec{k}))|^2\} \quad (43b)$$

Finally, the annihilation operators are now $(\hat{z}_1(\vec{k}))^\star$, $\hat{z}_2(\vec{k})$; the vacuum is given by $\Psi_0(z_1,z_2) = \delta(z_1(\vec{k})) \times \exp\,[\frac{1}{4\hbar}\int d^3k|k|^{-1}z_2(\vec{k})z_2(-\vec{k})]$; and, the normal ordered Hamiltonian is:

$$\hat{h} = -\hbar \int d^3k \,\Big(\,\hat{z}_1(\vec{k})\hat{z}_1(\vec{k})^\star + \hat{z}_2(\vec{k})^\star\hat{z}_2(\vec{k})\,\Big)$$
$$= -\hbar \int d^3k \Big[-z_1(\vec{k})(z_1(-\vec{k}) + \hbar|k|\frac{\delta}{\delta z_1(\vec{k})}) + (-z_2(-\vec{k}) + \hbar|k|\frac{\delta}{\delta z_2(\vec{k})})z_2(\vec{k})\Big] \quad (44).$$

This representation is unitarily equivalent to the Bargmann representation constructed in section 3 and hence also to the standard Fock representation for the spin-2 field.

The analysis of the last two sections has provided several important lessons to us. First, it is encouraging that the self dual representation does indeed exist in the linearized limit because the conceptual ingredients used in its construction are available also in the full theory. Second, the fact that there is a neat separation into helicity states is now a technical result. It arose naturally while implementing the quantization program; we did not feed it in at any stage. Indeed, we did not have to appeal anywhere to Poincaré invariance. It is the reality conditions that selected for us the inner product and the vacuum. Finally, we have learnt that the actual imposition of the reality condition can be quite an involved procedure. Even after the elementary classical variables are fixed, one may still have to invent new representations of the algebra of quantum operators to ensure the existence of a sufficiently large physical Hilbert space.

References

[1] A. Ashtekar and Joohan Lee (in preparation).

[2] A. Ashtekar, C. Rovelli and L. Smolin (in preparation).

[3] A. Ashtekar, J. Math. Phys. **27**, 824(1986).

12 CONNECTION REPRESENTATION: ISSUE OF TIME

1 Introduction

In this chapter we will first present a summary of the current status of the quantization program for the full theory in the connection representation and then discuss an important application. This representation has been very useful in quantum cosmology [1,2], where it has led to interesting solutions to all quantum constraints. Furthermore, key features of this representation –particularly the holomorphicity of the quantum wave functions– seem to play an important role in that discussion. However, in the full theory without any truncation, the representation has proved to be unmanageable so far. More precisely, in this representation, no one has yet been able to solve simultaneously all three quantum constraints of the full theory. In this respect, its status is comparable to that of the metric representation in geometrodynamics. However, it does differ from the metric representation in some key respects. Let me give two examples. First, because the connection representation arises also in Yang-Mills theory, it is easier to use while importing ideas from QCD into quantum gravity. Second, because of the detailed structure of the scalar constraint of general relativity, this representation is better suited to analyze the issue of time in quantum gravity. In this chapter we shall discuss the second of these issues; an illustration of the first application will be given in the next chapter.

Section 2 summarizes the status of the connection representation. Section 3 outlines the problem of time in quantum gravity and section 4 presents two strategies that one can adopt to address this problem. In section 5, we show that the problem can in fact be solved in a certain approximation if one uses the connection representation. (We also point out the reasons which make the metric (or the triad) representation is ill-suited for this task.) The end result is that there is a precise sense in which the familiar Schrödinger evolution that we use in everyday physics can be "derived" from the scalar constraint of quantum gravity (coupled with matter). Thus, as was emphasized in the Introduction to Part III, the connection representation appears to be well suited to probe avenues through which the standard space-time physics that we normally use can arise from the abstract framework of canonical quantum gravity.

2 Connection representation

We now wish to carry out the various steps in the quantization program, presented in the last section of chapter 10, for full general relativity.

Let us begin with the real classical phase space and introduce the $\star$-algebra of quantum operators as follows. Choose for the space S of elementary classical variables the complex vector space spanned by functions which are constant or linear[1] in the basic canonical variables A_a^i and $\widetilde{E}_i^a$. This space is obviously closed under the Poisson bracket and, at the same time, large enough so that any (sufficiently regular) function on the phase space can be expressed as (possibly a limit of) a sum of products of elements of S. To carry out the second step in the program, let us associate with each element of S a quantum operator. This is equivalent to introducing abstractly defined local operators[2] (or rather, operator-valued distributions) $A_a^i(x)$ and $\widetilde{E}_i^a(x)$ and considering their smeared out versions, $A(\widetilde{f}) := \int_\Sigma d^3x\, A_a^i \widetilde{f}_i^a$ and $\widetilde{E}(g) := \int_\Sigma d^3x \widetilde{E}_i^a g_a^i$. These are the elementary quantum operators. They are defined "abstractly" in the sense that *a priori*, there is no Hilbert space of states for them to act upon. To obtain the algebra $\mathcal{A}$ of quantum operators, we first construct the free algebra generated by these operators and subject it to the canonical commutation relations:

$$[A_a^i(x), A_b^j(y)] = 0, \qquad [\widetilde{E}_i^a(x), \widetilde{E}_j^b(y)] = 0,$$
$$[\widetilde{E}_i^a(x), A_b^j(y)] = \hbar\delta_b^a\delta_i^j\delta^3(x,y). \tag{1}$$

(Note that our space S is sufficiently small for the anti-commutation relations to be redundant.) This completes the second step. Our next task is to introduce the involution operation $\star$ on the algebra $\mathcal{A}$. As in chapter 10, this can be accomplished simply by requiring that the $\star$-operation be an involution and that if $\bar{F} = G$ for any two elementary classical variables F and G in S, then $(F)^\star = G$ in $\mathcal{A}$. This, in particular, implies that, *formally*, $\widetilde{E}_i^a\widetilde{E}^{bi}$ is its own $\star$ and that its time derivative (obtained by taking the commutator with the Hamiltonian) is also its own $\star$. (The conclusion is only formal because we have not regularized the products of operator-valued distributions involved.) This is the required $\star$-algebra $\mathcal{A}^{(\star)}$.

The next step is to construct a representation of $\mathcal{A}$. Since the variables $(\widetilde{E}_i^a, A_a^i)$ of general relativity are analogous to the pair (q, z) on the phase space of a harmonic

1 In this construction, for brevity we are implicitly fixing an origin in the affine space of connections.

2 In this and subsequent chapters, for notational simplicity, whenever there is no danger of confusion, we shall drop the hats on the quantum operators.

oscillator, it follows from section 3 of the last chapter that, in the connection representation, we should begin with the vector space V of (generalized) ***holomorphic*** functionals of A_a^i.[3] On this V, the elementary quantum operators are represented as:

$$\hat{A}_a^i(x) \cdot \Psi(A) := A_a^i(x)\Psi(A), \quad \text{and} \quad \hat{\tilde{E}}_i^a \cdot \Psi(A) := \hbar \frac{\delta\Psi(A)}{\delta A_a^i} \ . \tag{2}$$

The fifth step in the program is to express the quantum constraints as concrete (regularized) operators on V, and obtain wave functions $\Psi(A)$ which are in the kernel of these operators. These are the ***physical states*** of quantum gravity in the connection representation. The sixth and the last step is to introduce the Hermitian inner-product on the space V_{phy} of these physical states. It is at this stage that we would appeal to the $\star$-relations which have been ignored so far. As we saw in chapter 10, in simple examples, this requirement does suffice to single out an inner-product. The same is true of a number of model systems based on general relativity such as the linearized theory discussed in chapter 11, certain Bianchi models and 2+1 gravity.

However, as pointed out in chapter 2, the problem of finding physical states, and hence also of implementing the $\star$-relations, remains open in the connection representation. A large class of solutions to the quantum Gauss and scalar constraints [3] *have been* constructed. These are the Wilson loops, i.e., wave functions $\Psi_\gamma(A)$, labelled by loops γ, which are the traces (in the spin-$\frac{1}{2}$ representation of the internal-gauge group) of holonomies of connections A_a^i around the loop γ: $\Psi_\gamma(A) := \mathrm{tr}\, \mathcal{P}\, exp \oint_\gamma dS^a A_a$. One can also construct a large class of simultaneous solutions to the Gauss and the vector constraints as follows. Denote by S the space of connections A_a^i for which the magnetic field $\tilde{B}_i^a := \tilde{\eta}^{abc} F_{bci}$, regarded as a matrix, is non-degenerate at each point of the spatial 3-manifold. Given any connection in S, one can construct a "magnetic metric" $\tilde{\tilde{m}}^{ab}$ via $\tilde{\tilde{m}}^{ab} = \tilde{B}_i^a \tilde{B}^{bi}$. Using $\tilde{\tilde{m}}^{ab}$ in place of $\tilde{\tilde{q}}^{ab}$, one can construct invariants, such as the integral over Σ of the scalar curvature or of the Ricci tensor squared, etc., which, regarded as wave functions, are annihilated by the quantum Gauss and vector constraints. (All these wave functions have support only on S.) Thus, one has available large classes of solutions to the Gauss and the scalar constraints as well as the Gauss and the vector constraints. Unfortunately, however, very little known is about simultaneous solutions to *all* constraints. Consequently, as far as full quantum gravity is concerned, so

3 Kodama [1] has recently proposed a somewhat different approach in the context of quantum cosmology. There, the wave functions are holomorphic but not *entire* in the component of the connection that corresponds to time. The relation between the two approaches is not well-understood as yet.

far, the connection representation has not led to qualitatively new insight about the dynamics of the gravitational field in the Planck regime. Rather, at present its value lies primarily in the fact that it provides a general framework that is well-tailored to address certain conceptual issues of quantum gravity in a concrete way. An important example is the issue of time.

3 Problem of time

The issue of time in quantum gravity can be summarized as follows. Already in classical general relativity, because of the absence of a background geometry, the notions of time and dynamics are more subtle than in other field theories. In the canonical framework, one begins with just a 3-manifold Σ, and specifies the values of canonical variables on it. The infinite dimensional phase space Γ is constructed from these pairs. Given a lapse Y and a shift T^a, one can construct a Hamiltonian H_{T,T^a} and "evolve" any canonical pair *in the phase space* by following the integral curve of the Hamiltonian vector field X_{T,T^a}. The parameter along the integral curve is the affine parameter of the vector field X_{T,T^a}; *a priori*, it is not a time parameter in any space-time. Indeed, if the chosen canonical pair does not satisfy the constraints, the integral curves of the Hamiltonian vector fields for various choices of lapse and shift fail to provide us with any space-time whatsoever. When the constraints are satisfied, on the other hand, these curves are all mutually compatible and lead to a single space-time metric g_{ab} on the 4-manifold $M = \Sigma \times \mathbb{R}$. In this space-time, finally, one can reinterpret the phase space evolution as time evolution; the affine parameter along any Hamiltonian vector field can now be identified with a time parameter in the space-time that results from the evolution. Thus, finally, we do have a consistent picture of the evolution. However, the familiar notions of time and evolution can be introduced only *after* one has solved the field equations and singled out a solution. The notions do not refer to a background structure and are in this sense not "universal". They have to be introduced separately in each solution g_{ab} to the field equations.

In quantum theory, the problem gets more serious. For, now even the end product of evolution is not a space-time. For comparison, consider a particle in Euclidean space. A space-time metric, g_{ab}, the end product of the classical evolution, is analogous to the entire trajectory $\vec{x}(t)$ of the particle. A quantum state, on the other hand, is a wave function $\Psi(\vec{x})$ and the end product of the quantum evolution is therefore a solution, $\Psi(\vec{x}, t)$, to the Schrödinger equation. In a general solution, the particle trajectories $\vec{x}(t)$ play no role whatsoever. Similarly, one would expect that classical space-times will play no role whatsoever in a general solution to the equations of quantum gravity. Just as in particle dynamics one can focus on semi-classical approximations and consider solutions to the Schrödinger equation in this approximation to be built by a superposition of classical trajectories, so will quantum gravity presumably admit a semi-classical regime in which approximate solutions can be interpreted as being made out of a bunch of "nearby" space-time geometries. Even in this case, a single, classical space-time is not available and there is considerable difficulty in singling out a time variable. Beyond the semi-classical regime, the prospects of having a classical notion of time and dynamics seem very dim indeed. If there is no classical time, however, how can one do physics? After all, the business of physics is to make predictions. What can one predict if one does not have access to the familiar notion of time? This, broadly, is the issue of time in quantum gravity.

4 Two strategies

The problem of time has been discussed extensively in the literature[4] and I cannot possibly survey here the various ideas that have been proposed. Rather, I will just outline a couple of strategies. My own views on the subject have changed over the past decade or so. I will therefore begin with a summary of the approach I had adopted in the past, explain why I am now dissatisfied with it and then present my current viewpoint.

Let me return to the classical theory briefly. It is clear from chapters 7 and 9 that in the asymptotically flat context there is a clear distinction between constraints and Hamiltonians. Now, in Dirac's theory of constrained systems, first class constraints always generate gauge transformations. One can apply that interpretation to general relativity and conclude that the canonical transformations generated by all constraints should be regarded as gauge transformations. What transformations do these constraints generate? Recall from chapter 7 that these are the internal rotations and space-time diffeomorphisms *which tend to identity at spatial infinity.* A possible viewpoint then is to regard these transformations as gauge. The canonical transformations generated by the Hamiltonians, on the other hand, correspond to space-time diffeomorphisms which are asymptotic translations. These are to be interpreted as dynamical evolutions. Thus, one can adopt the attitude [5] that, in the asymptotically flat context, the overall situation is rather similar to that encountered in, say, non-Abelian gauge theories in Minkowski space.

4 For example, an entire Part of the Proceedings of the Osgood Hill conference on *Conceptual problems of quantum gravity* [4] is devoted to this issue. This chapter is based on my contribution to that volume.

One can therefore proceed in quantum theory following the route normally adopted in gauge theories. Thus, one may first complete the program presented in chapter 10, where the physical states are singled out by solving the quantum constraints, and then write down the Hamiltonian operator on the Hilbert space of physical states. Since the classical Hamiltonian weakly commutes with all constraints and is weakly real, one may expect that, with an appropriate choice of ordering, the quantum Hamiltonian would be a well-defined, self-adjoint operator on the physical Hilbert space. It would therefore generate a one-parameter family of unitary transformations. One could simply call this parameter "time" in quantum theory. Thus, it is possible to argue that it is straightforward to tackle the issue of time in the asymptotically flat context.

My present view is that, while it *is* consistent to adopt such an attitude, this strategy is not very useful in practice. To see this, let us again recall the situation in the classical theory. Given the classical phase space, one can pass to the reduced phase-space by factoring out the constraint surface by the orbits of the constraint vector fields. Classically, this is precisely the space of physical states. In the asymptotically flat context, there is nontrivial dynamics on this new symplectic manifold, and its generating functional is simply the projection of the Hamiltonian from the full to the reduced phase space. (Note that the reduced Hamiltonian depends *only* on the asymptotic values of the lapse and the shift.) One can construct the Hamiltonian vector field on the reduced phase space and call its affine parameter "time" in the classical theory. Although this procedure is "correct", without a suitable gauge-fixing prescription *to tie the value of the affine parameter to some geometric field in space-time*, the resulting notion of time is not very useful in practice. By itself, it has very little *direct* relation to the operational notions, such as the proper time measured by clocks, that we actually use in general relativity. In quantum theory, the notion of time as the parameter along the group generated by the Hamiltonian operator suffers from the same drawbacks. To make this notion useful, we have to make it more "concrete". In the quantum measurement theory, for example, the notion of time as an abstract parameter would not suffice. We need something like the quantum analog of the classical gauge-fixing procedure. However, in quantum theory the issue is substantially more complicated because, as was emphasized earlier in this section, the end result of solving all quantum equations is *not* a space-time. What we need, at least in some approximation, is "internal clocks": some of the mathematical degrees of freedom of the theory are to be the clocks, with respect to which other degrees –identified as the physical modes– "evolve".

For definiteness, let us first work in the traditional configuration representation of geometrodynamics in which the wave functions are functionals $\Psi(q)$ of the 3-metrics q_{ab}. Then, one can envisage the following procedure. First, we could try to split q_{ab} into two parts, $q^{(T)}$ representing "time", and $q^{(R)}$ representing the rest (i.e., the "physical degrees of freedom" contained in q_{ab}). The decomposition is to be achieved in such a way that the scalar constraint can be re-interpreted as a Schrödinger equation for wave functionals $\Psi(q) \equiv \Psi(q^{(R)}, q^{(T)})$ which tells us how Ψ "evolves" with respect to $q^{(T)}$ under the action of a Hamiltonian. The Hamiltonian operator itself should act only on the argument $q^{(R)}$ of the wave function. If this could be achieved, the situation would be similar to that in the case of the parametrized non-relativistic particle discussed in section 10.5. Dynamics would become manifest only when we deparametrize the theory by decomposing the argument q_{ab} of the wave functionals into time $q^{(T)}$, and the rest, $q^{(R)}$. Note that, as we saw in section 10.5, it *not* essential to single out time just to equip the space of physical states with an inner-product. The reality conditions seem well-suited to carry out that task. Rather, the decomposition is sought in order to be able to physically interpret the resulting mathematical framework in familiar terms.

These ideas are of course not new. They have been around for a long time now in the context of spatially compact space-times although, to my knowledge, the emphasis has been on the problem of singling out the correct inner product on physical states. Furthermore, the relevance of these considerations to the asymptotically flat case had not been spelled out in the early literature. In the spatially compact case, several proposals were put forward to accommodate these ideas using semi-classical techniques, particularly in the context of mini-superspaces. However, it has also been clear for about two decades that, in the traditional metric representation used above, the procedure of splitting q_{ab} in the desired fashion cannot be implemented even in the weak field limit [6]. I will now argue that the difficulty lies in the use of the metric representation; the problem is not inherently insoluble.

Let us begin by analyzing the source of the difficulty. Consider, again, the parametrized free particle with phase-space coordinates[5] $(q^1, ...q^n; p_1, ...p_n)$ subject to the constraint,

$$p_1 + H(q^2, ..q^n, p_2, ..p_n) = 0. \tag{3}$$

As we saw before, if we use the q-representation, the quantum constraint becomes $i\hbar\frac{\partial}{\partial q^1}\Psi(q^i) = \hat{H} \cdot \Psi(q^i)$. The form of this equation suggests that we identify q^1 with time and interpret the quantum constraint as the time-dependent Schrödinger equation. On the other hand, in the momentum representation, where p_1 is just a

5 To mimic the situation in general relativity, I am adopting the attitude that all we have is a Hamiltonian description of a system with the constraint given by equation (1); we did not begin with a particle in an (n-1)-dimensional space and *then* parametrize it. Thus, *a priori*, we do not know which of the variables is to play the role of time in quantum theory; this will be determined by the constraint. To emphasize this viewpoint, unlike in chapter 10, now I avoid the use of q_0.

multiplication operator, the quantum version of the classical constraint equation, $\hat{p}_1\Psi(p_i) = -\hat{H}\cdot\Psi(p_i)$, does *not* resemble the time-dependent Schrödinger equation. Thus, although the two representations are mathematically equivalent, one is better suited to single out "time" than another. I would like to argue that, in the gravitational case, the metric representation is ill-suited, much in the same way as the momentum representation is ill-suited in the above example. To see this, recall that, in terms of the traditional canonical variables, the scalar constraint can be rewritten as (see Eq.9.36):

$$\mathcal{E}(q) + H_T(q, \widetilde{p}) = 0 \qquad (4)$$

where $\mathcal{E}(q)$ is the ADM energy, $\widetilde{p}_{ab}$ is the momentum canonically conjugate to the 3-metric q_{ab} and $H_T(q,\widetilde{p})$ is the Hamiltonian corresponding to the lapse T. This constraint is very similar to the one in Eq.1, with $\mathcal{E}(q)$ playing the role of p_1. The similarity suggests that we should think of $\mathcal{E}(q)$ as the variable conjugate to "time". This idea is attractive especially because $\mathcal{E}(q)$ has the interpretation of the total energy of the system. Furthermore, the structure of (2) and the interpretation of $\mathcal{E}(q)$ is unaffected by inclusion of matter sources. Note, however, that the similarity also suggests that it would be very difficult to extract "time" in the q_{ab}-representation where $\hat{\mathcal{E}}(q)$ is just a multiplication operator.

What is the situation with new variables? The analysis of the parametrized particle suggests that the situation would be better in the A_a^i -representation: since $\mathcal{E}$ is now expressible as a functional of the triads $\widetilde{E}_i^a$ only, $\hat{\mathcal{E}}$ would be a differential operator in the A_a^i -representation. The "part of A_a^i conjugate to $\mathcal{E}$ " in the argument of the wave functional $\Psi(A)$ can now play the role of time. We shall now show that this idea can indeed be carried through in a weak field limit.

5 Truncated theory

For simplicity of presentation, in the main body of this section I will focus only on the scalar constraint, assume that Σ is topologically $\mathbb{R}^3$ and use the asymptotically flat boundary conditions of chapter 6. Then, at the end of the section, I will comment briefly on the other constraints and on the spatially compact case. The emphasis will be on conveying the general ideas; details will appear elsewhere.

We have already carried out the first four steps of the quantization program in the connection representation. The next step, as we saw, is to solve the quantum constraints. It is here that I would like to use the weak field truncation. Following the procedure used in chapter 11, let us introduce a background configuration $(\overset{\circ}{E}{}_i^a, \overset{\circ}{A}{}_a^i = 0)$, where $\overset{\circ}{E}{}_i^a$ is a flat triad, and expand out the operators that appear in the constraint equations in powers of deviations, $(\hat{E}^a_i - \overset{\circ}{E}{}^a_i)$ and $(\hat{A}^i_a - 0)$. Recall that, since we now have access to a background triad, we can convert the internal indices to vector indices, and using the flat metric $\overset{\circ}{q}_{ab}$ constructed from the background triad, carry out the standard decomposition of symmetric tensor fields into transverse-traceless, longitudinal and trace parts. With this machinery at hand, let us solve the scalar constraint order by order in the connection representation.

As we saw in chapter 11, if we keep terms just to *first order* in deviations, we find that the only non-zero part of $A_{ab} := A^i_a \overset{\circ}{E}{}_b{}^i$ is the symmetric, transverse-traceless part, A^{TT}_{ab}. That is, all other parts of A_{ab} are of at least second order. The idea now is to solve the scalar constraint *to second order*. The simplest way to do this is to first show that, using reality conditions, the scalar constraint of full, classical general relativity, can be written as:

$$-\epsilon^{abc}\partial_a \Gamma_{bc} = G^2(\overline{A}_a{}^b A_b{}^a - \overline{A}A)\ , \tag{5}$$

where $\Gamma_{ab} := \Gamma^i_a E_{bi}$, with Γ^i_a the spin connection of the triad E^a_i, ϵ^{abc} is the alternating tensor defined by the triad, $A_a{}^b := A^i_a E^b_i$, and, G is Newton's constant. It is because of this that the Hamiltonian of the full theory with unit lapse (i.e., $T = 1$ or $\underset{\sim}{T} = 1/\sqrt{q}$) can be written in the "quadratic form" of Eq.14 of chapter 7:

$$H(\widetilde{E}, A) = -G\int_\Sigma d^3x\,(E)\,((\overline{A}_a{}^b)(A_b{}^a) - \overline{A}A\,), \tag{6}$$

where $A = A_a{}^a$ and E, as before is the determinant of the co-triad. To impose the quantum constraint to second order, we have to keep in Eq.5 terms only to second order in deviations from the background and replace the classical fields by the corresponding quantum operators. It turns out that the left side of Eq.5 vanishes identically to first order because at this order the deviation in the 3-metric is also traceless and transverse. Thus, the left side is already of second order. Let us consider the right side. Since the background connection $\overset{\circ}{A}{}^i_a$ vanishes, it follows that the only non-zero contribution to second order comes from replacing the triad everywhere by its background value. Therefore, if one writes out the spin-connection in terms of the triad, the second-order truncated scalar constraint reduces to:

$$-\frac{1}{G}\Delta\, E^T(x) = G(A^{TT})^{ab}(x)\,(\overline{A^{TT}_{ab}})(x), \tag{7}$$

where Δ is the Laplacian with respect to the flat background metric, and E^T is the trace part of $(E^a_i - \overset{\circ}{E}{}^a_i)$. Consequently, the truncated quantum constraint is given by:

$$-\frac{1}{G}(\Delta\hat{E}^{T})\cdot\Psi(A^{TT},A^{L},A^{T},A^{A}) = G\hat{A}^{TTab}(x)(\hat{A}^{TT}_{ab}(x))^{\star}\cdot\Psi(A^{TT},A^{L},A^{T},A^{A}), \quad (8)$$

where, A^L, A^T and A^A are, respectively, the longitudinal, the trace and the anti-symmetric parts of $\hat{A}_{ab}$.

Let us set $\tau(x) = G(\Delta)^{-1}\cdot A^T(x)$. Then, $\tau(x)$ has the physical dimensions of time. Furthermore, it is canonically conjugate to $\frac{1}{G}(\Delta E^T)$, whence, in the connection representation, the operator $-\frac{1}{G}(\hat{E}^T)$ is represented simply by $-\hbar(\delta/\delta\tau(x))$. Therefore, the truncated quantum constraint becomes:

$$-\hbar\frac{\delta}{\delta\tau(x)}\cdot\Psi(A^{TT},A^{L},A^{T},A^{A}) = G(\hat{A}^{TT})^{ab}(x)(\hat{A}^{TT}_{ab}(x))^{\star}\cdot\Psi(A^{TT},A^{L},A^{T},A^{A}). \quad (9)$$

This equation constrains the functional dependence of $\Psi(A)$: the dependence of $\Psi(A)$ on A^T is completely determined by its dependence on A^{TT}. In this sense, it is a quantum *constraint* equation. On the other hand, if we simply integrate the equation over Σ using the volume element provided by the background $\mathring{q}_{ab}$, we obtain:

$$\begin{aligned}\hbar\left[\int d^3x\,\frac{\delta}{\delta\tau(x)}\right]&\cdot\Psi(A^{TT},A^{L},A^{T},A^{A})\\ &= -G\left[\int d^3x\,\hat{A}^{TT}_{ab}(x)(\hat{A}^{TTab}(x))^{\star}\right]\cdot\Psi(A^{TT},A^{L},A^{T},A^{A}) \quad (10)\\ &\equiv \hat{H}_{(2)}\cdot\Psi(A^{TT},A^{L},A^{T},A^{A}).\end{aligned}$$

Here, in the last step we have used the expression of the full Hamiltonian in Eq.6 to recognize that the operator in the second step is indeed the second order truncated Hamiltonian and can therefore be written as $\hat{H}_{(2)}$. Note that this $\hat{H}_{(2)}$ is precisely the Hamiltonian of the linearized theory we found in chapter 11 (see Eq.11.20). Finally, since $\Psi(A)$ is ***holomorphic***, we have:

$$\frac{\delta}{\delta(Im\tau)(x)}\cdot\Psi(A) = i\frac{\delta}{\delta\tau(x)}\cdot\Psi(A), \quad (11)$$

where $Im\tau(x)$ is the imaginary part of $\tau(x)$. Therefore, Eq.10 *can* indeed be interpreted as a Schrödinger equation provided we identify $-\int d^3x\,\frac{\delta}{\delta(Im\tau(x))}$ with the time translation operator[6] $\frac{\partial}{\partial t}$; then Eq.10 just reduces to:

$$i\hbar\frac{\partial}{\partial t}\Psi(A) = \hat{H}_{(2)} \cdot \Psi(A) \tag{12}$$

where $\hat{H}_{(2)}$ (independent of t) is the correct Hamiltonian operator in this approximation. Thus, the scalar constraint of general relativity does contain, at least approximately, a time dependent Schrödinger equation.

What is the situation if we bring in matter sources? Now, the right side of Eq.5 contains matter contributions – the energy density of matter– while the left side is unaffected. (This has the well-known consequence that, in full classical general relativity, the energy of matter fields just gets added to the expression of the Hamiltonian (of Eq.6); the expression of the ADM surface term is unaffected.) Consequently, in all subsequent equations, one needs only to add the appropriately truncated expression of the matter energy density to the right hand side. The left side is untouched. The final result is therefore a Schrödinger equation in which the definition of time is the same as before but the Hamiltonian on the right side is modified to allow for possible matter contributions. Thus, the emergence of the Schrödinger equation in the truncated approximation is not restricted to source-free general relativity.

Recall that, in canonical quantum gravity, we do *not* have access to a classical space-time. Even in the above truncation procedure, we worked only with the 3-manifold Σ and expanded operators around some background fields on Σ. Thus, we did not have access to a classically defined time variable. Rather, we were able to isolate, from among the mathematical variables contained in A_a^i , a preferred variable t which serves as an "internal clock" with respect to which the wave function "evolves". Put differently, by identifying time in the components of A_a^i we have *derived* the Schrödinger equation of linearized gravity without having access to a space-time metric –or indeed, even a 4-manifold. An immediate consequence is that the theory is unitary in this weak-field limit. Thus, from the viewpoint of canonical quantum gravity, the standard quantization of the spin-2 field on Minkowskian background is consistent because, to this approximation, there is a well-defined truncated theory, in which one can single out an internal clock.

Note that the quantum constraint plays a dual role. First of all, it is a constraint on the physically admissible states; it says that, to represent a physical state, the

6 The fact that 'time' arises as the imaginary part of a complex variable suggests a new way of performing the Wick rotation to pass to the 'Euclidean' regime. Unlike in Hawking's Euclidean program, this rotation would be only a computational tool. See [1] for one way of implementing this idea.

wave functional must have a specific dependence on the pair (A^T, A^{TT}). Solutions to the quantum constraint are thus analogous to classical solutions of Einstein's equation. They both present us with entire ***histories*** which are physically admissible. In neither case, there is any ***a priori*** dynamics. In classical general relativity, time evolution arises only ***after*** we slice our solution in to space and time and reinterpret the 4-dimensional, fully covariant metric in dynamical terms. Similarly, in the truncated quantum theory, to begin with we just have solutions to the scalar constraint equation. ***a priori***, nothing happens. "Happening" is forced into the picture by introducing a slicing of the configuration space of connections by surfaces with A^T=constant. Once this is done and A^T is identified with time, every physical state $\Psi(A^{TT}, A^L, A^T, A^A)$ can be reinterpreted as "evolving" with respect to A^T. Dynamics occurs only because we identify an internal clock. Note, however, that, now *the true arena for dynamics is not a space-time but the infinite dimensional configuration space of connections.* In simple cases, such as the one obtained here by truncating the theory to second order around the trivial initial data, the dynamics in the infinite dimensional space can be faithfully projected onto a 4-dimensional space-time. When this is possible, we can interpret the "happenings" which occur in the infinite dimensional space in space-time terms. Whether such a "projection" is possible in general is far from being clear. If it turns out not to be possible, we would be forced to describe dynamics in the infinite dimensional setting. In this case, the space-time Schrödinger evolution would only be an approximate notion and one may have to face the types of problems associated with the lack of exact unitarity that were, e.g., raised by Ted Jacobson in the context of quantum cosmology [4].

We conclude with a number of remarks.

i) Note that the procedure required that we keep terms to second order and that the construction breaks down if we set Newton's constant G equal to zero. The rough picture is that the linear gravitons, being transverse and traceless, themselves do not have a Coulombic degree of freedom. However, since they carry energy, they act as sources of a Coulombic gravitational field in a second order calculation. It is the second-order field that carries information about " time" (or, more precisely, $\frac{\partial}{\partial t}$).

ii) What is the interpretation of the above "time" variable in the classical theory? In terms of the more familiar variables, $-ImA^T$ is just the trace part, K^T, of the extrinsic curvature K_{ab}, when the Gauss law is satisfied. Thus, the "local" time function $t(x)$ of the truncated theory is given by $t(x) = \frac{1}{2}(\Delta)^{-1}K(x)$, where Δ is the Laplacian of the background metric $\mathring{q}_{ab}$ and $K(x)$ is the trace of the extrinsic curvature. It is easy to check that the resulting t is indeed an affine parameter of the Hamiltonian vector field in linearized gravity. Thus, what we have accomplished is to tie the affine parameter to a concrete, geometric field encoded in the components of the connection A_a^i . That this is a viable choice of the time function in the weak-field limit was pointed out by Kuchař already in 1970 [6]. However, while the representation used by Kuchař exists only in the weak-field limit, the A_a^i representation exists in the full theory.[7]

iii) What is the situation with respect to other constraints? The final picture is completely analogous to that sketched above for the scalar constraint. The action of the truncated Gauss constraint determines the dependence of the wave function $\Psi(A^{TT}, A^L, A^T, A^A)$ on the skew part, A^A, of A_{ab} while that of the truncated vector constraint determines the dependence on the longitudinal part, A^L. Thus, in the truncated theory under consideration, it is only the dependence of the wave function on the "physical modes" A^{TT} that is freely specifiable. Once we know this dependence, the dependence on all other parts of A_{ab} is *completely* determined by the truncated quantum constraints. This is the sense in which the quantum constraints "generate gauge". On the other hand, we can identify certain "displacement operators", $\frac{\partial}{\partial t}$ and its analogs for the Gauss and the vector constraints, and re-interpret these equations in space-time terms. Then, the quantum scalar constraint implies that the (truncated) Hamiltonian generates "time evolution", and the quantum vector equation implies that the (truncated) 3-momentum generates "space translations".

iv) Finally, let us briefly consider on the spatially compact case. In this case, $\mathcal{E}$ vanishes identically and we cannot repeat the above argument as is. However, since Eqs.8 and 9 continues to hold even in the compact case, it would seem that one would be able to extract a notion of "local time" $t(x)$ in a suitably truncated theory. However, it would not be possible to construct a preferred *global* operator $(\partial/\partial t)$.

References

[1] H. Kodama, Prog. Theo. Phys. **80**, 1024 (1988); Phys. Rev. **D42**, 2548 (1990).

[2] A. Ashtekar and J. Pullin, Proc. Phys. Soc. Israel **9**, 65 (1990).

[3] T. Jacobson and L. Smolin, Nucl. Phys. **B299**, 295 (1988).

---

7 Also, the close relation between Eqs.3, 4 and the role of energy $\mathcal{E}$ in the procedure seems to have been overlooked in the older treatments.

[4] *Conceptual Problems of Quantum Gravity*, edited by A. Ashtekar and J. Stachel (Birkhäuser, Boston, in press)

[5] A. Ashtekar and G.T. Horowitz, Phys. Rev. **D26**, 3342 (1982).

[6] K. Kuchař, J. Math. Phys. **11**, 3322 (1970).

13 CONNECTION REPRESENTATION: *CP* PROBLEM

1 Introduction

In chapter 12, we saw that, unlike the metric representation of geometrodynamics, the connection representation is well suited to tackle the issue of time in the weak field truncation. In this chapter, we shall see that the general framework provided by the connection representation is useful also in the analysis of another conceptual issue, about which there has been some confusion within the geometrodynamical approaches: the CP problem in quantum gravity.

As we saw in Part II, a striking feature of the new canonical framework for general relativity is that the basic variables $(A_a{}^i, \tilde{E}^a{}_i)$ are the same as those normally used in the canonical formulation of non-Abelian gauge theories: $A_a{}^i$ may be thought of as a Yang-Mills connection, and $\tilde{E}^a{}_i$ may be thought of as its conjugate momentum, the electric field.[1] It is this aspect that will play the key role in the present paper. Of specific interest to us is the fact that one of the constraint equations satisfied by the gravitational pair, $(\tilde{E}^a{}_i, A_a{}^i)$, is precisely the Gauss law, $\mathcal{D}_a \tilde{E}^a{}_i = 0$, where $\mathcal{D}$ is the gauge-covariant derivative operator defined by $A_a{}^i$. As in Yang-Mills theory, this constraint generates "internal gauge transformations," (which now have the interpretation of triad rotations.) Consequently, as in Yang-Mills theory, one is now naturally led to examine the (internal) topological aspects of the theory at the classical and semi-classical level; issues such as the difference between "small" and "large" gauge transformations, the presence of n-vacua, and of instantons tunneling between them. In the full quantum theory the use of new variables suggests that we work in a representation in which wave functions are functionals of the connection 1-forms $A_a{}^i$, in place of the traditional metric-representation in which they are functionals of the 3-metric q_{ab}, or of triads $\tilde{E}^a{}_i$. Our analysis will rely heavily on the availability of this $A_a{}^i$ representation. As in Yang-Mills theory, the quantum Gauss constraint will tell us that physical states be

1 The momentum conjugate to an unweighted field is always a density of weight one. In the flat space Yang-Mills theory, one generally divides $\tilde{E}^a{}_i$ by the square root of the background flat metric and deals with the resulting unweighted electric field.

represented by *gauge invariant* functionals of $A_a{}^i$. Now, the topology of the quotient of the space of connections by the action of the gauge group is again non-trivial and this, in essence, leads to a 1-parameter ambiguity in quantization of the theory. More precisely, (internal) topological considerations lead us to a 1-parameter family of quantum theories –labelled by θ– and that there is a P, and hence also CP, violation in every sector other than that corresponding to $\theta = 0$ [1].

We begin, in section 2, with a review of the CP problem in Yang-Mills theory. Since our focus in gravity is on canonical quantization, we recall here the strong CP problem in the canonical framework where its origin lies in the failure of the Hamiltonian operator to be CP-invariant for non-zero values of θ. In section 3, we turn to gravity. We recall that, although the kinematics of general relativity now resembles that of Yang-Mills theory in many ways, there are also some differences. In spite of these, we are able to carry over the essential steps, used in the Yang-Mills context in section 2, to show that the Hamiltonian constraint of general relativity fails to be CP invariant in sectors labelled by non-zero values of θ. Since physical states in quantum theory are built out of solutions to *all* (first-class) constraints, there is *no* operator on the Hilbert space of physical states which can even implement the CP transformation, except in the $\theta = 0$ (or π) sector. That quantum gravity can lead to CP-violation is intuitively clear because one can always add to the Einstein-Hilbert action a CP violating topological term, $\theta \int \mathrm{tr}\, R \wedge R$. However, in quantum geometrodynamics, where one uses the metric –or the triad– representation, the addition of this term *does not* lead to the prediction of CP-violation at least in any obvious way. This is because the topologically distinct triad configurations belong to *disconnected* pieces of the configuration space of geometrodynamics. Thus, there are no continuous paths in this configuration space –let alone instantons– connecting topologically distinct triads. The space of connection 1-forms, on the other hand, *is* connected. Not only does it admit continuous paths connecting topologically distinct "vacua", but there also exist instantons if one allows for a cosmological constant [2]. We conclude this chapter by clarifying this issue and correcting some errors that have been made in the literature.

The analysis leading to the main result of this chapter depends only on the qualitative feature of general relativity that it admits a representation in which $A_a{}^i$ is diagonal; it is surprisingly insensitive to the precise manner in which the canonical quantization program may be eventually completed. In particular, our analysis is insensitive to the details of regularization schemes that may be employed to solve the difficult scalar and vector constraints of the theory and to the choice of the inner product that will be made on the space of physical states. Indeed, even the details of dynamics are inessential. If one were to replace general relativity by supergravity, for example, our basic results would continue to hold because supergravity also admits a representation in which $A_a{}^i$ is diagonal [3]. Thus, we are dealing here with a qualitative, non-perturbative feature shared by many local field theories of gravity.

2 CP problem in Yang-Mills Theory

As a prelude to the discussion of the CP problem in gravity, in this section we shall summarize the situation in Yang-Mills theory. (For details, see, e.g. [4]).

Consider $SU(N)$ source-free Yang-Mills theory. Let $\mathcal{C}$ denote the the space of all connection 1-forms $A_a{}^i$on $\mathbb{R}^3$ satisfying the boundary condition, $A_a{}^i = O(\frac{1}{r})$ at infinity. Using the quantization program of chapter 10, it is straightforward to construct the connection representation. The elements of the representation space V -i.e. the unconstrained wave functionals– are now maps from $\mathcal{C}$ to the space of complex numbers. There is only one constraint in the theory: the Gauss law. The physical subspace V_{phy} of V therefore consists of the wave functionals which are annihilated by this constraint. To spell out the consequences of this requirement, let us introduce some terminology. Denote by $G_0{}^0$ the connected component of identity, of the group G_0 of local $SU(N)$ gauge transformations $g(\vec{x})$ which are asymptotically identity, i.e., which satisfy $g(\vec{x}) = id + O(\frac{1}{r})$. $G_0{}^0$ is a normal subgroup of G_0 and its elements are referred to as *small gauge transformations*. Since the exponentiation of the Gauss law operator yields a representation precisely of $G_0{}^0$, wave functionals are annihilated by the Gauss law if and only if they are invariant under the (induced) action of $G_0{}^0$. Thus, physical states correspond to wave functionals which are invariant under all small gauge transformations.

The structure of $G_0/G_0{}^0$ plays an important role in what follows. Let us therefore make a mathematical detour to review it. Consider, first, the 3-sphere S^3 obtained by the standard one point compactification of $\mathbb{R}^3$. Because every element of G_0 tends to identity at infinity, it admits a continuous extension to S^3. Thus, every $g(\vec{x})$ in G_0 defines a (continuous) mapping from the 3-sphere S^3 to $SU(N)$. Therefore, we have:

$$\pi_0(G_0) = \pi_3(SU(N)) = Z \tag{1}$$

The integers characterizing the disconnected components of G_0 can be computed explicitly. To see this, note first that, associated with every element $g(\vec{x})$ of G_0, is an integer, $w(g)$ defined by

$$w(g) = \frac{1}{24\pi^2}\int_{\mathbb{R}^3} d^3x\, \mathrm{tr}(g^{-1}\partial_a g)(g^{-1}\partial_b g)(g^{-1}\partial_c g)\epsilon^{abc}, \tag{2}$$

which tells us how many times $g(\vec{x})$ winds around the non-contractible 3-sphere in the $SU(N)$ manifold as $\vec{x}$ ranges over the 3-sphere S^3. It is precisely the elements of ${G_0}^0$ that have a zero winding number. Elements of G_0 with non-zero winding numbers are referred to as *large gauge transformations*. Using the expression of $w(g)$ given in Eq.2, it is straightforward to show that the winding numbers have the following property: $w(g \cdot g') = w(g) + w(g')$. Therefore, the winding numbers have a well-defined projection on the quotient $G_0/{G_0}^0$; each disconnected component of G_0 is uniquely labelled by an integer. Furthermore, this projection defines an isomorphism from $G_0/{G_0}^0$ to the group Z of integers. Fix an element g_1 of G_0 with winding number 1. One can regard this g_1 as generating $G_0/{G_0}^0$ since $(g_1)^n$ has winding number n; one has $G_0/{G_0}^0 \approx \{(g^1)^n | n \in Z\}$.

Let us now return to the physical problem of quantization. As noted before, the presence of the Gauss constraint tells us only that the physical states should be invariant under the action of small gauge transformations; it provides no clue as to how the states are to behave under large gauge transformations. On the other hand, physical *observables* –functions of ${A_a}^i$and its canonical momentum $\widetilde{E}^a{}_i$, the electric field – are, by definition, invariant under the entire group G_0. In particular, they commute with the g_1, introduced above. Therefore, the group $G_0/{G_0}^0$, generated by g_1, defines a superselection rule and each of its unitary irreducible representations defines a distinct quantum theory. Since $G_0/{G_0}^0$ is isomorphic with Z, its unitary irreducible representations are all of the type $(g_1)^n \rightarrow \exp(in\theta)$, where θ is an angular parameter. Thus, there is a 1-parameter family of ambiguities in quantization. In the sector labelled by θ, the wave functions transform according to the rule:

$$\Psi(A^{g_1}) = e^{i\theta}\Psi(A), \quad \text{where,} \quad A^{g_1} = g_1 \cdot A \cdot {g_1}^{-1} + g_1 \cdot d{g_1}^{-1} \tag{3}$$

To make the ambiguity in quantization transparent and to obtain the Hamiltonian operator explicitly in θ-sectors, it is useful to examine in detail the structure of the effective configuration space. Since physical states are invariant under small gauge transformations, the effective configuration space on which physical wave functionals are defined is the quotient C' of C by the action of ${G_0}^0$. The quotient $G_0/{G_0}^0$ has a well-defined, non-trivial action on C'. To visualize this action, it is convenient to introduce the Chern-Simons functional [4] $Y(A)$ on C:

$$Y(A) := -\frac{1}{16\pi^2}\int_{\mathbb{R}^3} d^3x \,\mathrm{tr}\{F_{ab}A_c - \tfrac{2}{3}A_aA_bA_c\}\epsilon^{abc}, \tag{4}$$

where α is the Yang-Mills coupling constant. It is easy to verify that $Y(A)$ has the property:

$$Y(A^g) = Y(A) + w(g) \tag{5}$$

Therefore, $Y(A)$ may be regarded as a functional on C' Using it, one can divide

C' into "slabs" labelled by integers: An element A' of C' belongs to the n-th slab if and only if $n \leq Y(A') < n+1$. Then, under the action of g_1, the n-th slab is mapped to the $(n+1)$-th slab. Thus, the function $Y(A)$ serves as a coordinate on C' (with values ranging over the entire real line) which increases under the action of g_1. Let us now consider the physical states in the θ-sector. Eq.3 implies that these wave functionals on C' satisfy[2]

$$\Psi(A)|_{Y=n+1} = e^{i\theta}\Psi(A)|_{Y=n} \tag{6}$$

Thus, the wave functionals are completely determined by their value on the zero-th slab and the value of the parameter θ. It is therefore convenient to rescale them as follows:

$$\Psi'(A) := e^{-i\theta Y(A)}\Psi(A). \tag{7a}$$

Then, the new wave functions, $\Psi'(A)$, are periodic, $\Psi'(A)|_{Y=n+1} = \Psi'(A)|_{Y=n}$, so that the effective configuration space now reduces *just to the zero-th slab*. The price paid in this rescaling is that the expression of the momentum operator, $\widetilde{E}^a{}_i$, changes by the addition of a gradient:

$$\begin{aligned}\widetilde{E}'^a{}_i \cdot \Psi'(A) &= e^{-i\theta Y(A)}\, \widetilde{E}^a{}_i\, e^{i\theta Y(A)} \cdot \Psi'(A) \\ &= i\hbar\frac{\delta}{\delta A_a{}^i} \cdot \Psi'(A) - i\hbar\Psi'(A)\, e^{-i\theta Y(A)}\, \frac{\delta}{\delta A_a{}^i} \cdot e^{i\theta Y(A)} \\ &= -i\hbar(\frac{\delta}{\delta A_a{}^i} - \frac{i\theta}{8\pi^2}\widetilde{B}^a{}_i) \cdot \Psi'(A),\end{aligned} \tag{7b}$$

where, $\widetilde{B}^a{}_i := \epsilon^{abc}F_{bci}$ is the magnetic field defined by $A_a{}^i$. Consequently, if one uses the primed wave functionals to represent states, the space of states is the same in all θ sectors: physical states can be represented by wave functionals $\Psi'(A)$, defined only on the zero-th slab, satisfying (θ-independent) periodic boundary conditions. The θ-dependence is now transferred to operators. Thus, we have two equivalent procedures. Either one works with wave functionals $\Psi(A)$ satisfying θ-dependent boundary conditions of Eq.6 and the θ-independent expression, $\widetilde{E}^a{}_i = i\frac{\delta}{\delta A_a{}^i}$, of the momentum operator, or, with θ-independent states $\Psi'(A)$ and the θ-dependent momentum operator of Eq.7.*b*. Let us use the second strategy. Then, the Hamiltonian operator takes the form:

$$H' = \frac{1}{2}\int_{\mathbb{R}^3} d^3x\ \mathrm{tr}\{(i\hbar\frac{\delta}{\delta A_a{}^i} + \frac{\hbar\theta}{8\pi^2}\widetilde{B}^a{}_i)^2 + (\widetilde{B}^a{}_i)^2\}. \tag{8}$$

Since $\widetilde{B}^a{}_i$ is a pseudo-vector, the Hamiltonian operator H' fails to be invariant

2 For notational simplicity, we shall regard physical states as functionals $\Psi(A)$ on C which are invariant under $G_0{}^0$ rather than as functionals $\Psi(A')$ on C'.

under P, and hence also under CP, for any non-zero value of θ. In terms of the unprimed wave functions, the same problem arises in a different disguise; since the Chern-Simons functional, and hence the boundary condition of Eq.6, is not CP-invariant, the CP-operator fails to be well-defined in any θ-sector except the one corresponding to $\theta = 0$.

Finally, we recall that the re-definition of the momentum in Eq.7*b* –and the expression of the Hamiltonian in Eq.8– can also be understood as arising from the addition of a θ-dependent, "topological term," S_θ, to the action:

$$S_\theta = \frac{\theta}{16\pi^2} \int d^4x \, \epsilon^{abcd} \, \mathrm{tr} \, {}^4F_{ab} \, {}^4F_{cd}, \tag{9}$$

where ${}^4F_{ab}$ is the curvature of the 4-connection. Since the integrand of S_θ is a total divergence, its addition does not change the equations of motion. However, it does modify the definition of momentum canonically conjugate to $A_a{}^i$. In quantum theory, this modification translates precisely to Eq.7*b*. *Any* addition of a total divergence to the action has the effect of modifying the formal expressions of the conjugate momenta (provided the additional term depends on the time derivative of field variables). However, normally, this modification of the expression of the momentum operator can be "undone" by a simple rescaling of the wave functions by a phase factor, which leaves the definition of the domain of the the momentum operator unchanged. In the present case, on the other hand, the required rescaling, $\Psi'(A) \to \Psi(A)$, changes the domain of the momentum operator since it destroys the periodicity of the wave functionals. That is, it is precisely because the term S_θ is of topological origin that different choices for values of θ lead to *inequivalent* representations of the quantum canonical commutation relations.

3 CP problem in quantum gravity

Fix a 3-manifold Σ. For simplicity, we shall assume here that Σ is topologically $\mathbb{R}^3$ and return briefly to more general topologies in section 4. Let us recall from Part II some features of the new canonical formulation in order to facilitate comparison with Yang-Mills theory. The new configuration variable $A_a{}^i$ is a complex field: It is related to the real, geometrodynamical canonical coordinates $(\widetilde{E}_i^a, \Pi_a^i)$ on the phase space via:

$$A_a{}^i = \Gamma_a{}^i - i\Pi_a{}^i, \tag{10}$$

where G is Newton's constant and where $\Gamma_a{}^i$ are the connection 1-forms on internal indices $i, j, \ldots$ compatible with the triad $\widetilde{E}^a{}_i$. (Note incidently that since Γ has the same physical dimensions as the Yang-Mills connection, ${A_a}^i$ does not.) As we saw in chapter 6, the transformation $(\tilde{E}^a{}_i, {\Pi_a}^i) \to (\tilde{E}^a{}_i, {A_a}^i)$ is canonical, whence the only non-zero Poisson brackets between $\tilde{E}^a{}_i$ and ${A_a}^i$ are:

$$\{A^i_a(x), \tilde{E}^b_j(y)\} = i\delta^b_a\, \delta^i_j\, \delta^3(x-y) \tag{11}$$

It is because of these relations that we regard the hybrid pair, consisting of the real triad $\tilde{E}^a{}_i$ and the complex connection ${A_a}^i$ as being "canonically conjugate". Finally, let us recall the expressions of constraints. To facilitate comparison with Yang-Mills theory, let us express the constraints in terms of $\tilde{E}^a{}_i$, and the magnetic field $\tilde{B}^a_i := \tilde{\eta}^{abc}\, F_{bci}$ of the connection ${A_a}^i$. We have:

$$\mathcal{G}_i \equiv \mathcal{D}_a \tilde{E}^a{}_i = 0, \tag{12}$$

$$\mathcal{V} \equiv B_i \times \tilde{E}^i = 0 \tag{13}$$

$$\text{and} \quad \mathcal{S} \equiv \epsilon^{ijk}\, B_i \times \tilde{E}_j \cdot \tilde{E}_k = 0\,, \tag{14}$$

where, for notational simplicity, we have dropped the space indices and used instead the vector cross and dot product notation. Of particular interest to us in this chapter is the first of these constraints, which is completely analogous to the Gauss constraint of Yang-Mills theory.

Let us now turn to the problem of quantization. Let us use the connection representation introduced in the last chapter. The resulting quantum description is formally analogous to that of the Yang-Mills theory discussed in section 2. However, there are two important differences: The connection is now *complex-valued* and the wave functions are *holomorphic*, rather than arbitrary, complex-valued functionals of this connection. It turns out, however, that the two differences, so to speak, "compensate each other". To see this, note that since the wave functionals $\Psi(A)$ are holomorphic, they are completely determined by their restriction to the real part of ${A_a}^i$, which, is an $SO(3)$-valued connection. Therefore, there is a 1-1 correspondence between the quantum states of gravity in the ${A_a}^i$ representation and those of $SO(3)$-Yang-Mills theory in the representation in which the Yang-Mills connection is diagonal. It is because of this that one can carry over the non-trivial topological features associated with the Gauss law constraint from Yang-Mills theory to gravity.

We can now essentially repeat the discussion of section 2. The configuration space $\mathcal{C}$ now consists of connections ${A_a}^i$ on Σ which satisfy the boundary conditions specified in chapter 6. (Note that these require that the connection should go to zero as $1/r^2$ rather than $1/r$ as in Yang-Mills theory. However, for topological considerations, this stronger rate of fall-off at infinity is irrelevant.) Even though the connection 1-forms are complex, the relevant gauge group is still $SO(3)$ rather than a complexification thereof since it must map real triads to real triads. As before, denote by G_0 the group of local $SO(3)$ gauge transformations which are asymptotically identity and by ${G_0}^0$ the connected component of G_0. The quotient, ${G_0}^0/G_0$, is isomorphic to the additive group Z of integers. The Gauss law constraint tells us again that physical states are represented by wave functionals $\Psi(A)$ which are invariant under the (induced) action of ${G_0}^0$. Thus, the effective configuration space C' is again the quotient of C by the orbits of ${G_0}^0$. The quotient $G_0/{G_0}^0$ has a well-defined and non-trivial action on C'. To study this action, one can introduce the Chern-Simons functional, $Y(A)$, defined by

$$Y(A) := -\frac{1}{16\pi^2}\int_{\mathbb{R}^3} d^3x\{F_{ab}{}^i A_{ci} - \tfrac{2}{3}\epsilon_{ijk}A_a{}^i A_b{}^j A_c{}^k\}\epsilon^{abc}. \tag{15}$$

However, now the functional is complex-valued.

Large gauge transformations, on the other hand, continue to have real winding number since the gauge group is real. Therefore, we are now led to divide C' into "slabs" as follows: the n-th slab now consists of all elements A' of C' satisfying $n \leq ReY(A') < n+1$, where Re stands for "real part of." Then, it is again true that a large gauge transformation, g_1, with winding number one, maps the n-th slab to the $(n+1)$-th slab. For reasons spelled out in section 2, physical states satisfy the boundary condition (see Eq.6):

$$\Psi(A)|_{ReY=n+1} = e^{i\theta}\Psi(A)|_{ReY=n} \tag{16}$$

Hence, as before, the restriction of the wave functional to the 0-th slab and the value of θ suffice to determine any physical state. We can absorb the θ-dependence in the boundary condition of Eq.16 by rescaling the physical states Ψ by a θ-dependent phase factor as in Eq.7.*a*. However, then the operator expression of the canonical momentum, $\widetilde{E}'^a{}_i$, acquires the θ-dependence as in Eq.7.*b*. Thus, as in Yang-Mills theory, we have a 1-parameter ambiguity in quantizing the theory.

Let us now return to the vector and scalar constraints (Eqs.13 and 14). Physical states have to be annihilated by these constraints as well. Let us work with wave functions Ψ' satisfying the periodic boundary conditions and the corresponding momentum operator $\widetilde{E}'^a{}_i$ Then, the vector constraint becomes:

$$\widetilde{B}^i \times (\frac{\delta}{\delta A^i} + \frac{i\theta}{8\pi^2}\widetilde{B}_i)\cdot\Psi' \equiv \widetilde{B}^i \times \frac{\delta}{\delta A^i}\cdot\Psi' = 0 \tag{17}$$

Thus, the θ-dependence drops out of this constraint operator: We can simultaneously solve for the vector constraint in *all* θ-sectors. This is analogous to the fact that the expression of the Gauss law constraint in Yang-Mills theory is θ-independent. The situation is quite different with respect to the Hamiltonian constraint. Its formal operator expression becomes:

$$\epsilon^{ijk}\widetilde{B}_i \times \left(\frac{\delta}{\delta A^j} + i\frac{\theta}{8\pi^2}\widetilde{B}_j\right) \cdot \left(\frac{\delta}{\delta A^k} + \frac{\theta}{8\pi^2}\widetilde{B}_k\right) \cdot \Psi' = 0 \tag{18}$$

Note that, as in the expression of the Yang-Mills Hamiltonian, the term linear in θ fails to be P and hence CP invariant. Denote by V_θ the vector space of solutions to all constraints for a fixed value of θ. Since V_θ is mapped to $V_{-\theta}$ by P , it follows that CP can be a well-defined operator *only* on the $\theta = 0$ sector. In particular, in $\theta \neq 0$ sectors, the ground state cannot be CP-invariant.

To conclude this section, we note that the quantization ambiguity can be traced back to the possibility of adding a topological term to the gravitational action. In terms of new variables, the gravitational action is a functional of tetrads, $e^a{}_I$, and (complex, self dual Lorentz) connections, ${}^4A_a{}^{IJ}$, in space-time. As we saw in chapters 3 and 4, the (self dual) action which leads to Einstein's vacuum equations is:

$$S(e, {}^4A) = \int d^4x\, (e)\, e^a_I\, e^b_J\, {}^4F_{ab}{}^{IJ}\ , \tag{19a}$$

where ${}^4F_{ab}{}^{IJ}$ is the curvature of the 4-connection. As in Yang-Mills theory, we can add to it a purely topological term S_θ:

$$S_\theta(e, {}^4A) := \frac{\theta}{16\pi^2}\int d^4x\, \mathrm{tr}({}^4F_{ab}\, {}^4F_{cd})\, \underset{\sim}{\eta}^{abcd}\ . \tag{19b}$$

(If the factors of G are restored, while the factor of G does not appear explicitly in the self-dual action, it does appear in the expression of S_θ: the right side of Eq.19*b* must be divided by the Planck density.) Note that S_θ is independent of the tetrads and that, as in Yang-Mills theory, it is a total divergence. The effect of this term is only to modify the definition of momenta and this modification leads to the 1-parameter quantization ambiguities.

4 Discussion

Both in the Yang-Mills and the gravitational case, we saw that the existence of θ-sectors in the $A_a{}^i$-representation is closely related to the possibility of adding a CP-violating, topological term S_θ to the action. In the gravitational case, this term depends only on ${}^4A_a$; it is independent of the tetrads. Therefore, *a priori*, there is no relation between this term and the Pontryagin index

$$\int d^4x(g)^{\frac{1}{2}}\epsilon^{abcd}\, R_{abm}{}^{n} R_{cdn}{}^{m} \tag{20}$$

defined by the space-time metric g_{ab}. However, as we saw in chapters 3 and 4, when the field equations are satisfied, the connection 1-form ${}^4A_a$ is completely determined by the tetrad: In any solution to the field equations, $A_a{}^i$ is the self dual (on internal indices) part of the space-time connection compatible with the tetrad. Therefore, in any *solution*, one can evaluate S_θ of Eq.19.*b* in terms of the tetrad. Then, it turns out that S_θ reduces precisely to (20), where g_{ab} is the space-time metric determined by the tetrad. Note, however, that the agreement holds only on solutions to field equations.

The CP-violating term (20) has received some attention in the literature on quantum gravity based on geometrodynamical variables. In these frameworks, one restricts oneself to *non-degenerate* tetrads on space-time and hence also triads on 3-manifolds Σ; here too, there exist again non-trivial vacua. They arise as follows. Let us suppose that Σ is topologically $\mathbb{R}^3$. Any two non-degenerate triads on Σ are related to one another by a local $GL(3)$ transformation g. A classical vacuum is a triad E^a_i on Σ for which the 3-metric $q^{ab} = E^a_i E^{bi}$ is flat. We will say that two triads are equivalent if the winding number of the element g of $GL(3)$ relating them is zero. Since $SO(3)$ admits elements with non-zero winding numbers and since under any local $SO(3)$ transformation the 3-metric does not change, it follows that there exist *inequivalent* flat triads. These have been called the non-trivial vacua. It has often been argued (see, in particular [5]) that the close formal similarity of the expression (20) with the Yang-Mills θ-term (Eq.9) implies that instantons with a non-trivial Pontryagin index can be used to calculate the transition amplitude between these topologically distinct gravitational vacua. However, note that now the configuration space is *disconnected*; only those triads which are equivalent to one another via the relation given above, belong to the same connected component of the configuration space. Consequently, there are no continuous paths whatsoever, let alone instantons, that can interpolate between them. Thus, the situation is very different from Yang-Mills theory or from gravity with the configuration space of self dual connections. Therefore, it is difficult to motivate the addition of expression (20) to the Einstein-Hilbert action. One can of course argue [6] that one is free to add any surface term to the action, add (20) and study the consequences using, say, the path integral approach. However, as is pointed out in [6], the physical significance of the results is then somewhat obscure. The use of the $A_a{}^i$ -representation is likely to clarify these results.

We conclude with three remarks.

i) It may seem surprising at first that the connection representation used in this paper and the traditional triad representation used in geometrodynamics lead to such different results. Shouldn't it be possible to pass from one to the other by a 'Fourier transform'? ([7], see also the discussion surrounding Eq.10 of chapter 10). In Yang-Mills theory, the use of this transform does indeed let one translate the results in the connection (configuration) representation to the electric-field (momentum) representation [4]. In the geometrodynamical approach to quantum gravity, on the other hand, the triads are assumed to be *non-degenerate* whence the required Fourier transforms do not exist even formally. Furthermore, it is impossible to relax this condition: the constraint equations of the theory in the canonical variables in geometrodynamics cease to be meaningful if one lets the triad be degenerate. This is why the geometrodynamical triad representation in gravity is quite different from the electric field representation in Yang-Mills theory. In terms of new variables, on the other hand, the constraints and the Hamiltonian continue to be meaningful even when the triad is allowed to become degenerate. It is therefore natural to allow this possibility in the quantum theory. Our representation of the operator $\widetilde{E}^a{}_i$ by $\delta/\delta A_a{}^i$ exploits this possibility. In the new framework, therefore, one *can* translate the results from the connection to the triad representation following the procedure followed in Yang-Mills theory.

ii) In absence of spinor fields, one can avoid entirely the introduction of tetrads and associated internal rotations and work only with metrics. It may therefore seem surprising that there is a possibility of CP violation at all in pure gravity. Our viewpoint on this issue is the following. Unambiguous physical effects associated with the θ-sectors will arise only in presence of fermionic matter. If we take fermions away, these effects leave imprints on those descriptions of pure gravity which carry some remnant structure left behind by fermions. The appropriate remnant appears to be the availability of the freedom to make internal rotations. That the full theory can leave such an imprint when the essential ingredients, the fermions, are taken away is a feature not unique to gravity. An analogous situation occurs in QCD where quark confinement of the full theory leaves its marks on *pure* QCD without any fermions.

iii) For simplicity, in this chapter we have restricted ourselves to 3-manifolds Σ which are topologically $\mathbb{R}^3$. If we consider more complicated spaces, the space of mappings from the (compactified) 3-space to the gauge group acquires a more interesting structure [8] and this, in turn, leads to additional ambiguities in quantization. If we allow space to have non-trivial topology, we also open up another, qualitatively different, source of ambiguities. These arise as follows. In the new phase-space formulation, the kinematic "gauge group" is the semi-direct product, $G_0 \circledS D_0$, of the group G_0 of (asymptotically identity) internal triad rotations with the group D_0 of (asymptotically identity) diffeomorphisms of Σ. If Σ is topologically non-trivial, the quotient, $D_0/{D_0}^0$, of D_0 by its connected component can be non-trivial. The irreducible unitary representations of this quotient –i.e., of the mapping class group of Σ– provide quite a different type of 'θ -sector' in quantum gravity [9]. To understand the quantization freedom completely, we need to consider the unitary irreducible representations of the quotient of $G_0 \circledS D_0$ by its connected component.

References

[1] A. Ashtekar, A. P. Balachandran and S. Jo, Intl. J. Mod. Phys. **A4**, 1493 (1989).

[2] J. Samuel, Class. & Quant. Grav. **5**, L123 (1988).

[3] T. Jacobson, Class. & Quant. Grav. **5**, 923 (1988).

[4] R. Jackiw, in *Relativity Groups and Topology II*, edited by B. Dewitt and R. Stora, (North Holland, Amsterdam, 1984); J. Goldstone and R. Jackiw, Phys. Lett. **74B**, 81 (1978).

[5] S. W. Hawking, in *Recent Developments in Gravitation*, edited by M. Levi and S. Deser (Plenum Press, N.Y., 1979).

[6] S. Deser, M. Duff and C. J. Isham, Phys. Lett. **B93**, 419 (1980).

[7] H. Kodama, Phys. Rev. bf D42, 2548 (1990).

[8] C.J. Isham, in *Old and New Questions in Physics, Cosmology, Philosophy and Theoretical Biology*, edited by A. Van Der Merwe; C. J. Isham and G. Kunstatter, J. Math. Phys. **23** 1668 (1982).

[9] J. Friedman and D. W. Witt, in *Mathematics and General Relativity*, edited by J. Isenberg (American Mathematical Society, Providence, 1988); D. W. Witt, J. Math. Phys. **27**, 573 (1986).

14 LOOP REPRESENTATION: MAXWELL THEORY

1 Introduction

Since the ideas used in the construction of the loop representation in quantum gravity are somewhat unconventional, in this chapter, we will illustrate them with a simpler example: the quantum Maxwell field in Minkowski space.[1] Chronologically, however, the Maxwell case was worked out *after* general relativity; the main ideas can be traced back to the work by Rovelli and Smolin in [2] on the loop representation for gravity. The fact that free photons can be adequately handled in the loop representation has, in turn, provided considerable support for some of the formal constructions presented in [2].

The key question we want to address is the following. Wave functionals in the connection representation, $\Psi(A)$, can be thought of as the expansion coefficients of the quantum state $|\,\Psi\,\rangle$ in the eigenbasis of the connection operator: $\Psi(A) = \langle\, A\,|\,\Psi\,\rangle$. In the loop representation, on the other hand, states arise as functionals $\Psi(\gamma)$ of closed loops γ. If we were to think of them as the expansions coefficients $\langle\,\gamma\,|\,\Psi\,\rangle$, how is one to interpret the basis $|\,\gamma\,\rangle$? To see the answer, let us return to the example of the hydrogen atom referred to in the introduction to this Part. The analogous states there are $\Psi(n,l,m)$ and *a priori* the basis $\{|\,n,l,m\,\rangle\}$, labelled just by three integers, has no counterpart on the classical phase space and seems equally mysterious. Indeed, the meaning of the basis becomes clear only after we specify that they are the eigenstates of the Hamiltonian $\hat{H}$, the total angular momentum $\hat{L}^2$ and the third component of the angular momentum $\hat{L}_3$ with eigenvalues $(-13.6\mathrm{eV}/n^2,\ l(l+1)\hbar^2,\ m\hbar)$ respectively. Similarly, to understand the interpretation of the loop basis $|\,\gamma\,\rangle$, we must specify the operators which it diagonalizes. In this chapter we carry out this task, using the standard Fock representation as the point of departure. We will find that the loop states are eigenstates of the positive

1 This chapter is based on joint work with Carlo Rovelli [1].

frequency electric field operator.[2]

In section 2 we briefly recall the Bargmann representation for photons and construct an explicit transform to pass to the loop representation. Section 3 then shows how the resulting description can be arrived at directly through the quantization program of chapter 10. This strategy forms the basis of loop quantization of gravity discussed in the next two chapters.

2 Rovelli-Smolin transform

Let us then begin by recalling the standard Fock-space description of photons in the Bargmann representation. The phase space Γ of the Maxwell field consists of a vector potential $A_a(\vec{x})$ and the electric field $E^b(\vec{x})$ satisfying the canonical Poisson bracket relations. The system has one first class constraint: $D_aE^a = 0$. For brevity, let us pass to the reduced phase space $\hat{\Gamma}$ by fixing the gauge, $D^aA_a = 0$. $\hat{\Gamma}$ is thus coordinatized by a pair of divergence-free (i.e. transverse) fields, $A_a^T(\vec{x}), E_a^T(\vec{x})$, on a space-like plane Σ. As in chapter 11, it is convenient to work in the momentum space. Then, the dynamical variables are $q_j(\vec{k})$, $p_j(\vec{k})$ with $j = 1,2$:

$$
\begin{aligned}
A_a^T(\vec{x}) &= \frac{1}{(2\pi)^{3/2}} \int d^3\vec{k}\, e^{i\vec{k}\cdot\vec{x}} \left(q_1(\vec{k})m_a(\vec{k}) + q_2(\vec{k})\overline{m}_a(\vec{k})\right) \\
E_a^T(\vec{x}) &= -\frac{1}{(2\pi)^{3/2}} \int d^3\vec{k}\, e^{i\vec{k}\cdot\vec{x}} \left(p_1(\vec{k})m_a(\vec{k}) + p_2(\vec{k})\overline{m}_a(\vec{k})\right).
\end{aligned} \tag{1}
$$

The negative sign in front of the right side of the second equation ensures that the $q_j(\vec{k})$ and $p_j(\vec{k})$ satisfy the canonical commutation relations with correct sign: $\{q_i(-\vec{k}), p_j(\vec{k}')\} = +\delta_{i,j}\delta^3(\vec{k},\vec{k}')$. (The necessity of this sign can be traced back

2 Note that the loop representation is *not* unique. Just as one can introduce, in non-relativistic quantum mechanics, a mathematical Hilbert space H of square integrable functions $\psi(\lambda)$ of a real variable λ with operators $\psi(\lambda) \to \lambda\psi(\lambda)$ and $\psi(\lambda) \to i\hbar d\psi/d\lambda$ satisfying the canonical commutation relations, one can introduce the space of suitably regular functions on the loop space and operators on it. The physical interpretation of states $\psi(\lambda)$ depends crucially on the meaning we associate with the two canonically conjugate operators. If the multiplication operator is taken to be the position observable, for example, the "states" $\delta(\lambda)$ are the eigenstates of the position operator. On the other hand, if the multiplication operator is interpreted as the momentum observable, "states" $\exp(ix\lambda)$ are the position eigenstates. Similarly, the interpretation of the state $|\gamma\rangle$ depends on the meaning we attach to the mathematically defined operators on the space of functions on the loop space. Thus, the states $|\gamma\rangle$ can be the eigenstates of the positive frequency electric field operator or the real electric field operator, or the self-dual part of the electric field operator, depending on one's framework. In this chapter, we have just made one specific choice for definiteness.

to the property $m_a(-\vec{k}) = -\overline{m}_a(\vec{k})$ of the "polarization vectors" $m_a(k)$. In the case of the linearized gravitational field, expansions analogous to Eq.1 involve only products $m_a m_b$ which are bilinear in m_a and the negative sign is unnecessary.) The positive frequency connection,

$$\begin{aligned} {}^{(+)}A_a^T(\vec{x}) &\equiv \tfrac{1}{\sqrt{2}}(A_a^T(\vec{x}) + i\Delta^{-\frac{1}{2}}E_a^T(\vec{x})) \\ &= \frac{1}{(2\pi)^{3/2}}\int \frac{d^3\vec{k}}{|k|}\, e^{i\vec{k}\cdot\vec{x}}\,(\varsigma_1(\vec{k})m_a(\vec{k}) + \varsigma_2(\vec{k})\overline{m}_a(\vec{k})), \end{aligned} \tag{2a}$$

is then canonically conjugate to the negative frequency electric field:

$$\begin{aligned} {}^{(-)}E_a^T(\vec{x}) &\equiv \tfrac{1}{\sqrt{2}}(E_a^T(\vec{x}) + i\Delta^{\frac{1}{2}}A_a^T(\vec{x})) \\ &= \frac{-i}{(2\pi)^{3/2}}\int d^3\vec{k}\, e^{i\vec{k}\cdot\vec{x}}\,(\overline{\varsigma}_1(-\vec{k})m_a(\vec{k}) + \overline{\varsigma}_2(-\vec{k})\overline{m}_a(\vec{k})). \end{aligned} \tag{2b}$$

(Here Δ is the Laplacian on the spatial 3-plane.) $\varsigma_j(\vec{k})$ are the two radiative modes of the Maxwell field corresponding to the two helicities. In terms of $q_j(\vec{k})$ and $p_j(\vec{k})$, they can be expressed as

$$\varsigma_j(\vec{k}) = \tfrac{1}{\sqrt{2}}(|k|q_j(\vec{k}) - ip_j(\vec{k})), \tag{2c}$$

and therefore satisfy the following Poisson bracket relations:

$$\{\varsigma_i(\vec{k}),\, \overline{\varsigma}_j(\vec{k}')\} = i|k|\delta_{ij}\delta^3(\vec{k},\vec{k}'). \tag{3}$$

Note that this final canonical description of the radiative modes is identical to the one we obtained in section 11.3 for the linearized gravitational field. The application of the quantization program can therefore be carried out exactly as before. Thus, once the elementary variables are taken to be $\varsigma_j(\vec{k}), \overline{\varsigma}_j(\vec{k})$, all steps in the program can be carried out in a straightforward way and the the reality conditions provide us with the correct, Poincaré invariant inner product. The final result is the following. The Hilbert space is the space of holomorphic functionals $\Psi(\varsigma_j(\vec{k}))$ with the inner product:

$$\langle\, \Psi(\varsigma_1,\varsigma_2) \,|\, \Psi'(\varsigma_1,\varsigma_2)\,\rangle = \int \Pi_j \mathrm{d\!I}\varsigma_j(\vec{k}) \wedge \mathrm{d\!I}\overline{\varsigma}_j(\vec{k})\; \mu(\varsigma_j,\overline{\varsigma}_j)\; \overline{\Psi(\varsigma_1,\varsigma_2)}\Psi'(\varsigma_1,\varsigma_2), \tag{4a}$$

where, as before, dI denotes the infinite dimensional exterior derivative operator, and where the measure $\mu(\varsigma_j, \bar{\varsigma}_j)$ is the Gaussian:

$$\mu(\varsigma_j, \bar{\varsigma}_j) = \exp\left[-\frac{1}{2\hbar} \int \frac{d^3k}{|k|}(|\varsigma_1(\vec{k})|^2 + |\varsigma_2(\vec{k})|^2)\right]. \tag{4b}$$

The "multiplication" operators $\hat{\varsigma}_j$ are the creators and the "derivative" operators $\hat{\bar{\varsigma}}_j(\vec{k}) \equiv (\hat{\varsigma}_j(\vec{k}))^*$ are the annihilation operators. The vacuum is the unit function $\Psi_0(\varsigma_1, \varsigma_2) = 1$. (For further details, see Eqs.11.24 -11.28.)

We can now carry out the Rovelli-Smolin transform. Their initial idea was to repeat the procedure one uses to pass from, say, the position to the momentum representation to construct the loop representation starting from the connection representation. To make such a transform, however, one needs a kernel, the analog of $\exp(i\vec{k} \cdot \vec{x})$ that one uses in the Fourier transform, and one needs a measure on the space of connections. The obvious candidate for the kernel is the trace of the holonomy which associates with each pair, (A, γ), consisting of a connection A and a closed loop γ, a number. However, in full 3+1 general relativity, it is hard to find the required measure on the space of connections. Therefore, in that case, the transform has remained a formal device. In the case of Maxwell (or, linearized gravitational) fields, on the other hand, we do have a satisfactory measure, given by Eq.4*b*. and we can make the transform rigorous. Thus, given a quantum state $\Psi(\varsigma_1, \varsigma_2)$ –i.e., a holomorphic function of a positive frequency connection ${}^{(+)}A_a^T(\vec{x})$, let us define a function $\psi(\gamma)$ on the space of closed loops via:

$$\psi(\gamma) = \int \Pi_j d\!I\varsigma_j(\vec{k}) \wedge d\!I\bar{\varsigma}_j(\vec{k})\, \mu(\varsigma_j, \bar{\varsigma}_j)\overline{H(\gamma, {}^{(+)}A^T)}\Psi(\varsigma_1, \varsigma_2), \tag{5a}$$

where the holonomy $H(\gamma, {}^{(+)}A^T)$ is given by:

$$H(\gamma, {}^{(+)}A^T) = \exp\left[\oint_\gamma dS^a \; {}^{(+)}A_a^T\right]. \tag{5b}$$

(Note that since we are dealing with Abelian connections, the trace operation is redundant and since the positive frequency connections are necessarily complex, there is no need to put a factor of i in front of the integral in the exponent.) We will now show that the integral exists for all states $\Psi(\varsigma_1, \varsigma_2)$ in the Fock space. Furthermore, the result can be written out "explicitly".

Let us begin with the expression of the holonomy. We have:

$$\begin{aligned}\int_\gamma dS^{a(+)}A_a^T &= \oint_\gamma dS^a \frac{1}{(2\pi)^{3/2}} \oint \frac{d^3\vec{k}}{|k|}\, e^{i\vec{k}\cdot\vec{x}}\, (\varsigma_1(\vec{k})m_a(\vec{k}) + \varsigma_2(\vec{k})\overline{m}_a(\vec{k})) \\ &= \int \frac{d^3k}{2\hbar|k|} \sum_j \varsigma_j(\vec{k})\overline{F}_j(\gamma, \vec{k})\end{aligned} \tag{6a}$$

where $F_j(\gamma, \vec{k})$, called the *form factors* of the loop γ, are given by:

$$F_1(\gamma,\vec{k}) = \frac{2\hbar}{(2\pi)^{\frac{3}{2}}} \oint_\gamma dS^a e^{-i\vec{k}\cdot\vec{x}}\, \overline{m}_a(\vec{k}),$$
$$\text{and} \quad F_2(\gamma,\vec{k}) = \frac{2\hbar}{(2\pi)^{\frac{3}{2}}} \oint_\gamma dS^a\, e^{-i\vec{k}\cdot\vec{x}}\, m_a(\vec{k}), \tag{6b}$$

In the last step in Eq.6a, we have achieved a clean separation of the dependence of the holonomy on its two arguments: the radiative modes $\varsigma_j(\vec{k})$ depends only on the connection ${}^{+}A^T_a$, while the form factors $F_j(\vec{k})$ depend only on the loop γ. We can now evaluate the Rovelli-Smolin transform. Substituting Eq.6a in the integral on the right side of Eq.5a, we find that $\Psi(\gamma)$ is given by the inner product:

$$\psi(\gamma) = \langle\, C_F(\varsigma_1,\varsigma_2)\,|\,\Psi(\varsigma_1,\varsigma_2)\,\rangle, \tag{7a}$$

where the "state" $C_F(\varsigma_1,\varsigma_2)$ has the form of a coherent state defined by the form factors $F_j(\varsigma_1,\varsigma_2)$,[3]

$$C_F(\varsigma_1,\varsigma_2) = \exp[\int \frac{d^3k}{2\hbar|k|} \sum_j \varsigma_j(\vec{k})\overline{F_j(\vec{k})}]\,. \tag{7b}$$

Thus, $\psi(\gamma)$ have an appealing interpretation: they can be regarded as the expansion coefficients of the state $\Psi(\varsigma_1,\varsigma_2)$ in the (overcomplete) coherent state basis provided by the form factors associated with closed loops.

To perform the Gaussian integration needed to evaluate the right side of Eq.7a, let us consider a system with a single degree of freedom ς. The integral in question is the infinite dimensional analog of:

$$\psi(F) := \int \frac{i d\varsigma \wedge d\bar{\varsigma}}{4\pi\hbar}\, e^{-\varsigma\bar{\varsigma}}\, e^{F\bar{\varsigma}}\, \Psi(\varsigma)\,. \tag{8}$$

It is straightforward to verify (e.g. using the fact that ς^n form an orthogonal basis in the Bargmann representation for the harmonic oscillator) that the integral always exists and its value is in fact given just by $\psi(F) = \Psi(F)$. Thus, the transform of the

3 We have used quotes in the label "state" here because $C_F(\varsigma_1,\varsigma_2)$ does not in general belong to the Fock space. As we shall see, the "inner-product" in Eq.7a is, however, well-defined. This is analogous to the fact that although $\exp(i\vec{k}\cdot\vec{x})$ does not belong to the Hilbert space $L^2(\mathbb{R}^3, d^3x)$, its "inner-product" with every element of $L^2(\mathbb{R}^3, d^3x)$ is always a well-defined element of $L^2(\mathbb{R}^3, d^3k)$.

function $\Psi(\varsigma) = \varsigma^n$ is just $\psi(F) = (F)^n$! In particular, the vacuum is represented by the unit function in both representations. Denote by Λ the mapping from the wave functions of ς to the wave functions of F: $\psi(F) = \Lambda \cdot \Psi(\varsigma)$. We can use this isomorphism between the two spaces to push various operators from the ς representation to the F representation in a straightforward fashion. One finds:

$$\Lambda^{-1} \circ \hat{\varsigma} \circ \Lambda = \hat{F}, \quad \text{and} \quad \Lambda^{-1} \circ \hat{\bar{\varsigma}} \circ \Lambda = \frac{\partial}{\partial F}. \tag{10}$$

Using the standard method of performing the Gaussian integrals over infinite dimensional (Hilbert) spaces, we can now evaluate the Rovelli-Smolin transform from the Bargmann to the loop representation of photons. Again, we have the result that the image $\psi(\gamma)$ of $\Psi(\varsigma_1, \varsigma_2)$ is given simply by: $\psi(\gamma) = \Psi(F_1, F_2)$, where F_j are the form factors of the loop γ. The vacuum state is represented by the function $\psi(\gamma) = 1$ on the loop space. A one photon state with helicity $+1$ and momentum k_0 is represented by the function $\psi(\gamma) = F_1(k_0)$. An n-photon state in which the photons have momenta $k_1, ...k_n$ and helicities $j_1, ...j_n$ (i.e., the ket $| k_1, j_1; ...k_n, j_n \rangle$ in the standard Fock space notation) is represented by the function $\psi(\gamma) = F_{j_1}(k_1)...F_{j_n}(k_n)$. Thus, a general element of the Fock space is represented in the loop representation by a function of closed loops γ of the type:

$$\psi(\gamma) = \sum_n \int \frac{d^3k_1}{2\hbar|k_1|} \cdots \frac{d^3k_n}{2\hbar|k_n|}\, g(k_1, ...k_n) F_{j_1}(k_1)...F_{j_n}(k_n)\,, \tag{11a}$$

where $g(k_1, ...k_n)$ is some function of n-momenta with

$$\int \frac{d^3k_1}{2\hbar|k_1|} \cdots \int \frac{d^3k_n}{|k_n|} |g(k_1, ...k_n)|^2 < \infty. \tag{11b}$$

Using the transform, it is straightforward to translate the creation and annihilation operators, the Hamiltonian etc. from the Bargmann to the loop representation.

To summarize, the Fock states are transformed to the loop representation via Eq.5. In this representation, physical states are essentially polynomials of the form factors of loops. In particular, photons with momentum $\vec{k}_0$ and helicity $j(= \pm 1)$ correspond to functions $\psi(\gamma) = F_j(\vec{k}_0)$. The operator $\psi(\gamma) \mapsto F_j(\vec{k})$ creates a photon in the state $|\vec{k}, j>$ and the operator $\hbar|k|(\delta/\delta F_j(\vec{k})$ annihilates the photon on that state.

3 Loop quantization

In the last section, we used the Rovelli-Smolin transform to construct the loop representation. We now show how the same representation can be constructed directly using the quantization program of chapter 10, without any reference to the Fock space or the Bargmann representation. It is this construction that we will extend to gravity in the next two chapters.

We begin by noting that there is a complete set of operators whose action on the loop states of section 2 can be given without any reference to form factors. Consider, on the reduced phase space, the functions $T^0[\gamma](A)$ parametrized by loops γ constructed from the *negative frequency parts* of the connection A^T_a as follows: $T^0[\gamma](A) := \exp \oint_\gamma dS^a\, {}^{(-)}A^T_a(\vec{x})$. Thus, T^0 are just the holonomies of the negative frequency connections. Their quantum analogs are therefore exponentials of the annihilation operators:

$$\hat{T}^0[\alpha] = \exp \int \frac{d^3k}{2\hbar|k|} \sum_j F_j(\alpha, \vec{k})(\hat{\varsigma}_j(\vec{k}))^\star. \tag{12a}$$

(The notation T^0 may seem a bit strange; its motivation will become clear in the next chapter.) The action of the operators $\hat{T}^0[\gamma]$ in the connection representation is straightforward to obtain. Let us translate their action to the loop representation via Eq.5. If $\Psi'(\varsigma_1, \varsigma_2) = \hat{T}^0[\alpha] \circ \Psi(\varsigma_1, \varsigma_2)$, we have:

$$\begin{aligned}
\psi'(\gamma) &:= \int [\mathcal{D}\mu]\, \overline{H(\gamma,\, {}^{(+)}A^T)}\Psi'(\varsigma_1, \varsigma_2) \\
&= \int [\mathcal{D}\mu]\, \overline{H(\gamma,\, {}^{(+)}A^T)} \exp \int \frac{d^3k}{2\hbar|k|} \sum_j [F_i(\alpha, \vec{k}) + F_i(\gamma, \vec{k})]\bar{(\varsigma)}_i(\vec{k}))\; \Psi(\varsigma_1, \varsigma_2) \\
&= \int [\mathcal{D}\mu]\, \overline{H(\gamma\#\alpha)}\Psi(\varsigma_1, \varsigma_2),
\end{aligned} \tag{12b}$$

where $[\mathcal{D}\mu]$ stands for the volume element $\Pi_j\, d\!\!I\varsigma_j(\vec{k}) \wedge d\!\!I\bar{\varsigma}_j\, \mu(\varsigma_1, \varsigma_2)$. The loop $\alpha\#\gamma$ is an "eye-glass loop" obtained by picking any curve β joining a point p of α to a point p' of γ, and moving along the following path: start at p, go to p' along β, go around γ, return to p along β^{-1} and then go around α. Thus, $\alpha\#\gamma = \alpha \circ \beta^{-1} \circ \gamma \circ \beta$. (The construction is actually much simpler than it seems from this wordy description!) Thus, in the loop representation, we have the following simple expression of the operator $\hat{T}[\alpha]$:

$$(\hat{T}^0[\alpha] \circ \psi)[\gamma] = \psi(\alpha\#\gamma), \tag{12c}$$

Another set of operators is constructed from *positive frequency* electric fields.

Consider on the reduced phase space the function $T^1[f](E)$ parametrized by equivalence classes of real valued 1-forms $f_a(\vec{x})$, where two are regarded as being equivalent if they differ by a gradient, defined as follows: $T^1[f](E) := \int d^3x\, f_a(\vec{x})(E^T)^a(\vec{x})$. In the connection representation, the corresponding operator is given by:

$$\begin{aligned}\hat{T}^1[f]\circ\Psi(\varsigma_1,\varsigma_2) &= \int d^3k\, f_a(\vec{x})^{(+)}\hat{E}^a(\vec{x})\\ &= -i\int d^3k[\overline{f_a(\vec{k})}m^a(\vec{k})\hat{\varsigma}_1(\vec{k}) + \overline{f_a(\vec{k})}\bar{m}^a(\vec{k})\hat{\varsigma}_2(\vec{k})].\end{aligned} \tag{13a}$$

As with $\hat{T}^0[\alpha]$, we can translate the action of $\hat{T}^1[f]$ to the loop representation via Eq.5. Setting $\Psi'[\varsigma_1,\varsigma_2] = \hat{T}^1[f]\circ\Psi[\varsigma_1,\varsigma_2]$, we have:

$$\begin{aligned}\psi'[\gamma] &= \frac{i}{2}\int [\mathcal{D}\mu]\overline{H(\gamma,{}^{(+)}A^T)}\;\Psi'[\varsigma_1,\varsigma_2]\\ &= (\int d^3k[f_a(\vec{k})\bar{m}^a(\vec{k})F_1(\gamma,\vec{k}) + f_a(\vec{k})m^a(\vec{k})F_2(\gamma,\vec{k})])\;\psi(\gamma).\end{aligned} \tag{13b}$$

Consequently, in the loop representation, the operator $\hat{T}^1[f]$ is given simply by:

$$(\hat{T}^1[f]\circ\psi)[\gamma] := \left(i\hbar\oint_\gamma dS^a f_a\right)\psi(\gamma). \tag{13c}$$

Again, the definition of the operator depends only on the structure naturally available on the loop space. The idea therefore, is to *introduce* the loop representation via Eqs.12 and 13.

Let us then choose $T^0[\gamma](A)$ and $T^1[f](E)$ as the elementary classical variables and carry out the quantization program. (Here we shall only sketch the final results; details will be given in [1].) One can check that these variables are closed under Poisson brackets: The only non-vanishing bracket is

$$\{T^0[\gamma](A),\, T^1[f](E)\} = \left(\oint_\gamma dS^a\, f_a\right)T^0[\gamma](A). \tag{14}$$

The T^0 are the configuration variables and the T^1, the momentum variables. It is easy to check that they form a complete set on the reduced phase space $\hat{\Gamma}$. In essence, the information in $\varsigma_j(\vec{k})$ is coded in $T^1[f]$ and that in $\bar{\varsigma}_j(\vec{k})$ in $T^0[\gamma]$. Following the first three steps in the quantization program we construct the algebra $\mathcal{A}$ of quantum operators, generated by $\hat{T}^0[\gamma]$ and $\hat{T}^1[f]$. Next, we must find a linear representation of this algebra. This is straightforward to accomplish. Choose for V the space of functions on the loop space spanned by functions of the type:

$$\oint_{\gamma_1} dS^{a_1} \ldots \oint_{\gamma_n} dS^{a_n}\, g_{a_1 \ldots a_n}(s_1, \ldots s_n)\,, \tag{15}$$

where s_i are points on the i-th loop, the index a_i refers to the tangent space at the point s_i on the i-th loop, and where g is totally symmetric in the obvious sense: $g_{a_1 a_2}(s_1, s_2) = g_{a_2, a_1}(s_2, s_1)$, etc. On this V, define the representation of $\mathcal{A}$ through Eqs.12 and 13. It is straightforward to check that the commutation relations required by Eq.14 are satisfied. To find the inner product, we can again use the reality conditions. The final picture is the following: The function $\psi(\gamma) = 1$ is the vacuum state; $\psi(\gamma) = \oint_\gamma dS^a f_a(\vec{x})$ is the 1-particle state with norm $\int (2\hbar|k|)^{-1} d^3k\, |f_a(\vec{k})|^2$, etc. (The two helicities are coded in the two transverse components of $f_a(\vec{x})$ that characterize the equivalence class to which this 1-form belongs.) Finally, the norm of the general state given in Eq.15 is:

$$\|\psi(\gamma)\|^2 = \int \frac{d^3k_1}{2\hbar|k_1|} \cdots \int \frac{d^3k_n}{2\hbar|k_n|}\, |(g_{a_1} \ldots g_{a_n}(\vec{k})|^2. \tag{16}$$

One can also write down physically interesting operators such as the total Hamiltonian, *directly* on this loop representation.

We can now return to the question we raised in the introduction. What is the interpretation of the state $|\,\gamma_0\,\rangle$? In the loop description, this state is represented by the characteristic function of γ_0: $\psi_0(\gamma) = 1$ if $\gamma = \gamma_0$, and 0 otherwise. Clearly, this is the eigenstate of all positive frequency electric field operators $\hat{T}^1[f]$ of Eq.13c: $\hat{T}^1[f] \circ \psi_0(\gamma) = (\oint_{\gamma_0} dS^a\, f_a)\psi_0(\gamma)$. Since the positive frequency operators are creators, it is clear that the state cannot have finite norm. It is analogous to the plane wave or the delta-distribution states of non-relativistic quantum mechanics. However, as a generalized state, it has a simple interpretation: It represents the simplest excitation of the positive frequency electric field operator. Indeed, since the electric field is divergence-free, it cannot admit eigenstates which are localized at a point; its simplest excitation is along a closed loop. Thus, the loop basis represents *the quantum Faraday lines* for the positive frequency part of the electric field operator.

In the next two chapters, we extend the loop representation to general relativity in 3+1 dimensions. In this case, the Rovelli-Smolin transform is only formal because we do not know the analog of the Gaussian measure $\mu(\varsigma_j, \bar{\varsigma}_j)$. Therefore, although one *can* arrive at the loop representation using the method used in section 2 of this chapter, the construction lacks rigor. We will therefore use the approach of section 3 instead: we will begin by introducing functions analogous to T^0 and T^1 on the phase space of general relativity and use them as the elementary variables in the quantization program of chapter 10. Strictly speaking, there is an intermediate step between the material presented in this chapter and that presented in the next two. As was pointed out in chapters 11 and 12, in full general relativity, we do not know the analog of the positive frequency connections; we are forced to work just with the self dual ones. Therefore, strictly, we should first carry out quantization of the Maxwell field as well in the self dual representation and show that the corresponding loop representation also exists. However, using the construction of the self dual representation for linear gravity presented in chapter 11 and the Rovelli-Smolin transform of section 2, it is straightforward to carry out these steps for Maxwell theory. In the resulting loop picture, states are again represented by functionals $\psi(\gamma)$ on the loop space and the elementary quantum operators, $\hat{T}^0[\gamma]$ and $\hat{T}^1[f]$ are again defined by Eqs.12*c* and 13*c*. However, their interpretation is now different. $T^0[\gamma]$ is now the exponential the of integral of the *anti-self dual* (rather than negative frequency) connection and $T^1[f]$ is the smeared out self-dual (rather than positive frequency) electric field. The corresponding operators therefore contain mixtures of the standard creation and annihilation operators. I urge the reader to construct this loop representation as a prelude to 3+1 gravity.

References

[1] A. Ashtekar and C. Rovelli, Quantum Faraday lines: Loop representation of the Quantum Maxwell field, Syracuse University pre-print (1990).

[2] C. Rovelli and L. Smolin, Phys. Rev. Lett. **61**, 1155 (1988); Nucl. Phys. **B331**, 80 (1990).

15 LOOP REPRESENTATION: CLASSICAL THEORY

1 Introduction

In the last chapter we made a brief detour into Maxwell theory to introduce the basic ideas of the loop representation. Let us now return to general relativity.

In the chapters 11-13, we constructed the connection representation for gravity and used its general framework to probe two conceptual issues. However, we were able to complete the quantization program only in a weak field truncation of quantum general relativity. In the full theory, we could only complete the first four steps. The fifth step –the singling out of physical states by solving the quantum constraints– remains incomplete. In this and the following chapter, we will show that this step can be completed, at least in part, in the loop representation. The result is an infinite dimensional space of physical states of quantum gravity.

In the geometrodynamical approach, the problem of finding solutions to quantum constraints has been investigated, off and on, for about three decades. Yet, outside the restricted context of quantum cosmology, little progress was made. Indeed, in the full theory, one still does not know a single solution to the quantum scalar constraint. Why then was it possible to find an infinite dimensional space of solutions to all constraints in the new approach? The reasons are two-fold. The first is technical: the constraints now have a much simpler form and are therefore manageable. The second reason is conceptual: because the canonical framework is based on connection dynamics, one can now introduce in quantum gravity a number of brand new concepts which were simply not available in geometrodynamics. The loop representation is, in fact, *based* on such ideas.

In the loop representation, one begins with the observation that the gauge invariant information in the connection is contained in the holonomies it defines around closed loops and introduces on the phase space of general relativity certain functions constructed from them. These will be the new elementary classical variables. As in Maxwell theory discussed in the last chapter, although these variables are *non-local* in their dependence on the 3-space, they have a number of compensating, attractive features. The elementary quantum operators are now labelled by closed loops (and selected points on these loops). The algebraic structure of these operators can now be specified using just the geometrical properties of closed loops. To commute two of these operators for example, one just breaks and joins the loops labelling them! Put differently, the symplectic structure on the phase space of general relativity is now coded in the geometry of the loop space of the spatial 3-manifold. To find the representation of the quantum algebra $\mathcal{A}$, it is natural to use, for the carrier space V, the space of (suitably restricted) functions on the loop space. To extract the space V_{phy} of physical states, one has to solve the quantum constraints. The *physical* states turn out to be functions which have support on certain types of loops and whose values do not change if the loop is replaced by another loop related by a diffeomorphism. Recall, however, that the diffeomorphism classes of loops (multiloops) are essentially knots (links). Thus, physical states of quantum gravity turn out to be functions with restricted support on the (generalized) knot and link classes of 3-space. Note that the space of knots and links is in fact discrete. There are concrete indications that this discreteness will simplify the analysis of several technical problems that still remain in the program.

The loop representation was introduced and extensively studied by Carlo Rovelli and Lee Smolin [1,2]. Their point of departure was a paper by Ted Jacobson and Lee Smolin [3] in which it was shown that the Wilson-loop functions of the self-dual connection –i.e., the traces of path ordered exponentials of $A_{aA}{}^{B}$ along closed loops– automatically solve the scalar constraint of general relativity in the connection representation. More recently, somewhat different approaches were developed by Miles Blencowe [4] and by Rudolpho Gambini and his collaborators [5]. In these notes, I will follow closely the original treatment of Rovelli and Smolin.[1] In section 2 we introduce the loop variables on the classical phase space. These consist of certain configuration variables $T^0[\gamma](A)$ and certain momentum variables $T^1[\gamma](A, \tilde{\sigma})$, both labelled by closed loops γ. In section 3, we show that they are closed under the Poisson bracket. This is the *small* $\mathcal{T}$ *algebra* of loop variables. Its elements will serve as the elementary variables in the classical theory. It turns out, however, that the constraint functionals of general relativity cannot be expressed in a convenient way as elements of this algebra. Therefore, in section 4, we enlarge this algebra by introducing functions, also labelled by closed loops, which are of higher order in momenta. In section 5, we investigate various properties of these T variables with an emphasis on the relations which are to yield the anti-commutation relations in the quantum theory. This machinery will be used in the next chapter to implement the quantization program based on the T variables.

---

1 I am grateful to them for their kind permission to use material from [2]. I would also like to thank Bernd Brügmann for correcting several numerical errors in the first version of chapters 15 and 16.

2 Classical loop variables

Consider continuous, piecewise smooth, nondegenerate mappings $\alpha : S_1 \to \Sigma$ from a circle to the (spatial) 3-manifold Σ. Each such mapping gives us a parametrized closed curve on Σ. We will call such curves loops, and we denote them by greek letters $\alpha(s), \beta(s), \gamma(s), \ldots$, where the loop parameter s will always be considered modulo 2π; $s = s + 2\pi$. The inverse of a curve α is defined to be the curve $(\alpha^{-1})(s) = \alpha(2\pi - s)$.

We begin by introducing the new configuration variables $T^0[\gamma]$. For this, let us first recall the notion of non-Abelian holonomy. If γ is a parametrized loop on Σ, then given a $SL(2,C)$ connection[2] $A_{aA}{}^B$ on Σ and any two points $\gamma(s)$ and $\gamma(t)$ on the loop, we acquire an element $U_\gamma(s,t)$ of $SL(2,C)$ which maps spinors at $\gamma(s)$ to those at $\gamma(t)$:

$$U_\gamma(s,t) := \mathcal{P} \exp\left(\int_{\gamma(s)}^{\gamma(t)} dS^a\, A_a\right), \tag{1}$$

where $\mathcal{P}$ stands for "path ordered". We shall denote the holonomy, or the parallel transport around the loop, by $U_\gamma(s) (\equiv U_\gamma(s,s))$. This $U_\gamma(s)_{AB}$ is an element of $SL(2,C)$, and satisfies the identity

$$U_\gamma(s)_{AB} = -U_{\gamma^{-1}}(s)_{BA}\,. \tag{2}$$

The function $T^0[\gamma](A)$ on the gravitational phase-space, labelled by a loop γ, is defined to be simply the trace of the holonomy of the self-dual connection along γ

$$T^0[\gamma] := \mathrm{tr}\ U_\gamma(s) = \mathrm{tr}\ \mathcal{P}\ e^{\oint_\gamma A}. \tag{3}$$

The set of $T^0[\gamma]$ are now the analogs of the complex variable z on the phase space of the harmonic oscillator; they will replace the linear functions of $A_{aA}{}^B$ that we used in chapter 12 to construct the connection representation. Note, however that $T^0[\gamma]$ are neither adapted to the affine space structure of the space of connections, nor are they independent. In fact, as we will see, they are overcomplete. Therefore, unlike in the connection representation, now we will have non-trivial anti-commutation relations of the type discussed in chapter 10. The key advantage of these variables is that, in contrast to the elementary variables $A_{aA}{}^B$ underlying the connection representation, the $T^0[\gamma]$ are manifestly gauge invariant.

---

2 In part I we often regarded the connection $A_{aA}{}^B$ as a complex valued 1-form that takes values in the Lie algebra of $SU(2)$. One can equivalently regard $A_{aA}{}^B$ as a 1-form that takes values in $SL(2,C)$. We shall adopt this convention in the discussion of the loop representation.

The $T^0[\gamma]$ are functions only of the connection $A_{aA}{}^B$. They do not depend on the momenta $\tilde{\sigma}^a{}_A{}^B$. We now introduce the appropriate momentum variables. We want these variables to be $SU(2)$ gauge invariant. However, since they are to be linear in momentum, they cannot contain products of more than one $\tilde{\sigma}^a{}_A{}^B$. The simplest way of constructing such a variable is to insert a $\tilde{\sigma}^a{}_A{}^B$ inside the trace of the parallel transport around loops. Thus, we define a second set of functions $T^a[\gamma](s)$ as follows. For any loop γ, and loop parameter s, $T^a[\gamma](s)$ is given by inserting $\tilde{\sigma}^a(x)$ along the holonomy of γ, at the point $x = \gamma(s)$. That is, we have:[3]

$$T^a[\gamma](s) := \sqrt{2}\,\mathrm{tr}\ U_\gamma(s)\tilde{\sigma}^a(\gamma(s))\ , \tag{4}$$

where we have inserted a factor of $\sqrt{2}$ to simplify the Poisson algebra. Each $T^0[\gamma]$ is invariant under reparametrization of γ; this is also true if the reparametrization changes the orientation, since the trace of an $SL(2,C)$ matrix is equal to the trace of its inverse. $T^a[\gamma](s)$, on the other hand, is not reparametrization invariant since it depends on a preferred value of the loop parameter, s. However, it is reparametrization covariant. Let γ' be a reparametrization of γ with the same orientation, $\gamma'(s) = \gamma(f(s))$, $\frac{df(s)}{ds} > 0$. Then clearly

$$T^a[\gamma'](s) = T^a[\gamma](f(s))\ , \tag{5}$$

so that we can identify these two objects. Note, however that $T^a[\gamma](s)$ does change sign under reparametrizations of the loop that reverse its orientation. This follows from Eq.2 and from the fact that $\tilde{\sigma}^a{}_A{}^B$is trace-free in the spinor indices. For notational simplicity, we shall sometimes refer to these momentum variables simply as T^1, where the superscript 1 is a reminder that the function depends *linearly* on the momentum $\tilde{\sigma}^a$. Note that the T^1 variables depend on *oriented unparametrized loops with a preferred point.* At times, at intermediate steps in calculations, we will associate an orientation with the loop on which T^0 depends. In such cases, the orientation can be chosen arbitrarily and final results will not depend on the choice.

Thus, we now have gauge invariant functions on the phase space which can serve as the configuration and momentum variables. We will see later in this chapter that they do form a complete set in an appropriate sense and are also closed under the Poisson brackets. Therefore, we will be able to use then as our elementary classical

3 If the internal group had been $U(1)$, the right side of Eq. 4 would have reduced to a simple product: $T^1[\gamma](s) = \sqrt{2}(U_\gamma)\ \tilde{E}^a(\vec{x}) = \sqrt{2}(T^0[\gamma])\ \tilde{E}^a(\vec{x})$, where $\vec{x} = \gamma(s)$. Since $T^0[\gamma]$ just factors out, it is now easier to use $\tilde{E}^a(\vec{x})$ in place of $T^a[\gamma](s)$. This is why in chapter 14 we defined $T^1[f]$ to be simply $\int d^3x f_a \tilde{E}^a$. In the non-Abelian case, the analog of this operator is inconvenient to use because it does not yield a closed Poisson algebra with the variables $T^0[\gamma]$.

variables. We conclude this section by showing that certain smeared out versions of the momentum variables can in fact be "derived" from the configuration variables. This fact will simplify a number of calculations later in this chapter. (This relation between the two sets of variables was first noticed in 2+1 gravity [6] where the structures involved are considerably simpler. For details, see, chapter 17.)

We note first that the configuration space $\mathcal{C}$ of general relativity –the space of connections $A_{aA}{}^{B}$ – is equipped with certain pre-Poisson structures, i.e., second-rank, anti-symmetric, contravariant tensor fields which are constant with respect to the affine structure of $\mathcal{C}$. These can be used to define (generalized) Poisson brackets between functions on $\mathcal{C}$. Since $\mathcal{C}$ is just the configuration space –and not the phase space– it is somewhat surprising that such structures exist. They arise as follows. Given a vector field v^a on the spatial 3-manifold Σ, we define a pre-Poisson structure ${}^{v}\omega$ via its action on any two cotangent vectors $(\delta\tilde{\sigma}^a)$ and $(\delta\tilde{\sigma}^a)'$ at any point of $\mathcal{C}$:

$$ {}^{v}\omega((\delta\sigma),(\delta\sigma)') := \sqrt{2}\int_{\Sigma} d^3x\, \underset{\sim}{\eta}_{abc} v^a \operatorname{tr}\,(\delta\tilde{\sigma}^b)(\delta\tilde{\sigma}^c)'. \tag{6} $$

By inspection, this tensor field has the desired properties stated above. Each of these pre-Poisson structures is degenerate. For example, given a vector field v^a, the right side of Eq.6 vanishes for all $(\delta\tilde{\sigma}^c)'$ if $(\delta\tilde{\sigma}^a)$ is chosen to be proportional to v^a. (This is why we use the suffix "pre" in pre-Poisson structures.) However, the *collection* of pre-Poisson structures as a whole *is* non-degenerate: if the integral on the right vanishes for all vector fields v^a and all $(\delta\tilde{\sigma}^c)'$, then $(\delta\tilde{\sigma}^a)$ must be zero. Now, given a pre-poisson structure, we can use it in place of the inverse of a symplectic structure to convert functions into vector fields: Given any function f on $\mathcal{C}$ we acquire vector fields ${}^{v}X_{(f)}$ on $\mathcal{C}$ via: ${}^{v}X_{(f)} := {}^{v}\omega \circ \mathrm{d}\!\mathrm{I} f$, where, as usual, $\mathrm{d}\!\mathrm{I}$ is the infinite dimensional exterior derivative. Recall, furthermore, that, given any vector field X on the configuration space $\mathcal{C}$, we acquire a function $P(X)$ on the *phase space*, linear in momentum, given by: $P(X)(A,\tilde{\sigma}) := \tilde{\sigma}\circ X$. Therefore, given a configuration variable f on $\mathcal{C}$, we can construct a momentum variable $P({}^{v}X_{(f)})$. Let us choose for f our gauge invariant configuration variables $T^0[\gamma]$. The resulting momentum variable is:

$$ T^1[\gamma, v^a](A_{aA}{}^{B},\tilde{\sigma}) = \sqrt{2}\oint_{\gamma} dS^a\, \underset{\sim}{\eta}_{abc}\, v^b \operatorname{tr}\tilde{\sigma}^c U_\gamma(A) \equiv \oint_{\gamma} dS^a\, \underset{\sim}{\eta}_{abc} v^b T^a[\gamma]. \tag{7} $$

This fact will enable us to establish a number of properties of the momentum variables $T^1[\gamma]$ starting from the properties of the configuration variables $T^0[\gamma]$.

3 The small $\mathcal{T}$ algebra

Let us now examine Poisson algebra of the functions $T^0[\gamma]$, $T^1[\gamma]$. We will find that the set is in fact closed under the Poisson brackets. From general considerations one does know that the space of all configuration and momentum variables on a cotangent bundle is necessarily closed under the Poisson bracket. Note, however, that we are dealing here with only a *subset* of all configuration and momentum variable s, namely the ones labelled by closed loops. Therefore, there is no a priori reason to expect the closure. In particular, had we replaced $T^1[\gamma](s)$ simply by $\tilde{\sigma}^a(\vec{x})$, we would not have attained this closure.

For any two loops α and β, we have:

$$\{T^0[\alpha],\, T^0[\beta]\} = 0 \tag{8}$$

simply because both functions are configuration variables, independent of momenta. To express the rest of the Poisson algebra we need to use loops that result from the breaking and joining of two loops that intersect at a point. Let us therefore first introduce some notation. Let α and β be two loops that intersect at the point $x = \alpha(\hat{s}) = \beta(\hat{t})$. We can construct a loop by starting from x, going first around α and then around β. We call this loop $\alpha\#_x\beta$. The subscript is needed because the two curves may intersect in more than one point. We will not write the subscript where the context is clear. We will also use $\alpha\#_s\beta$ for $\alpha\#_{\alpha(s)}\beta$. To complete the definition we need to specify how the loop $\alpha\#_x\beta$ is parametrized. We adopt the following convention. If $\hat{s}$ and $\hat{t}$ are the values of the parameters of α and β at the intersection, $\alpha(s+\hat{s})$ and $\beta(t+\hat{t})$ are parametrized loops that begin and end at the intersection (recall that the loop parameters are defined modulo 2π). We define

$$\alpha\#_x\beta\ (u) = \begin{cases} \alpha(2u+\hat{s}) & \text{for } 0 < u < \pi, \\ \beta(2u+\hat{t}) & \text{for } \pi < u < 2\pi. \end{cases} \tag{9}$$

We will use the notation $u'(s)$ and $u(t)$ for the values of the parameter of $\alpha\#\beta$ that correspond to the points $\alpha(s)$ and $\beta(t)$; the functions u' and u are defined by $\alpha\#\beta(u'(s)) = \alpha(s)$, $\alpha\#\beta(u(t)) = \beta(t)$ and may be easily computed from Eq.9.

Now we can compute the remaining Poisson brackets. Since T^0 is independent of momentum and T^1 is linear in momentum, their Poisson bracket is again independent of momentum (see Eq.1 in chapter 10). It turns out that the result is in fact a simple linear combination of the T^0 variables. By using the identity

$$\begin{aligned}\{\sqrt{2}\tilde{\sigma}^a_{AB}(x), U_\gamma^{CD}(0,s)\} &= -i\,\frac{\delta}{\delta A_a^{AB}(x)}\left[\mathcal{P}\exp\left(\int_0^s A_b(\gamma(u))\dot{\gamma}(u)^b du\right)\right]^{CD} \\ &= -i\int_0^s du\,\delta^3(\gamma(u),x)\dot{\gamma}^a(u)U_\gamma(0,u)^C{}_{(A}U_\gamma(u,s)_{B)}{}^D\,,\end{aligned} \tag{10}$$

we have:

$$\begin{aligned}\{T^a[\gamma](s), T^0[\eta]\} &= i\, U_\gamma(s)^{AB}\, \{\sqrt{2}\tilde{\sigma}^a_{AB}(\gamma(s)), T^0[\eta]\} \\ &= i\, U_\gamma(s)^{AB} \int dt\, \delta^3(\gamma(s), \eta(t))\dot{\eta}^a(t) U_\eta(t)_{(AB)} \\ &= \tfrac{1}{2i} \int dt\, \delta^3(\gamma(s), \eta(t))\ \dot{\eta}^a(t)\ [T^0[\gamma\#\eta] - T^0[\gamma\#\eta^{-1}]].\end{aligned} \tag{11}$$

Thus, the bracket is zero unless η intersects γ at the point s.

It is convenient to introduce a special notation for the distributional terms that appear in the commutators of the $\mathcal{T}$ algebra. Let us set

$$\Delta^a[\gamma, \eta](s) \equiv \tfrac{1}{2} \int dt\, \delta^3(\gamma(s), \eta(t))\ \dot{\eta}^a(t). \tag{12}$$

Then the right side of the Poisson bracket between the T^0 and T^1 variables is a linear combination of the T^0 variables with "singular" coefficients Δ^a. We will show below that these "singularities" are harmless and can be eliminated by interpreting the T variables as distributions. That is, the commutators are fully regular if the t-variables are smeared by suitable test functions. Thus, the situation is similar to the one encountered in the canonical commutation relations, $\{\phi(x), \pi(y)\} = \delta^3(x, y)$, of any classical field theory. Finally, for later convenience, let us introduce the notation

$$(\alpha\#\beta)^{><} = \alpha\#\beta, \quad (\alpha\#\beta)^{\vee\wedge} = \alpha\#\beta^{-1}, \tag{13}$$

and a symbol $\diamond$ that may take the value $><$ or $^{\vee}_{\wedge}$. The reasons for this notation will become clear at the end of this section where we introduce a diagramatic notation. Using this notation, the Poisson bracket of Eq.11 becomes

$$\{T^a[\gamma](s), T^0[\eta]\} = -i \sum_\diamond (-1)^{|\diamond|}\ \Delta^a[\gamma, \eta](s)\ T^0[(\gamma\#\eta)^\diamond], \tag{14}$$

where we have defined $|><| = 0$, $|^{\vee}_{\wedge}| = 1$.

Let us now compute the Poisson bracket between two T^1-variables. Since each T^1 is linear in momentum, so must the Poisson bracket be. It again turns out the bracket is in fact a simple linear combination of the T^1. We have:

$$\begin{aligned} i\{T^a[\gamma](s), T^b[\eta](t)\} &= 2U_\gamma(s)^{AB}\{\tilde\sigma^a_{AB}(\gamma(s)),\, U_\eta(t)^{CD}\}\,\tilde\sigma^b_{CD}(\eta(t)) \\ &+ 2\tilde\sigma^a_{AB}(\gamma(s))\,\{U_\gamma(s)^{AB},\, \tilde\sigma^b_{CD}(\eta(t))\}\,U_\eta(t)^{CD}] \\ &= \sqrt{2}\Delta^a[\gamma,\eta](s)\,\Big(\mathrm{tr}[U_\gamma(s)U_\eta(u,t)\tilde\sigma^b(\eta(t))\,U_\eta(t,u)] \\ &- \mathrm{tr}\,[U_{\gamma^{-1}}(s)U_\eta(u,t)\tilde\sigma^b(\eta(t))U_\eta(t,u)]\Big) \\ &- \sqrt{2}\Delta^b[\eta,\gamma](t)\,\Big(\mathrm{tr}[U_\eta(t)U_\gamma(v,s)\tilde\sigma^a(\gamma(s))U_\gamma(s,u)] \\ &- \mathrm{tr}\,[U_{\eta^{-1}}(t)U_\gamma(v,s)\tilde\sigma^a(\gamma(s))U_\gamma(s,u)]\Big) \end{aligned} \tag{15}$$

The result is then,

$$\begin{aligned} \{T^a[\gamma](s),\, T^b[\eta](t)\} =& i\sum_\diamond\,(-1)^{|\diamond|}\,\Delta^b[\eta,\gamma](t)\,T^a[(\eta\#_t\gamma)^\diamond](u(s)) \\ &- i\sum_\diamond\,(-1)^{|\diamond|}\,\Delta^a[\gamma,\eta](s)\,T^b[(\gamma\#_s\eta)^\diamond](u(t)). \end{aligned} \tag{16}$$

Following Rovelli and Smolin [1,2], we call the Poisson algebra of the T^0, T^1 variables, defined by Eqs.8, 11, and 16 the *small $\mathcal{T}$ algebra.* In section 4, we will introduce a larger algebra consisting of functions which are of higher order in momenta. That algebra will be called the *full $\mathcal{T}$ algebra.*

Before proceeding further, let us pause to verify that the "singularities" in the Poisson brackets are indeed harmless. To show this, we need to consider appropriately smeared out versions of the T variables. At the end of the last section, we presented one way to smear the T^1-variables. If one computes the Poisson brackets of those smeared out functions, $T^1[\gamma, v^a]$, one does find that the singularities are "softened". However, they do not entirely go away. The reason, roughly, is that in the definition of these functions, the smearing has occurred only in 1-dimension, along the loop γ. Therefore, to obtain a completely regular Poisson algebra, we must carry out a more appropriate smearing.

Let us consider, instead of single loops, *two parameter congruences* of loops $\gamma_\tau(s)$, $\tau = (\tau_1, \tau_2)$. We can then define the smeared T^1 variables to be

$$T^1[\gamma_\tau](f) = \int d^2\tau\; ds\; f_a(\gamma_\tau(s))\; T^a[\gamma_\tau](s), \tag{17}$$

where f is a smooth one form on Σ with compact support. The Poisson bracket of the smeared T^1 with a T^0 is given by

$$\{T^1[\gamma_\tau](f),\ T^0[\eta]\} = -i \oint_\eta dS^a\ f_a\ (T^0[\gamma_\tau\#\eta] - T^0[\gamma_\tau\#\eta^{-1}]), \tag{18}$$

where the τ in the right hand side is a function of the integration variable. Note that the singular coefficient Δ has been replaced by the line integral of the one form f along the grasped loop η. Thus, if we let the configuration variables, T^0, be associated with loops as before but smear-out momentum variables appropriately, the Poisson bracket is regularized. One can easily verify that the bracket between two smeared T^1-variables is also regular. Thus, the situation is better than what one might have naively anticipated; we do not have to smear-out T^0 at all.[4]

To conclude this section, we shall introduce some graphical notation which is well-suited to various calculations both in the classical and quantum theories. Let us denote a class of loops equivalent under reparametrization by a closed line. Since T^0 depends on single loops, but not on their parametrization, we can represent a $T^0[\gamma]$ simply by means of the corresponding equivalence class of parametrized loops, as in figure 1*a*.

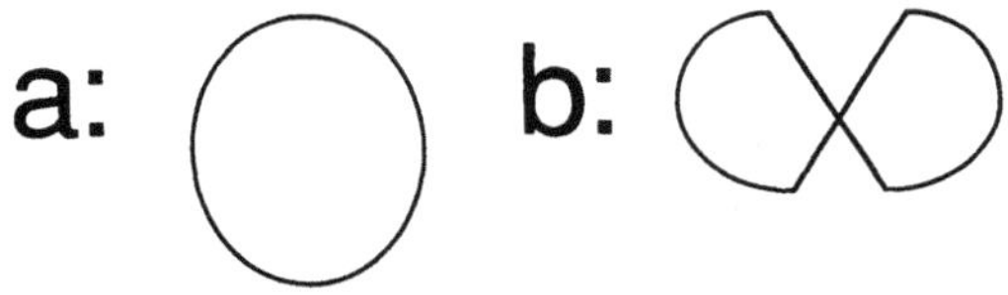

Figure 1: $T^0[\gamma]$

Our convention will be such that the intersections and the points of non-differentiability on the loops and their order (which are parametrization invariant concepts) are reproduced by the drawing. For instance a loop with one self intersection is denoted as in figure 1*b*. Then we can denote the equivalence classes of oriented loops with a particular selected point by putting a dot on the point and an arrow on the loop that fixes the orientation. Using this we denote the T^1 as in figure 2. We call the dots on the drawing of the loops which represent the insertion of a $\tilde{\sigma}^a{}_A{}^B$ "*hands*".

4 In fact, even with T^1, we can do with less; to regularize the Poisson algebra, it suffices to associate T^1 with 2-dimensional "strips", i.e., to use a 1-parameter congruence of loops in Eq.17. This fact may be significant to quantum theory.

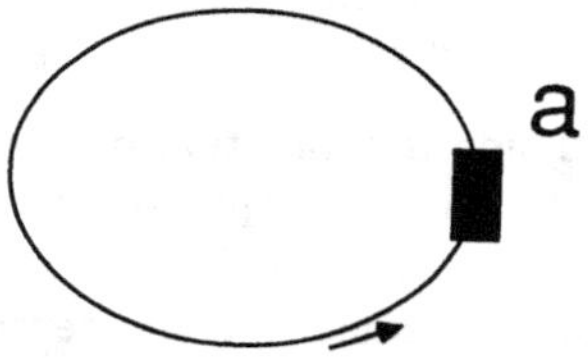

Figure 2: $T^a[\gamma](s)$

One can now perform the Poisson bracket calculations between the $T^0[\gamma]$ and $T^a[\gamma](s)$ using this diagrammatic notation. The Poisson bracket is an operator that sends two functions, each parametrized by a loop, to a linear combination of functions parametrized by loops that are related to the original ones by simple topological operations. The Poisson brackets of a T^0 and T^1 such that the hand indicating the presence of the $\tilde{\sigma}^a{}_A{}^B$ is at one of the intersections can be expressed as in figure 3.[5]

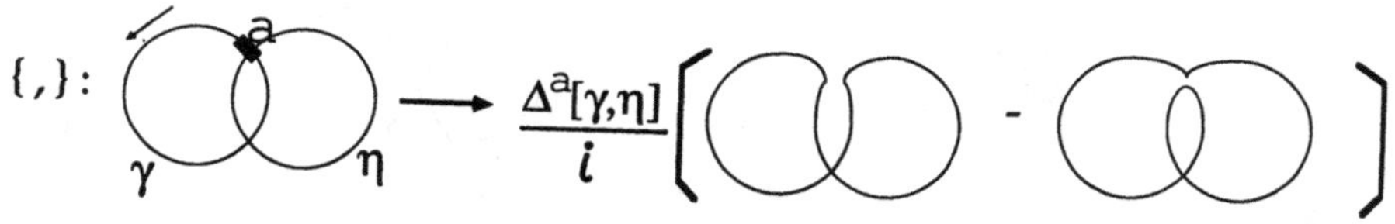

Figure 3: Poisson bracket of a T^0 and a T^1

Similarly, the Poisson brackets of two T^1s such that each of the two $\tilde{\sigma}^a(x)$ is at a point of intersection is drawn as in figure 4. It is clear that there is a simple graphical rule at work here. The { , } operator acts at each point where there is a hand. There is an elementary action at any hand, which we now describe.

5 In the figures we use the following convention to distinguish the two ways in which the four legs arriving at an intersection can be rearranged. We consider the rearrangement in which the two legs coming from the right (and the two coming from the left) are connected (the >< figure) to be the one in which the orientation of the two loops matches naturally. Note that in figure 3 the dotted loop is oriented, while the orientation of the undotted loop is arbitrary: by reversing its orientation, the role of the two resulting loops is inverted, but also the Δ changes sign, and the overall result is invariant.

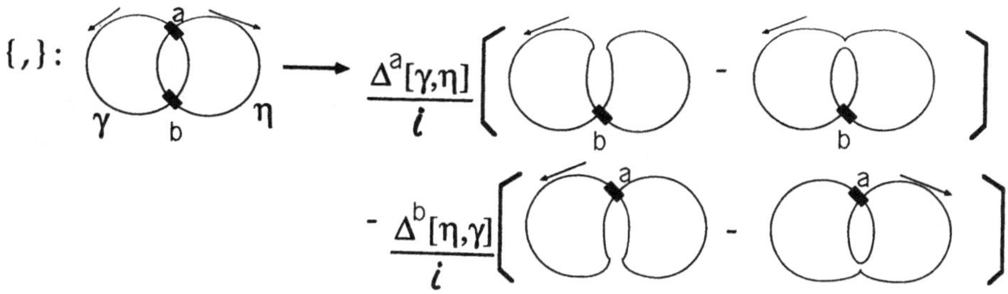

Figure 4: Poisson brackets of two T^1

If the hand is not at an intersection point of the two loops the result is zero. If the hand is on γ at a point of intersection with η and has index ^{a}, the action is the following. We obtain two new loops from γ and η by removing the hand, breaking each of them at the point of intersection, and rejoining each resulting leg of γ with one of η. In one of these two loops the orientations are consistent, in the other they clash so that we have to reverse the orientation of η before rejoining; we take the difference between the first loop and the second one and, finally, we multiply the result by $\Delta^a[\gamma,\eta]$.

This graphical rule expresses exactly the contents of Eq.10. It is summarized in figure 5.

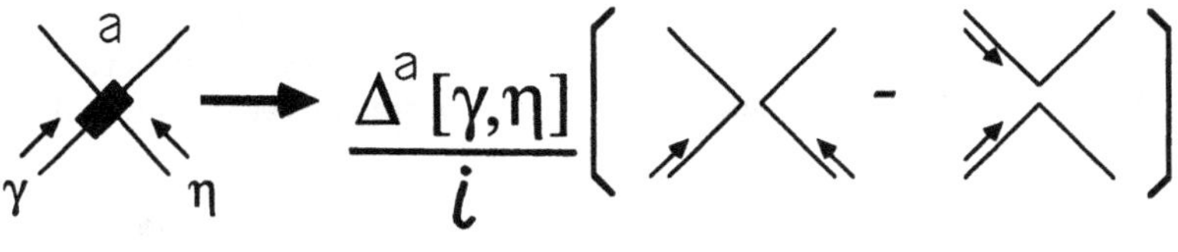

Figure 5: The action of the grasp operator

Now the total action of the operator $\{\,,\}$ is given by its action over the hand of the first loop (if it has a hand) minus its action over the hand of the the second one (if it has a hand).

For conciseness we will use the following terminology. We say that a hand on a loop γ "sees" a loop η if γ and η intersect at the location of the hand. We say that the hand "grasps" the loop η to indicate the operation described in figure 5, and we call the "grasp" of the hand over the loop η the result of the operation (the grasp is zero unless the hand sees η). Then we can re-express our result about the Poisson algebra in the following terms: The Poisson brackets of two handed loops α and β is given by the grasp of the hand of α over β, minus the the grasp of the hand of β over α. (Note that intersections which are not at the locations of the dots are not relevant, as, for instance, the second intersection of figure 3.) Finally, at this point, the pictorial meaning of the notation $><$, $\overset{\vee}{\wedge}$ should be clear. These symbols refer exactly to the pictorial description, as in figure 5, of the rearrangement of the legs at the intersection.

4 The full $\mathcal{T}$ algebra

We will see in the next section that the T^0 and the T^1 variables that form the small $\mathcal{T}$ algebra form a complete set of functions on the phase space in an appropriate sense. However, it is inconvenient to express the scalar constraint of general relativity directly in terms of them.[6] Since the scalar constraint is quadratic in momenta, it is therefore extremely useful to enlarge the small $\mathcal{T}$ algebra to include functions which are quadratic in momenta. It is remarkable that such an extension exists. As we noted in section 10.2, once we add to the space of elementary variables functions which are quadratic in momenta, in order to get closure under Poisson brackets, we are forced to include in the algebra functions which are of arbitrarily high order. It is because of this that we do not introduce just T^2 but include in the full algebra variables T^n for all n. From the point of view of the quantization program presented in chapter 10, the fact that such a larger sub-algebra of the Poisson algebra exists is, however, a luxury. However, since the structures involved are so natural, it is very likely that the existence of this larger algebra will be useful at some later stage.

The T^n are obtained by inserting more than one $\tilde{\sigma}^a{}_A{}^B$ along the holonomy of a loop. More precisely, we define T^n as a function of a loop and of n points on the loop fixed by loop parameter values $s_1, ..., s_n$, which satisfy $0 < s_1 < ... < s_n \leq 2\pi$.

---

6 The situation is somewhat analogous to the one in particle mechanics. Suppose we are given a system of particles whose phase space has the structure of a cotangent bundle. Then, as we saw in section 10.2, the configuration variables $f(q)$ and the momentum variables $V^a(q)p_a$ (where $V^a(q)$ are vector fields on the configuration space) are overcomplete. However, to express a physically interesting observable in terms of these variables, one often has to take a *limit* of a sum of their products. Algebraic operations do not suffice.

We set

$$T^{a_1 \dots a_n}_{ordered}[\gamma](s_1 \dots s_n) \equiv \\ \sqrt{2}\,\mathrm{tr}\,[\tilde{\sigma}^{a_1}(\gamma(s_1))U_\gamma(s_1, s_2)\tilde{\sigma}^{a_2}(\gamma(s_2)) \dots \tilde{\sigma}^{a_n}(\gamma(s_n))U_\gamma(s_n, s_1)] \tag{19}$$

For later convenience we adopt the convention that the order in which the s_i (and related a_i) are written as arguments of a T^n is irrelevant. If $P = (i, j, .., p, q)$ is a permutation of the first n natural numbers we define T^n (without the subscript *ordered*) to be

$$T^{a_1 \dots a_n}[\gamma](s_1 \dots s_n) = \sqrt{2} \sum_P \, \theta(s_q - s_p) \dots \theta(s_j - s_i) \, T^{a_i \dots a_p}_{ordered}[\gamma](s_i \dots s_p), \tag{20}$$

where $\theta(t) = 1$ for $t > 0$ and zero otherwise; T^n may be graphically represented as an oriented curve with n hands.

The main property of the T^n's is that they form a closed graded Poisson algebra, which has the structure

$$\{T^n, T^m\} \sim T^{n+m-1}. \tag{21}$$

More precisely

$$\begin{aligned} &\{\, T^{a_1 \dots a_n}[\gamma(s_1 \dots s_n)] \,,\, T^{b_1 \dots b_m}[\eta](t_1 \dots t_m) \,\} \\ &= -i \sum_{k=1}^{n} \sum_{\diamond} (-1)^{|\diamond|} \, \Delta^{a_k}[\gamma, \eta](s_k) \; T^{a_1 \dots \not{a}_k \dots a_n, b_1 \dots b_n}[(\gamma \#_{s_k} \eta)^\diamond] \\ &\qquad \times (u'(s_1) \dots \not{u}'(s_k) \dots u'(s_n), u(t_1) \dots u(t_m)) \\ &+ i \sum_{k=1}^{m} \sum_{\diamond} (-1)^{\diamond} \, \Delta^{a_k}[\eta, \gamma](t_k) \; T^{b_1 \dots \not{b}_k \dots b_m, a_1 \dots a_n}[(\eta \#_{t_k} \gamma)^\diamond] \\ &\qquad \times (u'(t_1) \dots \not{u}'(t_k) \dots u(t_m), u(s_1) \dots u(s_n)) \,. \end{aligned} \tag{22}$$

The slash over a term means that that term is not present: $(a_1 \dots \not{a}_k \dots a_m) = (a_1 \dots a_{k-1} a_{k+1} \dots a_n)$. Each of the two terms contains a sum over the hands of one of the loops, and, for every hand, the sum on the two ways to rearrange the legs. This result is obtained after some algebra, or much more easily by the graphical calculus described in the previous section. The full Poisson algebra can then be expressed in the following form. The Poisson bracket of two handed loops α and β is given by the sum of all the grasps of the hands of α, over β, minus the sum of all the grasps of the hands of β over α. For instance the Poisson bracket of a T^2 and a T^3 that intersect at two points, such that there is a hand of the T^2 at each intersection, is given in figure 6.

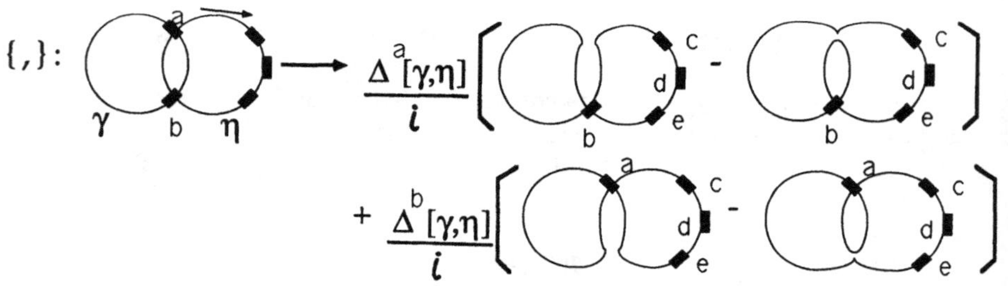

Figure 6: Poisson bracket of one T^2 and one T^3 with some hands that do not see the other loop.

To summarize, we have shown that the T variables form a graded Poisson algebra. As we already said we call this algebra $\mathcal{T}$. This algebra is entirely expressible in term of breaking and rejoining of handed loops and, remarkably, the full symplectic structure of the phase space of general relativity is coded in this algebra.

To conclude this section, we show that the distributions appearing in the Poisson brackets of the full $\mathcal{T}$ algebra can also be eliminated by smearing the T^n-variables appropriately. Since we already considered the case $n = 1$ in the last section, let us now suppose we have $n \geq 2$. Then, we have to introduce an independent three dimensional smearing for every hand. That is we consider $2n$-parameter families of loops

$$T^n[\gamma_{\tau_1...\tau_n}](f) = \int d^2\tau_1...d^2\tau_n \int ds_1...ds_n \\ f^{(1)}_{a_1}(\gamma_{\tau_1...\tau_n}(s_1))...f^{(n)}_{a_n}(\gamma_{\tau_1...\tau_n}(s_n))\; T^{a_1...a_n}[\gamma_{\tau_1...\tau_n}](s_1...s_n). \tag{23}$$

It is straightforward to verify that all the singularities of the $\mathcal{T}$ algebra are eliminated in this way.

We assume that the dependence of the loop on each pair of parameters τ_i is nondegenerate only in an interval around one of the hands (that is on the support of the corresponding f_a) and that these intervals do not overlap. In this way we keep the smearing of each hand separate from the others, and the regularized loop appears as one which is 'fattened' in a neighborhood of each hand. Note that, in the classical theory, it is not necessary to smear each hand independently. However, this will be essential in the discussion of the quantum theory in the next chapter. We will return to this point then.

5 Properties of T variables

It is a standard result in the theory of connections that from the holonomies of a connection around arbitrary loops in the manifold Σ, one can reconstruct that connection upto a (local) gauge transformations. Thus, the holonomies contain the entire "gauge invariant information" that a connection has. Our configuration variables are the traces of holonomies, $T^0[\gamma](A)$. One might therefore expect these variables to provide an (overcomplete) coordinatization of the "gauge invariant configuration space", i.e., of the quotient, $\hat{C}$, of the space C of connections by the local gauge group. This expectation is essentially correct. In fact, had the gauge group been $SO(3)$, the traces of holonomies, in turn, would have enables us to reconstruct the holonomies –and therefore the connection upto gauge– rigorously. In the present case, the gauge group is $SL(2,\mathbb{C})$ rather than $SO(3)$ and we need to be more careful. Now, some information *is* in general lost in taking the traces. Fortunately, however, the failure occurs only on a "set of measure zero" on $\hat{C}$. More precisely, the gradients of $T^0[\gamma](A)$ do span the full cotangent space at points of $\hat{C}$ almost everywhere.[7] We shall take this result to mean that the T^0 provide us with "sufficiently many" configuration variables. There is, of course, a great amount of redundancy in this choice since the gradients are far from being linearly independent; they "overspan" the cotangent space of $\hat{C}$. Thus, the choice we are making is somewhere "between" choosing for configuration variables *arbitrary* functions of connections and only those functions which are linear in connections.

What is the situation with respect to momenta T^1? Recall that $T^1[\gamma, v]$ can be "constructed" from $T^0[\gamma]$ using the pre-Poisson structures introduced at the end of section 2. Now, since the gradients of $T^0[\gamma]$ (over)span the cotangent space of $\hat{C}$ almost everywhere, and since the collection of the pre-Poisson structures as a whole is non-degenerate, the Hamiltonian vector fields ${}^{v}X_{T^0[\gamma]}$ constructed from them (over)span the tangent space of $\hat{C}$ almost everywhere. Hence, the $T^1[\gamma, v]$, obtained by "contracting" these Hamiltonian vector fields by the momenta $\tilde{\sigma}^a{}_A{}^B$ provide us with an overcomplete set of momentum variables on the gauge invariant phase space. Therefore, the complex vector space generated by $(T^0[\gamma], T^1[\gamma, v])$ can indeed be used as the space S of elementary variables in the quantization program. (Consequently, it would suffice and perhaps even be better to use $T^1[\gamma, v]$ in place of $T^a[\gamma](s)$ in the quantization procedure. In these notes, for simplicity, I have continued to use the initial choice, $T^a[\gamma](s)$, made by Rovelli and Smolin.)

---

7 Details will appear elsewhere. The overall situation is similar to that in 2+1 gravity discussed in chapter 17. As in the 2+1 case and in the in the third example discussed in section 10.5, the failure of the T^0 to provide us with a coordinatization of $\hat{C}$ *everywhere* is probably a signal that a superselection occurs in the quantum theory.

Recall, however, from chapter 10 that if the elementary classical variables are over-complete, it is important to spell out the algebraic relations between them because these relations have to be taken over to the quantum theory. Let us therefore collect the algebraic relations between the T^0 and the T^1 variables.

The $SL(2, C)$ algebra in which the self-dual connection assumes values is characterized by the following algebraic identity which plays an important role in the loop formalism

$$\delta_A^B \delta_C^D + \epsilon_{AC} \epsilon^{DB} = \delta_A^D \delta_C^B . \tag{24}$$

By inserting this identity between the four legs of the loops arriving at one intersection we obtain the identity expressed in figure 7.

Figure 7: $\delta_A^B \delta_C^D + \epsilon_{AC} \epsilon^{DB} = \delta_A^D \delta_C^B$.

For instance, by using this identity we have the relation

$$T^0[\alpha \# \beta] + T^0[\alpha \# \beta^{-1}] = T^0[\alpha] T^0[\beta], \tag{25a}$$

which is graphically expressed in figure 8.

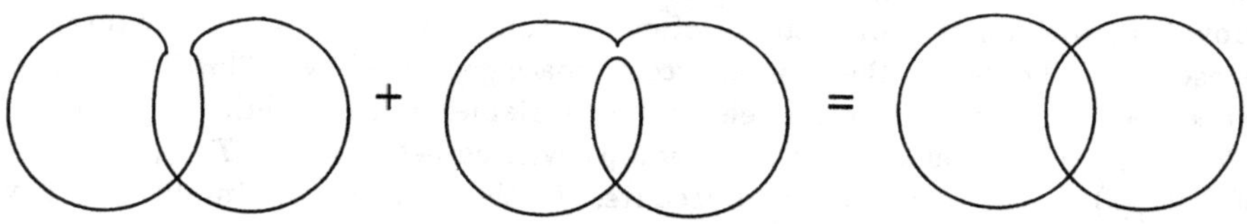

Figure 8: The fundamental two spinor identity as a condition on T^0

Therefore we have another interpretation of the two symbols $><$ and $^{\vee}_{\wedge}$. They refer respectively to $\delta_A^B \delta_C^D$ and $\epsilon_{AC} \epsilon^{BD}$. Their difference $\delta_A^D \delta_C^B$ represents the crossing of the two loops and may be represented with the symbol $\times$. In this spirit we will also use $(\alpha\#\beta)^{\times} = \alpha \cup \beta$. Note that in the notation of figure 7 the identity in Eq.24 is expressed in a way that is similar to that used in Penrose's spin network formalism [6].

The T variables also satisfy a second identity related to retracing over portions of a loop. If α is a loop and η is a segment of a loop, as in Figure 9,

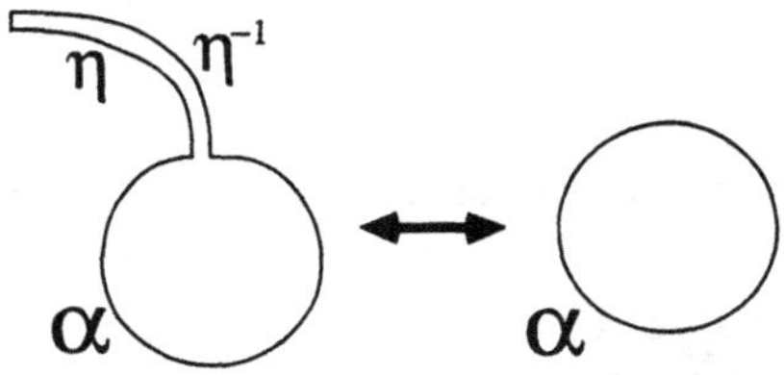

Figure 9: $\alpha\#\eta\#\eta^{-1}$

then we have,

$$T^0[\alpha\#\eta\#\eta^{-1}] = T^0[\alpha] \tag{26a}$$

Eqs.25 and 26 imply the following identity, which we will need later. Let α and β be any two loops, and let η be a segment of a curve joining some point of α to some point of β. Then we have,

$$T^0[\alpha]T^0[\beta] = T^0[\alpha\#\eta\#\beta\#\eta^{-1}] - T^0[\alpha\#\eta\#\beta^{-1}\#\eta^{-1}] \tag{27}$$

We will call loops of the form $\alpha\#\eta\#\beta\#\eta^{-1}$ "eyeglass" loops.

Another nontrivial identity following from Eqs.25 and 26 is the following. Let α be any smooth loop made of segments α_1 and α_2 which are joined smoothly at points p_1 and p_2. Then we have

$$T^0[\alpha] = T^0[\alpha_1\#\eta]T^0[\alpha_2\#\eta^{-1}] + T^0[\alpha_1\#\eta\#\alpha_2^{-1}\#\eta], \tag{28}$$

where η is a segment of a curve joining the points p_1 and p_2. This case is completely analogous to the first case, Eq.27, and will also be referred to as an eyeglass loop. One can show that all non-trivial identities which follow from a combination of Eqs.25 and 26 are of one of these two types.

The identities of Eqs.25 and 26 allow a complete characterization of T^0. In fact it is possible to argue [7], that given any function on the space of unparametrized loops which satisfies these identities, there should exist an $SL(2,C)$ connection such that the given function is the holonomy of that connection. Thus, we have exhausted the algebraic relations between the T^0.

Let us next consider the identities satisfied by the T^1-variables. Perhaps the simplest way to obtain them is to use the "derivation" of the T^1 variables from the T^0, presented at the end of section 2. Then, Eqs.25a and 26a imply the following relations:

$$T^1[\alpha\#\beta, v] + T^1[\alpha\#\beta^{-1}, v] = T^0[\alpha]T^1[\beta, v] + T^1[\alpha, v]T^0[\beta] \qquad (25b)$$
$$T^1[\alpha\#\eta\#\eta^{-1}, v] = T^1[\alpha, v], \qquad (26b)$$

for all vector fields v^a. The identities contained in the subsequent equations, 27 and 18, also go over to the T^1 variables in an obvious way. This concludes the discussion of the algebraic relations between the elementary variables.

Let us now turn to the last part of this subsection and show how the variables of physical interest can be expressed directly in terms of the T^n. It turns out that most of the familiar, local variables of general relativity lie at the "border" of the T set: they can be obtained as the limit of certain linear combinations of Ts, as the loop shrinks down to a point.

Let us begin with the three-metric. It is related to the T^2 variables in the following way. Consider a loop $\gamma(x)$ which starts and ends at the point x and the sequence $\gamma^\delta(x)$ obtained by shrinking $\gamma(x)$ down to the point x: $\gamma^1(x) \equiv \gamma(x)$ and $\gamma^0(x) \equiv 0$. Then we have that

$$\tilde{\tilde{q}}^{ab}(x) = -\frac{1}{\sqrt{2}} \lim_{\delta \to 0} T^{ab}[\gamma^\delta(x)](s_1, s_2). \qquad (29)$$

In the limit $\delta \to 0$ the Us in this expression go to the identity, so that only the $\tilde{\sigma}^a(x)$'s survive in the trace.

The second local variable that will be of interest in the next chapter is the scalar constraint. It can be defined in terms of T^2s in a similar way. To define it we consider a fixed coordinate system in the neighborhood of a point x. Given these coordinates let us define a coordinate circle of radius δ beginning at x which lies in the a-b coordinate plane. We call this $\gamma^\delta_{ab}(x)$. Then we have

$$U_{\gamma^\delta_{ab}(x)}(s) = \mathbf{I} + \delta^2 F_{ab}(x) + o(\delta^2). \qquad (30)$$

In order to get the scalar constraint we have to start from a combination of T^2's such that in the limit the first term that contains the identity cancels. If we take

$$S^\delta(x) = T^{[ab]}[\gamma^\delta_{ab}(x)](\delta^2, 2\pi), \qquad (31)$$

it follows that the scalar constraint $S(\vec{x})$ is given by:

$$S(x) = \lim_{\delta \to 0} \frac{1}{\delta^2} C^\delta(x). \tag{32}$$

To see this, note that

$$C^\delta(x) = \sqrt{2}\,\mathrm{tr}[\tilde{\sigma}^{[a}(\gamma(\delta^2))U_\gamma(\delta^2, 2\pi)\tilde{\sigma}^{b]}(\gamma(2\pi))U_\gamma(2\pi, \delta^2)], \tag{33}$$

where, for simplicity we have used γ for $\gamma^\delta_{ab}(x)$. Now, we can neglect the difference between $\tilde{\sigma}^(x) = \tilde{\sigma}^a(\gamma(2\pi))$ and $\tilde{\sigma}^a(\gamma(\delta^2))$ since it is of order $O(\delta^3)$, and for the same reason we can substitute $U_\gamma(2\pi, \delta^2)$ by identity and $U_\gamma(\delta^2, 2\pi)$ by $U_\gamma(0)$. Then we have

$$C^\delta(x) = \sqrt{2}\,\mathrm{tr}[\tilde{\sigma}^{[a}(x)\tilde{\sigma}^{b]}(x)\, U_{\gamma^\delta_{ab}(x)}(0)] + o(\delta^2). \tag{34}$$

Let us now expand U as in Eq.30. The leading term is zero because of the antisymmetrization, whence we have:

$$C^\delta(x) = \sqrt{2}\delta^2\,\mathrm{tr}\,[\tilde{\sigma}^a(x)\tilde{\sigma}^b(x)F_{ab}(x)] + o(\delta^2), \tag{35}$$

which implies Eq.32.

One can similarly analyze the expressions of other constraints. However, We will not need them in the next chapter. Rather we will need to know the transformation properties of the Ts under the transformation generated by these constraints in the phase space. Since these constraints generate the kinematical symmetry group of spinorial general relativity, it will turn out that, in quantum theory, to enforce the transformation properties of $\hat{T}$ operators is essentially equivalent to imposing these constraints. Let us begin with the Gauss constraint. Under the $SU(2)$ gauge transformations it generates, the T variables are invariant. Under a diffeomorphism ϕ generated by the remaining constraint, they transform as scalars as far as their dependence on the loop γ is concerned and as vector densities of weight 1 at the points $\gamma(s_i)$ with index a_i. That is in the new coordinate system $x'^a = \phi^a(x)$ they are given by

$$\begin{aligned} T'^{a_1 \dots a_n}[\gamma](s_1 \dots s_n) &= J^{-1}(\gamma(s_1)) \dots J^{-1}(\gamma(s_n)) \\ &\frac{\partial \phi^{a_1}(\gamma(s_1))}{\partial x^{b_1}} \dots \frac{\partial \phi^{a_n}(\gamma(s_n))}{\partial x^{b_n}}\, T^{b_1 \dots b_n}[\phi \cdot \gamma](s_1 \dots s_n), \end{aligned} \tag{36}$$

where $(\phi \cdot \gamma)^a(s) = \phi^a(\gamma(s))$ and $J(x)$ is the Jacobian of the coordinate transformation ϕ.

Let us conclude with a summary of the results obtained in this chapter. We have shown that general relativity may be formulated in terms of the T variables, defined in Eqs.3 and 4 for the small $\mathcal{T}$ algebra and Eqs.19 and 20 for the full $\mathcal{T}$ algebra. The symplectic structure is given by the Poisson bracket structure of the small $\mathcal{T}$ as coded in Eqs.8, 11 and 16. The Hamiltonian constraint is given by Eq.32. The transformation properties of the T variables under the gauge and diffeomorphism constraints are well defined. The spatial metric can be expressed in terms of the T variables using Eq.29.

References

[1] C.Rovelli and L.Smolin, Phys. Rev. Lett. **61**, 1155 (1988).

[2] C.Rovelli and L.Smolin, Nucl. Phys. **B331**, 80 (1990).

[3] T.Jacobson and L.Smolin, Nucl. Phys. **B299**, 295 (1988).

[4] M.P.Blencowe, Nucl. Phys. **B341**, 213 (1990).

[5] R. Gambini, "Loop space representation of quantum general relativity and the group of loops" (Montevideo pre-print, 1990).

[6] R.Penrose, "Application of negative dimensional tensors", in *Combinatorial mathematics and its applications*, D.J.A.Welch (Ed.), Academic Press, London (1971).

[7] F.Gliozzi and M.A.Virasoro, Nucl. Phys. **B164**, 141 (1980).

16 LOOP REPRESENTATION: QUANTUM THEORY

1 Introduction

Let us now carry out the first five steps of the quantization program of chapter 10 using as elementary variables the elements of the small T-algebra constructed in the last chapter. (Many of the notions used in this chapter were introduced in the last chapter and will not be recalled again.) This quantization will lea us to the loop representation.

How is this representation related to the connection representation used in chapters 11-13? Already in their initial work [1,2] Rovelli and Smolin introduced a formal transform to pass from the connection to the loop picture. This transform is useful in many ways: it provides the much needed intuition for the various definitions and constructions in the loop representation. Furthermore,as we saw in chapter 14, for Maxwell fields the transform exists rigorously and can in fact be used to *introduce* the loop representation. We will see in the next chapter that the same is true in 2+1 general relativity as well. In the 3+1 theory, however, the transform has remained a formal device since we do not know if the measure on the space of connections needed in its definition exists. Therefore, in these notes, I will not use the transform. Rather, we will arrive at the loop representation directly in the process of "quantizing the classical T-algebra."

In the next section, we use $T^0[\gamma]$ and $T^a[\gamma](s)$ as the elementary variables and construct the quantum algebra by imposing the appropriate commutation and anti-commutation relations. We then find a representation of it on a linear vector space V. As expected from the discussion of the loop representation for the Maxwell field in chapter 14, the space V consists of suitable functionals on the loop space of the "spatial" 3-manifold Σ. In section 3, we write out the quantum constraints in terms of $\hat{T}$-operators and find a large class of physical states. More precisely, we will be able to present the *general* solution to the diffeomorphism constraint of general relativity and an infinite dimensional space of solutions to the Hamiltonian constraint. These solutions strongly suggest that quantum geometry at the Planck length is governed by discrete structures and is therefore *very* different from the continuum picture underlying perturbative treatments. These discrete structures have an eerie similarity with the ones, such as spin networks, that are governed by combinatorial techniques.

Although these developments represent significant progress, important mathematical issues still remain. The space of states found so far is incomplete; issues of regularization are not fully settled; and, the problem of finding the correct inner-product is completely open. To gain further insight into these questions, in the next chapter we will discuss a simpler system –2+1-dimensional general relativity– where they can be fully analysed.

2 Quantization ignoring constraints

We now wish to use the machinery constructed in the last chapter to carry out the quantization program of chapter 10. In this section, we shall complete the first four steps.

2.1 Algebra of quantum operators

Using the small $\mathcal{T}$ algebra as the space $\mathcal{S}$ of elementary classical variables in the quantization program, it is straightforward to construct the algebra $\mathcal{A}$ of quantum operators. Elementary quantum operators are now linear combinations of $\hat{T}^0[\gamma]$ and $\hat{T}^1[\gamma](s)$. Consider the free algebra they generate and impose on it commutation relations via the Dirac prescription: $[\hat{f}, \hat{g}] = i\hbar\widehat{\{f,g\}}$ for all elementary variables f and g. Thus, we require:

$$\begin{aligned}
[\hat{T}^0[\alpha],\, \hat{T}^0[\beta]] &= 0 \\
[\hat{T}^0[\alpha],\, \hat{T}^a[\beta](s)] &= -\hbar \sum_{\diamond} (-1)^{|\diamond|}\, \Delta^a[\beta,\alpha](s) \hat{T}^0[(\beta\#\alpha)^\diamond)] \\
[\hat{T}^a[\gamma](s),\, \hat{T}^b[\eta](t)] &= \hbar \sum_{\diamond} (-1)^{|\diamond|}\, \Delta^a[\gamma,\eta](s)\, \hat{T}^b[(\eta\#_s\gamma)^\diamond](u(t)) \\
&\quad -\hbar \sum_{\diamond} (-1)^{|\diamond|} \Delta^b[\eta,\gamma](t)\, \hat{T}^a[(\gamma\#_t\eta)^\diamond](u(s)),
\end{aligned} \tag{1a}$$

and the algebraic relations:

$$\begin{aligned}\hat{T}^0[\alpha]\hat{T}^0[\beta] &= \hat{T}^0[\alpha\#\beta] + \hat{T}^0[\alpha\#\beta^{-1}]\\ \hat{T}^0[\alpha\circ\eta\circ\eta^{-1}] &= \hat{T}^0[\alpha], \quad \text{and} \quad \hat{T}^0[0] = 2\\ \{\hat{T}^1[\alpha,v],\ \hat{T}^0[\beta]\} &= 2\hat{T}^1[\alpha\#\beta,v] + 2\hat{T}^1[\alpha\#\beta^{-1},v]\\ \hat{T}^1[\alpha\circ\eta\circ\eta^{-1},v] &= \hat{T}^1[\alpha,v], \quad \text{and} \quad \hat{T}^1[0,v] = 0\,,\end{aligned} \tag{1b}$$

for all vector fields v^a on the "spatial' 3-manifold Σ, where $\{\ ,\ \}$ denotes the anti-commutator. (The last relation of $\hat{T}^0$ and $\hat{T}^1$ need to be imposed if we allow zero loops, i.e., loops of the type $\gamma(s) = p \in \Sigma$.) Unlike in the connection representation, non-trivial anti-commutation relations arise here because the set of elementary classical variables is now overcomplete. Denote the resulting algebra by $\mathcal{A}$. The next step is to introduce the $\star$-relations on $\mathcal{A}$ and construct $\mathcal{A}^{(\star)}$. However, since we will not discuss the issue of selecting the inner-product, these relations will not play any role in the subsequent discussion. Through out this chapter, therefore, we will work just with the associative algebra $\mathcal{A}$.

2.2 Representation of the CCR

We now construct a vector space $\overline{V}$ and represent the $\hat{T}$ operators by concrete linear operators on $\overline{V}$ such that the commutation relations of Eq.1a are faithfully represented. In the next sub-section, we will show that one must restrict oneself to a suitable sub-space V of $\overline{V}$, in order that the anti-commutation relations of Eq.1b can also be captured faithfully. Thus, it is V that will carry a linear representation of the algebra $\mathcal{A}$ constructed above.

Let us begin by introducing a new notion that is needed to construct $\overline{V}$. A *multiple loop* is a set containing a finite number of (continuous, piecewise smooth) loops. We will use the notation $\{\eta\}$ to denote a multiple loop formed by single loops $\eta_1, ...\eta_n$ and assume that the multiple loop has a unique parameter, $s \in (0, 2\pi n)$, which can "jump" from one point to another. Thus, we can use the notation $\int_{\{\eta\}} = \int_{\eta_1} + ... + \int_{\eta_n}$. Let $\mathcal{M}$ be the space of multiple loops. Then, $\mathcal{M}$ is a graded space; the direct sum of the space of single loops, of two loops, of three loops and so on. We assume that $\mathcal{M}$ also carries in it the zero loop, represented just by a point.

We are now ready to define the vector space $\overline{V}$: it will be the space of complex-valued functionals on $\mathcal{M}$. Let us denote the elements of $\overline{V}$ by $\Psi[\{\eta\}]$. Any $\Psi[\{\eta\}]$ may be expressed in terms of its "components" $\Psi_n[\eta_1, ..., \eta_n]$ on the subspaces of $\mathcal{M}$ of the sets composed of n loops. Thus $\Psi[\{\eta\}]$ has the form

$$\Psi = \{\Psi_0,\ \Psi_1[\alpha],\ \Psi_2[\beta_1,\beta_2],\ \Psi_3[\gamma_1,\gamma_2,\gamma_3], \dots\} \tag{2}$$

Note that these amplitudes depend on *unordered* sets of loops. Thus, in Eq.2, the

Ψ_n are symmetric in their entries.

The quantum operator $\hat{T}^0[\gamma]$ corresponding to the variable $T^0[\gamma]$ will be represented on $\overline{V}$ as:

$$\hat{T}^0[\gamma]\,\Psi[\{\eta\}] := \Psi[\gamma \cup \{\eta\}], \tag{3}$$

where $\gamma \cup \{\eta\}$ is the collection of loops which is the union of γ and $\{\eta\}$. It follows immediately from the definition that if the initial state Ψ has support only on sets of n loops then $\hat{T}^0[\gamma]\Psi$ defined by Eq.3 has support only on sets of $n-1$ loops; $\hat{T}^0[\gamma]$ is again a "lowering operator". Note that the definition of the operator is essentially the same as the one we used in the loop representation of Maxwell theory in chapter 14. (Indeed, it would have been exactly the same had we represented quantum states by functions of multi-loops also in the Maxwell case.) The interpretation of the operator in gravity is, of course, different from that in the Maxwell case.

Let us now define the operator corresponding to the variable T^1:

$$\hat{T}^a[\gamma](s)\,\Psi[\{\eta\}] := -\hbar \sum_{\diamond} (-1)^{|\diamond|}\,\Delta^a[\gamma,\{\eta\}](s)\;\Psi[(\gamma\#\{\eta\})^\diamond]. \tag{4}$$

Note that this action can be defined in terms of the elementary grasp operations of figure 4 of the last chapter. Here, however, the grasp is on the argument of the loop functional and the coefficient $(-1)^{|\diamond|}\,\Delta^a[\gamma,\{\eta\}](s)$ multiplies the value of the loop functional. In the quantum context we will use the expression "grasp" in this sense. Put differently, we express the content of Eq.4 by saying that the action of the quantum operators corresponding to the loop γ with a hand is given by the grasp of the hand on the loop functional.

Let us compute the commutator algebra of these operators. First of all it is straightforward to show that the $\hat{T}^0$ operator commutes with itself. We have,

$$\hat{T}^0[\alpha]\hat{T}^0[\beta]\Psi[\{\eta\}] = \Psi[\alpha \cup \beta \cup \{\eta\}] = \hat{T}^0[\beta]\hat{T}^0[\alpha]\Psi[\{\eta\}], \tag{5}$$

because the action of $\cup$ is commutative. The commutator of $\hat{T}^0$ and $\hat{T}^1$ is given by

$$\begin{aligned}\left[\hat{T}^0[\alpha],\,\hat{T}^a[\beta](s)\right]\Psi[\{\eta\}] &= \hbar\sum_{\diamond}(-1)^{|\diamond|}\,\Delta^a[\beta,\{\eta\}](s)\;\Psi[\alpha\cup(\beta\#\{\eta\})^\diamond]\\ &\quad -\hbar\sum_{\diamond}(-1)^{|\diamond|}\Delta^a[\beta,\alpha\cup\{\eta\}](s)\;\Psi[(\beta\#(\alpha\cup\{\eta\}))^\diamond]\end{aligned} \tag{6}$$

To simplify this expression note first that

$$\Delta^a[\beta,\alpha\cup\eta](s) = \Delta^a[\beta,\alpha](s) + \Delta^a[\beta,\eta](s)\,, \tag{7}$$

where the first (second) term is different from zero only if the hand of β sees α

(respectively, η). Thus the second term in the commutator may be decomposed as

$$\hbar \sum_{\diamond} (-1)^{|\diamond|} \Delta^a[\beta, \alpha \cup \{\eta\}](s)\ \Psi[(\beta \# (\alpha \cup \{\eta\}))^\diamond]$$

$$= \hbar \sum_{\diamond} (-1)^{|\diamond|} \Delta^a[\beta, \alpha](s) \Psi[(\beta\#\alpha)^\diamond \cup \{\eta\}] + (-1)^{|\diamond|} \Delta^a[\beta, \{\eta\}](s) \Psi[\alpha \cup (\beta \# \{\eta\})^\diamond] \tag{8}$$

Now, the last term in Eq.8 cancels with the first term of the commutator and we obtain the final result

$$[\hat{T}^0[\alpha], \hat{T}^a[\beta]](s) = -\hbar \sum_{\diamond} (-1)^{|\diamond|} \Delta^a[\beta, \alpha](s) \hat{T}^0[(\beta\#\alpha)^\diamond] \tag{9}$$

Thus, the difference between first adding a loop α to the argument of the functional and then grasping it by an operator, and first grasping it and then adding the loop α is the same as adding the grasp of α. This completes the verification of the first two of the three commutation relations. It only remains to check the third. Using the same techniques, we have:

$$\begin{aligned}[\hat{T}^a[\alpha](s), \hat{T}^b[\beta](t)] = \hbar \sum_{\diamond} (-1)^{|\diamond|} \Delta^a[\alpha, \beta](s)\ \hat{T}^b[(\beta \#_s \alpha)^\diamond](u(t)) \\ -\hbar \sum_{\diamond} (-1)^{|\diamond|} \Delta^b[\alpha, \beta](t)\ \hat{T}^a[(\alpha \#_t \beta)^\diamond](u(s)).\end{aligned} \tag{10}$$

These commutators faithfully represent the commutation relations 1*a* on $\mathcal{A}$ which themselves reflect the Poisson brackets between the classical T^0 and T^1 variables we found in the last chapter. Thus, we have obtained a linear representation of the canonical commutation relations. There is still the question of whether the anti-commutation and other relations between the elementary operators are correctly represented. In fact, as matters stand, they are not. To capture them, one must restrict oneself to an appropriate subspace V of $\overline{V}$. Thus, it is V that will carry a linear representation of the algebra $\mathcal{A}$.

2.3 Representation of $\mathcal{A}$

Let us first recall the relations satisfied by the elementary classical variables. The first set of these relations follows from the invariance of the holonomy under reparametrizations. We may implement these relations in the quantum theory simply by demanding that the loop functionals $\Psi[\{\eta\}]$ in V be *invariant under reparametrizations*, i.e. they depend only on the unparametrized multiple loops. It is straightforward to verify that Eqs.2-10 are consistent with this restriction.

The second set consists of the anti-commutation relations, Eq.1*b*. The idea is to realize these relations faithfully by imposing directly a corresponding condition on the states. Let us begin with the first of these equations. The key point in its implementation is that although the identity itself is a *nonlinear* condition on the loop operators $\hat{T}^0[\gamma]$, it can be realized by imposing a *linear* restriction on states which are functionals of *multiple* loops. We require, for any pair of loops γ and η and any intersection point involving them, that elements $\Psi[\{\eta\}]$ of V should satisfy:

$$\Psi[\gamma\#\eta,...] + \Psi[\gamma^{-1}\#\eta,...] - \Psi[\gamma\cup\eta,...] = 0, \tag{11}$$

where ... refers to any other loops that may be present in the argument of the loop functional. Using the $\times$ symbol introduced in section 15.3, we can rewrite this condition on permissible states as:

$$\Psi[(\gamma\#\eta)^{\times}] - \Psi[(\gamma\#\eta)^{><}] + \Psi[(\gamma\#\eta)^{\overset{\vee}{\wedge}}] = 0 \tag{12}$$

It is straightforward to demonstrate that on the space of the states that satisfy Eq.12, the loop operators of Eq.3 satisfy the first anti-commutation relation of Eq.1*b*. To ensure that the second anti-commutation relation is also satisfied, we just restrict ourselves to wave functionals satisfying:

$$\Psi[\alpha\#\eta\#\eta^{-1},...] = \Psi[\alpha,...]. \tag{13}$$

Finally, the last of the anti-commutation relations involving $\hat{T}^0[\gamma]$ requires that our wave functions satisfy:

$$\Psi[0\#\{\eta\}] = 2\Psi[\{\eta\}]. \tag{14}$$

Finally, we note that Eqs.12 and 13 imply that the wave functionals must satisfy additional conditions which are the counterparts of Eqs.15.27 and 15.28. For example, the states automatically fulfill the condition: $\Psi[\alpha\#\eta\#\beta\#\eta^{-1}] - \Psi[\alpha\#\eta\#\beta^{-1}\#\eta^{-1}] = \Psi[\alpha\cup\beta]$, for any two loops α and β and any segment η joining them.

Before going on to consider the anti-commutation relations involving the $\hat{T}^1[\gamma, v]$, let us note an important consequence of of the restrictions on the wave functionals we have obtained so far. Eqs.12 and 13 imply that *any state Ψ in V is completely fixed by its value on single loops*: its value on any multiple loop can be determined by its value on suitable "eyeglass (single) loops" using these identities. Thus, as in the Maxwell theory, we could have avoided the introduction of multiple loops. However, in practice, it is generally more convenient to work with the multiple loop states since the action of the $\hat{T}$ operators is easier to express in terms of them.

Finally, let us consider the last three anti-commutation relations involving the $\hat{T}^1[\gamma, v]$ operators. It turns out that on those states which satisfy the restrictions contained in Eqs.12, 13 and 14, these anti-commutation relations are identically satisfied. Hence there are no further restrictions. Thus, V is the sub-space of $\overline{V}$ on which Eqs. 12-14 hold. It carries a representation of the algebra $\mathcal{A}$.

2.4 Higher order operators

It is convenient, for the analysis of constraints, to represent on V also the higher order operators $\hat{T}^n$. In order to carry out this task, we need to generalize our notation on the breaking and rejoining of loops at intersections to the case in which the breaking and rejoining happens *simultaneously* at two or more intersections. This is illustrated by figure 10.

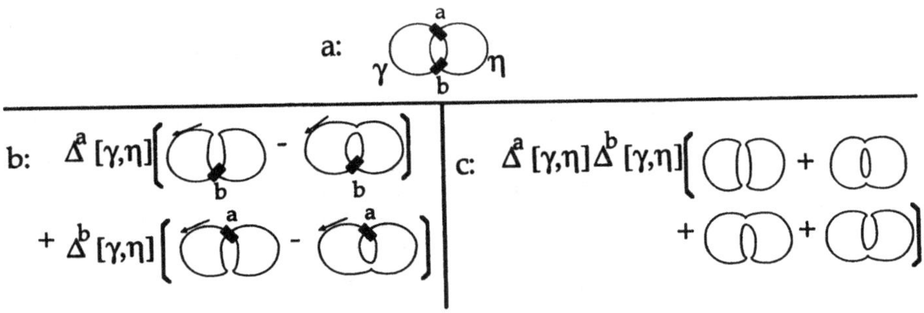

Figure 10: Two hands grasp, one at a time and simultaneously

In figure 10*a* we have two loops, one with two hands and the other with no hands. The loops intersect twice, and there is a hand at each intersection. In figure

10*b* we draw the result of grasping with the hands *one at a time* (as in the classical Poisson bracket), while figure 10c shows the result of grasping with the two hands *simultaneously*. In general the grasp of n hands one at a time is the sum of the n grasps of the n hands; i.e. it is the sum of the $2n$ terms obtained by breaking and rejoining at each intersection one at a time. The simultaneous grasp of n hands is the sum of 2^n terms obtained by breaking and rejoining *at all the hands* at the same time.

Given a loop α with n hands at the points $s_1 ... s_n$ and a loop β which intersects α at each of these points we denote by

$$(\alpha \#_{s_1 ... s_n} \beta)^{(\diamond_1 ... \diamond_n)}$$

the loop obtained by substituting the n intersections with the n alternatives $(\diamond_1 ... \diamond_n)$. We do not need to specify how the resulting loop is parametrized, since the parametrization will not play any role in the future. We do maintain the notation, introduced in the previous section, that $u'(s)$ and $u(t)$ indicate the parameters of the composed loop that correspond to the points $\alpha(s)$ and $\beta(t)$. Our notation for single loops generalizes in a straightforward way to multiple loops. Note incidentally that the result of a simultaneous grasp will often be a multiple loop even when the two original loops are single loops.

We can now introduce the quantum loop operators with more than one hand; the analogs of the classical variables T^n. Let us define

$$\begin{aligned} \hat{T}^{a_1 ... a_n}[\gamma](s_1 ... s_n)\Psi[\{\eta\}] := -\hbar^n \sum_{\diamond_1} ... \sum_{\diamond_n} (-1)^r \\ \Delta^{a_1}[\gamma, \{\eta\}](s_1) ... \Delta^{a_n}[\gamma, \{\eta\}](s_n)\ \Psi[(\gamma \#_{s_1 ... s_n} \{\eta\})^{(\diamond_1 ... \diamond_n)}]. \end{aligned} \quad (15)$$

where the value of r –which determines the sign of each term in the sum– depends on the orientation of the two loops as follows. Each of the two original loops has (or can be given) an orientation. Each term of the grasp is a (multiple) loop made out of segments of the original loops, whence each of these segments has an orientation. In order to assign an orientation consistently to each of the loops some of the orientations of the segments may have to be reversed. r –or more properly $r((\gamma \#_{s_1 ... s_n} \{\eta\})^{(\diamond_1 ... \diamond_n)})$– is defined as the number of these segments whose orientations must be reversed.[1]

1 In case of grasps of a single hand we have $r((\gamma \# \{\eta\})^{\diamond}) = |\diamond|$, therefore the previous way to keep track of the sign is a particular case of the present one.

Thus the action of a many handed loop quantum operator is proportional to the simultaneous grasp of all its hands. That is, it is given by the sum of all the 2^n possible ways in which one can simultaneously substitute all of the intersections at which there is a hand with one of the two ways to join the legs of the two intersecting loops to each other. Note that the result of the grasp is zero unless *all* the hands of γ see $\{\eta\}$.

It turns out, actually, that these higher order operators also have rather simple commutation relations. As one might expect of variables which are of quadratic and higher order in momenta, the rule $[\hat{F}, \hat{G}] = i\hbar\widehat{\{F, G\}}$ no longer applies. Nonetheless, the higher order $\hat{T}$ are again closed under the commutator bracket [1,2]. We do not write out this algebra because we will not need the commutators in this chapter.

2.5 Smearing in the quantum theory

In sections 15.3 and 15.4 we showed that the distributional singularities contained in the Poisson algebra of the loop variables may be eliminated by smearing over a space of test functions. This may also be done for the quantum algebra. The situation is more complicated because the simultaneous grasps produce terms containing products of delta functions. These are prevented from coinciding by the requirement that the loop parameters of the hands must be unequal. To construct a completely non-singular algebra we must be careful to define suitable spaces of congruences and test functions in such a way that these simultaneous delta function singularities are always completely eliminated by the integrations over the test functions.

By smearing the $\hat{T}^n$ operators in the same way as the classical variables, the delta distributions in the Δ functions only appear inside the corresponding integrals; however an inspection of the integrals shows that if two of the integration regions overlap some degeneracy may appear such that some of the delta distributions which result are not compensated by a corresponding integration. In order to avoid this problem we have to assume that the integration regions corresponding to different hands on the same loop do not overlap. This requirement can be satisfied as long as there is an open interval in the loop parameter separating each insertion of $\tilde{\sigma}^a$ in $\hat{T}^n$. We defined the quantum loop algebra so that this would be the case.

Intuitively we want to substitute, for a loop with n hands, one which has "fattenings" in n regions, then put each hand in a region and require that the regions not overlap. Each fattening is described by a (two dimensional) congruence of loops and each congruence depends on a different two dimensional parameter τ_i. We obtain this by considering only n-dimensional congruences in which the dependence on the i-th parameter τ_i is trivial everywhere except on an interval γ^i of the loop parameter. We require that the intervals not overlap and that in the interval the loop actually spans a congruence as the parameter varies.

More precisely let us consider mappings $\tilde{\gamma}^n$ from $R^{2n} \times S^1$ into Σ, which we denote $\gamma_{\tau_1...\tau_n}(s)$, with the following characteristics. We call γ^i the interval of S^1 in which the dependence from τ^i is non trivial. We assume that each γ^i is connected; and disconnected from the others γ^j, and that the restriction of $\tilde{\gamma}^n$ to any γ^i is injective. In terms of these objects it is now easy to define the smeared quantum operators. What we have to do is simply to make sure that any hand $\tilde{\tau}^a(\gamma(s))$ stays in its interval γ^i. We obtain this by inserting in the definition of the smeared operator a universal smooth function $R^n_{\tilde{\gamma}}(s_1...s_n)$ which is defined to be zero unless each s_i is in the corresponding γ^i. Finally the definition of the smeared quantum loop operator is

$$\begin{aligned} \hat{T}[\tilde{\gamma}^n](f) = \int d^2\tau_1...d^2\tau_n \int ds_1...ds_n R^n_{\tilde{\gamma}}(s_1...s_n) \\ f_{a_1..a_n}(\gamma_{\tau_1...\tau_n}(s_1)...\gamma_{\tau_1...\tau_n}(s_n)) \;\; \hat{T}^{a_1...a_n}[\gamma_{\tau_1...\tau_n}](s_1...s_n) \end{aligned} \tag{16}$$

Note that the overlap of two integration regions is still allowed if the two fattenings of the loop belong to the two different arms of a self intersection of the loop. An inspection of the commutators shows that this case is safe and introduces no additional singularities.

One can now explicitly compute the commutator of two of these operators and verify that it is given by a finite dimensional integral over smeared operators of the same form.

3 Quantum constraints

In this section, we shall investigate the action of the quantum constraints on the space V. We will be able to display the general solution to the diffeomorphism constraint and a large class of solutions to the scalar constraint.

3.1 The diffeomorphism constraint

There is a natural action of the diffeomorphism group of Σ on the loop space

$\mathcal{M}$. For any $\phi \in Diff(\Sigma)$ its action is given by

$$(\phi \cdot \{\gamma\})(s) = \{\phi(\gamma(s))\} \tag{17}$$

This action induces a natural linear representation, which we will call U, of the diffeomorphism group on the space of the loop functionals on $\mathcal{M}$, by

$$U(\phi)\Psi[\{\eta\}] = \Psi[\phi^{-1} \cdot \{\eta\}]. \tag{18a}$$

If ϕ_t is a one parameter group of diffeomorphisms generated by the vector field v on Σ, the generators $D(v)$ of the representation U are given by

$$D(v)\Psi[\{\eta\}] \equiv \frac{d}{dt}U(\phi_t)\Psi[\{\eta\}]|_{t=0}. \tag{18b}$$

The operators $D(v)$ are unbounded and we do not expect them to be defined on the entire space V, but only in some "dense" domain. This domain consists essentially of the loop functionals which are differentiable in the directions tangent to the orbits of the diffeomorphism group in $\mathcal{M}$. (Recall that we did not put any general differentiability requirement on the loop functionals.)

The $D(v)$'s defined in this way satisfy the algebra of the diffeomorphism group

$$[D(v), D(w)] = D([v, w]). \tag{19}$$

We are interested in the commutation relations of the quantum loop operator $\hat{T}$ with the $D(v)$ or, equivalently, in the transformation properties of the loop operators under the representation U. We use the notation

$$\phi \cdot \hat{T}^{a_1 \dots a_n}[\gamma](s_1 \dots s_n) \equiv U(\phi)\, \hat{T}^{a_1 \dots a_n}[\gamma](s_1 \dots s_n)\, U(\phi^{-1}). \tag{20}$$

It is straightforward to show that

$$\phi \cdot \hat{T}^0[\gamma] = \hat{T}^0[\phi \cdot \gamma]. \tag{21}$$

The calculation is:

$$\begin{aligned} \phi \cdot \hat{T}^0[\gamma]\Psi[\eta] &\equiv U(\phi)\, \hat{T}^0[\gamma]\, U(\phi^{-1})\Psi[\eta] = U(\phi)\, \Psi[\phi \cdot (\gamma \cup \eta)] \\ &= \Psi[(\phi \cdot \gamma) \cup \eta] = \hat{T}^0[\phi \cdot \gamma]\, \Psi[\eta], \end{aligned} \tag{22}$$

for all $\Psi[\{\eta\}]$. It is slightly more complicated to show that

$$\begin{aligned} \phi \cdot \hat{T}^{a_1 \dots a_n}[\gamma](s_1 \dots s_n) &= J^{-1}(\gamma(s_1)) \dots J^{-1}(\gamma(s_n)) \times \\ \frac{\partial \phi^{a_1}(\gamma(s_1))}{\partial x^{b_1}} \dots \frac{\partial \phi^{a_n}(\gamma(s_n))}{\partial x^{b_n}}\, &\hat{T}^{b_1 \dots b_n}[\phi \cdot \gamma](s_1 \dots s_n); \end{aligned} \tag{23}$$

(see Eq.15.34). Eqs.21 and 23 show that the quantum operators $\hat{T}$ transform under the representation U of the diffeomorphism group exactly as the corresponding classical variables transform under diffeomorphisms. Since the Poisson brackets of the diffeomorphism constraints generate infinitesimal diffeomorphisms on the classical phase space, it follows that the commutator of the $D(v)$ with all the operators of the theory reproduces exactly the Poisson brackets of the diffeomorphism constraints with the corresponding classical variables. Thus we can identify the $D(v)$ operators with the quantum diffeomorphism constraints, or more precisely, with the quantum operator corresponding to the smeared form, $\int_\Sigma d^3x\ v^a(x)C_a(x)$, of the diffeomorphism constraints.[2]

3.2 Diffeomorphism invariant states

To single out physical states, we now want to find the general solution to

$$D(v)\Psi[\{\eta\}] = 0 \tag{24}$$

Since $D(v)$ is the generator of the action of the diffeomorphism group on the loop space $\mathcal{M}$, Eq.24 is equivalent to the requirement that the state $\Psi[\{\eta\}]$ is constant along the orbits of this action. We denote by $K(\{\eta\})$ the orbit in which the multiple loop $\{\eta\}$ lies. Then, the general solution to Eq.24 is,

$$\Psi[\{\eta\}] = \Psi[K(\{\eta\})] \tag{25}$$

To extract information about the structure of these solutions, let us analyze the orbits of the diffeomorphism group.

First of all, as ϕ sends a multiple loop composed of n loops into a multiple loop composed of the same number of loops, a first characterization of the orbits is given by the number of loops. A second diffeomorphism invariant of the loop is the number of intersections. A third is the number of points of discontinuity of the first derivatives of each loop. Thus, each orbit carries a string of integers that codes these characteristics.

Finally, sets of smooth and nonintersecting loops fall into equivalence classes under the diffeomorphism group which are the knot and link classes of the loops.

2 Blencowe[3] has given an alternate treatment, involving area derivatives, of the diffeomorphism constraint. There, one does not seek to represent the action of the diffeomorphism group on V but rather regularizes the diffeomorphism constraint operator itself and *shows* that this regularized operator in fact has the interpretation of the generator of the diffeomorphisms. Note also that since the Gauss constraint is redundant in the loop representation, the distinction between the vector and the diffeomorphism constraints (see chapter 7) disappears here.

The knot classes of Σ are the equivalence classes of smooth loops in Σ under an operation known as *ambient isotopy* ([4]). It is then a standard theorem in knot theory that the equivalence classes of smooth loops with respect to diffeomorphisms are the same as the equivalence classes under ambient isotopy. The generalization of knot theory to include loops with corners is rather trivial. The generalization to include intersecting loops (or graphs) is less trivial, and has already been considered by some authors. (See, e.g., Smolin's contribution to [5].)

For the case of *multiple* smooth non self-intersecting loops there is, beside the knotting of each single loop, the additional phenomenon of the linking of different loops which is also invariant under diffeomorphisms. The orbits of the diffeomorphisms in the space of these multiple loops are called the *link classes* of the manifold. We use the notation L for the link classes of the manifold. Again, if one allows loops with corners or intersecting loops, one is led to consider generalized link classes. Let us therefore call the orbits in $\mathcal{M}$ of the diffeomorphism group of Σ the *generalized link classes* of Σ.

Functionals $\Psi[\{\eta\}]$ on the loop space that depend only on the generalized link class of the loops will then be solutions to the diffeomorphism constraints provided the condition that the states be constant under the orbits of the diffeomorphisms is compatible with the conditions we imposed on the states space in section 2. The first condition on the states (i.e. elements of V) is *reparametrization invariance*. Since the image of the loop is transformed by a diffeomorphism in a way that is independent of its parametrization it is clear that reparametrizations commute with the action of the diffeomorphisms, and therefore that the orbits of the diffeomorphisms are well defined on the space of unparametrized loops. The second restriction on elements of V is the first of anti-commutation relations, Eq.1*b*. The diffeomorphisms preserve this condition, because it is a topological relation, expressed by the breaking and rejoining at the intersections, and these operations are diffeomorphism invariant. Therefore the condition is well defined on the orbits K. The second anti-commutation relation, which expresses the *retracing identity*, is also preserved under diffeomorphisms, for the same reason. Finally, the last anti-commutation relation is trivially preserved since the zero loop is always mapped to itself under diffeomorphisms.

To summarize, then, the general solution to the diffeomorphism constraint is given by functionals of the form given in Eq.25, (with $K(\{\eta\})$ denoting the generalized link class of η in Σ), which, in addition, satisfy the conditions of Eqs.12, 13 and 14. It is important to stress the fact that the generalized link classes form a *denumerable* set. Thus, the loop space $\mathcal{M}$ is, on the one hand, rich enough that a complete algebra of quantum operators for general relativity can be defined on it and, on the other hand, simple enough that, using it, the complete solution to the diffeomorphism constraints can be given in terms of a denumerable basis. This fundamental discreteness in the description of quantum gravity is an essentially non-perturbative feature. It has a number of attractive facets.

3.3 The scalar constraint

Let us now consider the scalar constraint. Recall from Part II that in terms of the new variables, the constraint is purely quadratic in the momenta $\tilde{\sigma}^a_A{}^B$. In quantum theory, a direct translation of this expression involves products of operator-valued distributions which are ill-defined and need to be regularized. Thus, the constraint operator has to be expressed as the limit of a sequence of regulated operators. Therefore, in order to define the scalar constraint in the quantum theory additional structure must be introduced to allow us to define a suitable regularization procedure. From this point of view the situation is the same as in Minkowskian quantum field theories in which a regularization procedure must be introduced in order to define the Hamiltonian. What is different is that we must introduce a suitable regularization in the absence of a background metric, in a way that does not destroy the diffeomorphism invariance of the theory at least as far as the final results are concerned.

The loop formulation provides a natural solution to this problem. In fact the classical scalar constraint is already defined, in Eqs.15.31 and 15.32, as the limit of the functions $C^\delta(x)$. In the quantization procedure introduced in the last section, each of these objects is represented by a well defined quantum operator,

$$\hat{C}^\delta(x) = \hat{T}^{[ab]}[\gamma^\delta_{ab}(x)](\delta^2, 2\pi) \, . \tag{26}$$

Thus the loop representation gives us directly a natural definition of a ***regulated*** quantum scalar constraint. The idea now is to impose this constraint in quantum theory in the form:

$$\lim_{\delta \to 0} \hat{C}^\delta \; \Psi[\{\eta\}] = 0. \tag{27}$$

In the next sub-section, we will study a class of solutions to this equation.

Before accepting this definition of the quantum scalar constraint, we must discuss its compatibility with the diffeomorphism constraints. Note that a non diffeomorphism invariant structure is necessary to define the regulated constraint operator since the local operator $C(x)$ has to be replaced by an extended object, defined in terms of some extended structures on space. This is unavoidable since regularization necessarily involves some form of point splitting or short distance cutoff. In our construction of the regulated scalar constraint, given by Eqs.15.31, 15.32 we make use of a specific set of loops, which are defined in terms of a particular coordinate system. This is the way in which our regularization of the scalar constraint breaks the diffeomorphism invariance. Therefore we must ensure that the space of solutions is nonetheless diffeomorphism invariant. This amounts to showing that to impose the constraint in one particular set of coordinates is equivalent to imposing it any other coordinate system. This will be true if the regulated scalar constraint is transformed linearly into itself under the representation U of the diffeomorphism group. This is indeed the case. To see this, let us fix a coordinate system O and define $\hat{C}^{\delta}(x)$ with respect to it. If we call $\phi \cdot \hat{C}^{\delta}(x)$ the regulated scalar constraint defined in the coordinate system $\phi \cdot O$, we need to show:

$$\phi \cdot \hat{C}^{\delta}(x) = U(\phi^{-1})\hat{C}^{\delta}(x)\, U(\phi)\,. \tag{28}$$

This relation follows directly from our definitions.

3.4 Solutions to the scalar constraint

We now show that the regulated scalar constraint has a nontrivial space of solutions. In particular, we will show that if Ψ has support only on smooth loops, without intersections, then it is in the kernel of the scalar constraint, in the sense that

$$\lim_{\delta \to 0} \frac{1}{\delta^2} \hat{C}^{\delta}(x)\, \Psi[\eta] = 0. \tag{29}$$

Thus, we will recover, in the loop representation, the results obtained in [6] using the connection representation. In these notes, I will only present the formal calculation. The issue of regularization *has* been discussed in detail in [2]. However, recently, the regularization procedure has been criticized by Blencowe in [3]. It is my understanding that the criticisms are valid but can be met by suitable modifications. However, this material has not been published, and the status of the issue is somewhat "fluid".

Let us begin with some notation. Denote by $\mathcal{M}_0$ the subspace of $\mathcal{M}$ formed by the multiple loops which are non-intersecting and smooth everywhere, and by $\mathcal{M}_0'$ the subspace formed by multiple loops which include at least one eyeglass loop, but are otherwise nonintersecting and smooth. Let us consider the regularized scalar constraint without the smearing described in section 2.4 and use it to obtain solutions by a formal calculation allowing the Δ distributions. The action of the regularized constraint on any loop functional $\Psi[\{\eta\}]$ is given by:

$$\hat{C}^{\delta}(x)\cdot\Psi[\{\eta\}]=\sum_{i=1}^{4}(2\pi)c_i\Delta^{[a}\left[\gamma^{\delta}_{ab}(x)\right](\delta^2)\Delta^{b]}\left[\gamma^{\delta}_{ab}(x)\right]\Psi[\eta_i], \tag{30}$$

where η_i are four loops and c_i four finite coefficients, which are given by the action of the T^2's of $\gamma^{\delta}_{ab}(x)$ on η. (Recall that $\gamma^{\delta}_{ab}(x)$ is defined in the discussion of the classical scalar constraint. See Eqs.15.31 and 15.33.)

We consider separately three cases. In the first case, the loop η is in $\mathcal{M}_0$. In the second η is in $\mathcal{M}'_0$, and in the third it has corners and/or intersections, but is not in $\mathcal{M}'_0$. On $\mathcal{M}_0$ it is not difficult to show that $\hat{C}^{\delta}(x)\Psi[\{\eta\}]$ is $O(\delta^3)$. Since a loop in $\mathcal{M}_0$ is smooth the first of the two factors of Δ may be expanded in the loop parameter around the origin. If η has no self-intersections, then the leading term is equal to that in the second Δ factor and the two cancel by the antisymmetrization in the vector indices. The next term is then $O(\delta^3)$ so that in the limit we have zero. Note that if the loop has a corner this expansion is not possible, since the $\dot{\eta}$ in Δ may be discontinuous in the region where we want to expand it.

Now let us consider the value of the right hand side of Eq.30 on multiloops $\{\eta\}$ in $\mathcal{M}'_0$. Let us impose on the values of Ψ evaluated at the eyeglass loops the conditions,

$$\Psi[\alpha\#\rho\#\beta\#\rho^{-1}]=-\Psi[\alpha\#\rho\#\beta^{-1}\#\rho^{-1}]=\frac{1}{2}\Psi[\alpha\cup\beta] \tag{31}$$

and

$$\Psi[\alpha_1\#\eta\#\alpha_2\#\eta^{-1}]=\Psi[\alpha_1\#\eta\#\alpha_2^{-1}\#\eta^{-1}]=\Psi[\alpha] \tag{32}$$

We will call any state that satisfies these two conditions an extendible state because when this condition holds the value of Ψ on $\mathcal{M}_0$ extends uniquely to $\mathcal{M}'_0$. It is then straightforward to show that

$$\hat{C}^{\delta}(x)\Psi[\alpha\#\rho\#\beta\#\rho^{-1}]=\hat{C}^{\delta}(x)\Psi[\alpha\cup\beta] \tag{33}$$

with a similar identity holding for the other kind of eyeglass loops.

Finally, consider the value of the right side of Eq.30 on loops η with intersections or corners. One may show that if α has a corner or an intersection, then $\alpha\#\beta$ also has a corner or an intersection for any β in $\mathcal{M}$, since the operation of breaking and joining cannot subtract a corner or intersection.

Thus we have the following result. If we apply the scalar constraint operator to an extendible state $\Psi[\{\eta\}]$ with support only on $\mathcal{M}'_0$, then $\lim_{\delta\to 0} \mathcal{C}^\delta(x)\Psi[\{\eta\}]$ is zero both inside $\mathcal{M}_0 \cup \mathcal{M}'_0$ and outside it. Thus, it vanishes everywhere in $\mathcal{M}$. Thus, we have shown that using the loop representation, the scalar constraint may be regularized in such a way that it has a nontrivial kernel, and that any loop functional with support on $\mathcal{M}_0$ is in this kernel. This calculation is, however, formal because the Δ's in the right hand side of Eq.30 are distributions. Rovelli and Smolin [2] have proposed a regularization scheme that extends the ideas of section 2.5. However, as I said earlier, the issue is not yet fully settled.

Let us conclude this discussion with three remarks.

i) As was pointed out in chapter 2, as far as the scalar constraint is concerned, the loop representation for quantum gravity is analogous to the momentum representation for the Klein-Gordon theory in Minkowski space. In the Klein-Gordon theory, the solutions are given by arbitrary functions on the mass shell in the momentum space. From this point of view $\mathcal{M}$ is analogous to momentum space and its subspace $\mathcal{M}_0$ is analogous to the mass shell.

ii) Blencowe [3] has obtained additional topological solutions to all constraints. However, the solutions available so far are far from being complete. In particular, although a careful analysis has not been done, they appear to solve the scalar constraint for *any* value of the cosmological constant. Therefore, they are likely to constitute only a "small" sector of all physically relevant solutions. Indeed, from the results of Jacobson and Smolin [6] and more recent work of Brügmann and Pullin [7], one knows that there exist solutions in which states $\Psi[\{\eta\}]$ have support on intersections. However, relatively little work has been done in this area so far.

iii) The limit $\delta \to 0$ needed here to define the scalar constraint is taken using the point wise topology in $\mathcal{M}$. If we had a Hilbert structure on the space of the unconstrained states, i.e., on all of V, we could have imposed a stronger requirement using the Hermitian inner-product. However, as discussed in chapter 10, for systems with first class constraints, in general the physical inner product is meaningful only on the space of *solutions* to the quantum constraints. However, it *is* possible that another, *non-physical* inner product could be used to give a better definition of this limit. This entire issue arises because we do not have a background geometry. So far it has received very little attention in the literature.

4 Discussion

Let us combine the results obtained in the last section to exhibit some physical states of the gravitational field.

We note first that if a loop functional $\Psi[\{\eta\}]$ is in the kernel of the scalar constraint, then so is its image $U_\phi \Psi[\{\eta\}]$ under a diffeomorphism. This means that the transformation properties of the scalar constraint under diffeomorphisms are correct; the commutator of the scalar constraint with the diffeomorphism constraint is proportional to the scalar constraint itself. Because of this property, the scalar constraint is well defined on the orbits of the diffeomorphism group. We also saw that if a state $\Psi[\{\eta\}]$ in V has support only on smooth, non-self intersecting loops, it solves the scalar constraint. We can combine these results to arrive at the following solutions of all quantum constraints. Let Ψ be an extendible element of V defined such that,

$$\Psi[\{\eta\}] = 0 \tag{34}$$

unless $\{\eta\}$ is in $\mathcal{M}_0 \cup \mathcal{M}'_0$, and, when it is in $\mathcal{M}_0$,

$$\Psi[\{\eta\}] = \Psi[L[\{\eta\}]]. \tag{35}$$

where $L[\{\eta\}]$ is any link class of Σ. (Recall that generalized link classes constructed from elements of $\mathcal{M}_0$ are just the ordinary link classes.) Then $\Psi[\{\eta\}]$ is a physical state of the gravitational field. There is one independent state of this kind for every link class L of Σ. Therefore, characteristic functions of link classes –functions which assume the value 1 on a given link class and zero elsewhere– provide a basis in this space of physical states.

Thus, there now exists an infinite dimensional space of physical states in non-perturbative canonical gravity. A basis for this space is in one to one correspondence with the link classes of Σ. Recent developments in knot theory are therefore of immediate relevance at least to the mathematical description of non-perturbative quantum gravity. These states may be written in closed and explicit form in the loop representation. Furthermore, the action on any of these states of any (regularized) operator in the operator algebra $\mathcal{A}$ may be computed explicitly. Using these operators, one should be able to develop a physical intuition for the these states.

From the viewpoint of the standard perturbation theory, it is at first quite surprising that the physical states are based on discrete structures: In Minkowskian field theories one is used to local fields and quantum states arise as functionals of the continuous field variables. However, from rather general considerations, the underlying discreteness seems unavoidable. To see this, note first that any theory of connection dynamics admits a loop representation in its quantum version.

Therefore, if one accepts the reformulation of general relativity presented in Part II as a valid point of departure for canonical quantization and if one grants that the physical states have to be invariant under spatial diffeomorphisms, one is naturally driven to the conclusion that, in the loop representation, states can only depend on (generalized) link classes on the 3-manifold. Thus, the discreteness underlying canonical quantum gravity seems to be inescapable. It is important to realize that we are led to this conclusion starting from rather conservative assumptions. We did not *postulate* discreteness. We did not *begin* by declaring what the fundamental structures underlying space-time geometry should be. Rather, we are led to the description merely by combining general relativity with the well-established principles of canonical quantization.

Thus, the picture of the micro-structure of space-time that the non-perturbative treatment provides is very different from the one used in the perturbative treatments. One of the immediate goals therefore is to try to understand the successes and failures of perturbation theory using this picture. I will conclude by summarizing these ideas briefly.[3] The idea is to replace Minkowski space by an exact solution to all quantum constraints and analyse perturbations around it. A candidate for this background state, Ψ_0, is already available. Fix a flat 3-metric q^0_{ab} on $\mathbb{R}^3$ and consider a grid of non-intersecting straight lines in the x, y and z directions separated by a distance a with respect to q^0_{ab}. (The lines thus resemble a grid of piano wires.) We will call this giant multiple loop a *weave*. Consider the state $\Psi[\{\eta\}]_0$ which takes value 1 on the knot class of this loop and zero elsewhere. This state solves all constraints and approximates the flat space we began with in a certain well-defined sense. Furthermore, if one evaluates the action of the operator corresponding to the smeared out 3-metric, one finds an interesting and unexpected result: the approximation is the best precisely when the "grid separation" a is the Planck length! It is not surprising that the approximation would be bad if the weave is too loose, i.e., if the grid size is too large. However, one would have naively expected the approximation to improve continuously as we shrink the grid size and take the continuum limit of the weave. This does not happen. In fact, in the limit, the action of the smeared metric operator on the weave diverges as $(L_P/a)^4$ where L_P is the Planck length. Substantial progress has already been made on translating linearized gravity as a theory of fluctuations around this state. The linearized graviton states link with the weave in a non-trivial fashion; they resemble an embroidery. The hope is that these calculations will enable one to pin-point the reason behind the breakdown of the perturbation theory.

---

3 The discussion that follows is based on unpublished work done in collaboration with Carlo Rovelli and Lee Smolin.

References

[1] C. Rovelli and L. Smolin, Phys. Rev. Lett. **61**, 1155 (1988).

[2] C. Rovelli and L. Smolin, Nucl. Phys. **B331**, 80 (1990).

[3] M.P. Blencowe, Nucl. Phys. **B341**, 213 (1990). Nucl. Phys., (to appear).

[4] L.H. Kauffman, *Formal knot theory* and *On knots*, Princeton University Press, Princeton (1987).

[5] A.Ashtekar (with invited contributions), *New Perspectives in Canonical Gravity* (Bibliopolis, Naples, 1988).

[6] T. Jacobson and L. Smolin, Nucl. Phys. **B299**, 295 (1988).

[7] B. Brügmann and J. Pullin, "Intersecting N-loop solutions to the Hamiltonian constraint equations of quantum gravity" (pre-print, 1990).

17 2+1 GRAVITY

1 Introduction

As we saw in the last six chapters, a number of steps of our quantization program have now been completed in the case of 3+1 dimensional general relativity. In particular, we have introduced two new representations of quantum states. The first is the ***connection representation***, in which the operator A_a^i is diagonal, while the second is the ***loop representation*** in which quantum states arise as functions on the loop space of the spatial 3-manifold. The former has enabled us to make contact with familiar physics dictated by the Schrödinger equation as well with several aspects of Yang-Mills theories. The latter has turned out to be especially convenient to solve the quantum constraints exactly and seems well suited for probing the Planck regime. These results are certainly encouraging. Nonetheless, a number of important problems are still open; even if one restricts oneself just to mathematical considerations, the program is incomplete in several respects.

The difficulties we face fall, broadly, into two categories: those arising from the presence of an infinite number of degrees of freedom and those associated with specific features –such as the diffeomorphism invariance– of the full, non-linear general relativity. In chapter 11, we saw that the program can be completed in the case of linearized gravity, which has an infinite number of degrees of freedom but none of the problems associated with non-linearity and absence of background geometry. In this chapter we will discuss a complementary case, a system which shares with 3+1 dimensional general relativity difficulties of the second kind. We will see that the quantization program can be carried out to completion in this model as well. Thus, at least in principle, the program is capable of handling both categories of problems.

The system we now consider is 2+1 dimensional general relativity. In 2+1 dimensions, the gravitational field has no local degrees of freedom; there are no gravitons. However, the theory is not "trivial": it does have "global" degrees of freedom and, furthermore, shares with the 3+1 theory all the technical and conceptual problems arising from non-linearity and diffeomorphism invariance. Indeed, the perturbation theory around 2+1 dimensional Minkowski space appears, at first sight, to be non-renormalizable essentially for the same reasons that hold in 3+1 dimensions. However, two years ago, Edward Witten [1] showed that 2+1 general relativity is essentially isomorphic to Chern-Simons theory for the Poincaré group, $ISO(2,1)$, and therefore exactly soluble. Witten's phase space formulation of this theory, as he pointed out, is closely related to the new canonical framework for 3+1 general relativity discussed in these notes. Hence, in this version, 2+1 gravity is an ideal toy model to test the viability of the 3+1 quantization program.

Let us therefore apply the program, step by step, to this model. Although, we will find some new results for 2+1 gravity, our primary interest lies rather in what this model can teach us about the 3+1 program. In particular, we are interested in the following types of questions: Can the program be *completed* in the 2+1 case where the theory is completely manageable? Or, are there some essential obstructions which have so far remained buried in the technical difficulties of the 3+1 theory discussed in chapters 12 and 16? Can one construct both the connection and the loop representations? Are they equivalent? Is it true, as one hopes in the 3+1 program, that the requirement that real classical observables become Hermitian operators –the "reality condition"– suffices to determine the inner product on the space of physical quantum states? Can one introduce this inner product in a way that could be generalized to the 3+1 theory? In the 3+1 analysis, while the loop representation is mathematically very convenient, it does not admit an obvious physical interpretation. Can one gain insight into the physical meaning of various objects that feature in the loop representation by investigating their analogs in the 2+1 theory? In particular, what is the physical interpretation of the basic operators that arise in the loop representation? Thus, the key questions that we are interested in are related more to the quantization program in the 3+1 theory than to the structure of 2+1 gravity itself. Therefore, we do not want to base our analysis on those features of the 2+1 theory which are specific to 3-dimensions, such as the equivalence of 2+1 general relativity to a Chern-Simons theory and the availability of a manageable reduced phase space description. Similarly, since the 3+1 program emphasizes connection dynamics, we will not use ideas from the extensive literature on 2+1 geometrodynamics due to Deser, Jackiw, 'tHooft and others (see, e.g., Ref. [2]). Instead, our constructions will be based on notions which do admit 3+1 analogs in the new canonical formulation.

The purpose of this chapter is to summarize the results [3,4] on 2+1 gravity obtained by application of our general quantization program, emphasizing aspects which are likely to be useful in 3+1 quantum general relativity. The material is divided as follows. Section 2 introduces the phase space description of 2+1 classical general relativity with emphasis on the mathematical similarities and differences between the 2+1 and the 3+1 theories. Section 3 contains the main results on 2+1 quantum gravity. Section 4 briefly summarizes the status of the 3+1 program and points out the conceptual as well as technical lessons we have learned from the 2+1 analysis.

2 Canonical formulation of 2+1 gravity

Since we are interested in a Hamiltonian description, we shall assume that the space-time manifold M is topologically $\Sigma \times \mathbb{R}$. For simplicity, we shall further assume that Σ is compact. As in the discussion of the 3+1 theory (see chapters 3,4) the idea is to use a first order framework *a la* Palatini. Thus, the basic variables will be co-triads, e_a^I, and $SO(2,1)$ connections A_a^I [1] To recover general relativity, one must assume that the co-triads are linearly independent. However, our framework itself is more general; it can accommodate degenerate metrics. In what follows, we shall consider this more general theory (except when we comment on the relation of this framework to 2+1 general relativity). The action is given by:

$$S(e, {}^3\!A) := \int_M d^3x \, \tilde{\eta}^{abc} e_a^I \, {}^3F_{bcI}, \tag{1}$$

where, $\tilde{\eta}^{abc}$ is the metric independent Levi-Civita density on M, and $F_{abI} := 2\partial_{[a}{}^3A_{b]I} + \epsilon_{IJK} \, {}^3A_a^J \, {}^3A_b^K$ is the field strength of the connection ${}^3A_a^I$. This action is in fact the 2+1 dimensional analog of the self dual action we used in 3+1 gravity, even though the connection here is real while that in the 3+1 theory is complex [5]. The classical equations of motion are:

$${}^3\mathcal{D}_{[a} e_{b]I} = 0 \qquad \text{and} \qquad {}^3F_{ab}^I = 0, \tag{2}$$

where ${}^3\mathcal{D}$ is the gauge covariant derivative operator determined by ${}^3A_a^I$. Note that as defined ${}^3\mathcal{D}$ acts only on internal indices. However, *if* we were to extend the action of ${}^3\mathcal{D}$ to space-time tensors, the first equation ensures that this extension is *torsion-free*. A choice of such an extension will then completely determine ${}^3\mathcal{D}$. For the purposes of calculations it is sometimes convenient to extend ${}^3\mathcal{D}$ to be compatible with the triad. However, all results are independent of the specific extension chosen. Together with the second equation (which implies that the connection is flat), this tells us that the 3-metric $g_{ab} := e_a^I e_{bI}$ constructed from the triad is flat, i.e., satisfies the 3-dimensional Einstein's equation.

1 In this chapter uppercase latin letters $I, J\ldots$ denote internal $SO(2,1)$ indices (and label the co-triads). As before, generally, stem letters will distinguish space-time fields from spatial fields. In case of ambiguity, a suffix 3 will denote a space-time field. Thus, ${}^3A_a^I$ is a space-time connection while A_a^I stands for its pull-back to a 2-dimensional spatial slice.

It is straightforward to perform the Legendre transform of this action. The resulting phase space description is as follows. The configuration space $\mathcal{C} \ni A_a^I(x)$ is the space of pull-backs of the connection ${}^3A_a^I$ to the 2-dimensional spatial slice Σ. The momentum $\tilde{E}_I^a$ conjugate to A_a^I is the dual of the pull-back of the co-triad e_b^I, $\tilde{E}_I^a := \tilde{\eta}^{ab} e_{bI}$, where $\tilde{\eta}^{ab}$ is the Levi-Civita density of weight 1 on Σ . Thus, the phase space Γ consists of pairs $(A_a^I, \tilde{E}_I^a)$ of fields on Σ; the basic (non-vanishing) Poisson brackets are simply

$$\{A_a^I(x),\ \tilde{E}_J^b(y)\} = \delta_a^b\, \delta_J^I\, \delta^2(x,y). \tag{3}$$

Note that although $\tilde{E}_I^a$ does determine the induced metric q_{ab} on the spatial 2-manifold Σ via $(q)\, q^{ab} = \tilde{E}_I^a \tilde{E}^{bI}$, the momenta $\tilde{E}_I^a$ are *not* dyads; the internal index still runs from 1 to 3. *Thus, we are not assuming that there is a global frame field on* Σ. As emphasized by Moncrief [6], such an assumption would have restricted the topology of Σ severely; our construction, on the other hand, places no such restriction.

The system has first class constraints. These can be obtained simply by pulling back the equations of motion (1) to the spatial slice:

$$\mathcal{D}_a \tilde{E}_I^a = 0 \qquad \text{and} \qquad F_{ab}^I = 0, \tag{4}$$

where F_{ab}^I is the curvature of A_a^I. The first equation is the Gauss constraint which ensures that the internal $SO(2,1)$-rotations are gauge transformations. The "time" component, on the internal index, of the second equation is equivalent to the usual scalar constraint of 2+1 general relativity while the "space" component gives us the vector constraint. As could have been expected, the Hamiltonian is a linear combination of constraints.

Let us compare this canonical description with that of the 3+1 theory discussed in Part II. The basic variables in the 3+1 theory are also certain connections A_a^i and their conjugate momenta $\tilde{E}_i^a$ (which, however, *can* be interpreted as spatial triads.) The theory has three constraints; the Gauss constraint, $\mathcal{D}_a \tilde{E}_i^a = 0$, a vector constraint, $\tilde{E}_i^a F_{ab}^i = 0$, and a scalar constraint, $\epsilon^{ijk}\, \tilde{E}_I^a \tilde{E}_j^b\, F_{abk} = 0$. The constraints of the 2+1 theory can also be written in this form. However, in this case, since F_{ab}^I has only three independent components, a simple counting argument tells us that the satisfaction of both the scalar and the vector constraints is equivalent to the vanishing of the curvature F_{ab}^I itself. This is why the 2+1 theory has no local degrees of freedom. This is also the reason why, in this theory, the system of constraints is so simple; the constraints are either linear in momenta or independent of momenta. In the 3+1 theory on the other hand, as one should expect, the same counting argument shows that the satisfaction of the vector and the scalar constraints does *not* imply that the curvature itself has to vanish. This, from the canonical viewpoint, is the reason why the theory now has local degrees of freedom. Furthermore, in 3+1 theory there is always a constraint quadratic in momenta which makes both the construction of a reduced phase space in the classical theory and the problem of quantization highly nontrivial.

Since the constraints of the 2+1 theory are so simple, it is possible to write down explicitly a complete set of Dirac observables, i.e., functions on the phase space Γ whose Poisson brackets with constraints vanish weakly. Furthermore, these observables are the 2+1 analogs of the T variables introduced in chapter 15. Thus, each of these observables is associated with a closed loop γ on Σ. The configuration observables, $T^0[\gamma](A)$, are independent of the momentum $\widetilde{E}^a_I$ while the momentum observables $T^1[\gamma](A, \widetilde{E})$ are linear in $\widetilde{E}^a_I$. As in chapters 14 and 15, the superscript 0 or 1 refers to the order of the momentum dependence. One can also construct observables $T^n[\gamma]$ which are higher order in momenta. However, they turn out to be redundant in 2+1 dimensions. The configuration and the momentum observables are given by[2] :

$$T^0[\gamma](A) := \mathrm{tr}\, U_\gamma \qquad \text{and} \qquad T^1[\gamma](A, \widetilde{E}) := \oint_\gamma dS^a \ \mathrm{tr}\, E_a U_\gamma, \tag{5}$$

where $U_\gamma(s) := \mathcal{P} \exp \int_\gamma A$ is the holonomy of A^I_a around γ, evaluated at the point $\gamma(s)$, $E_a := \underset{\sim}{\eta}_{ab} \widetilde{E}^b$, and where we have used the 2-dimensional representation of $SO(2,1)$ to take the trace. (These observables are closely related to the ones introduced by Steven Martin [7]. In the case when Σ is non-compact, with the topology of a punctured 2-plane, T^0 is essentially the mass defined by Deser, Jackiw and 'tHooft [2] and T^1, the angular momentum, of the "particle" at the puncture enclosed by the loop γ.) It is straightforward to verify that the Poisson brackets of these observables with the constraints vanish weakly. Furthermore, these functions are themselves closed under the Poisson bracket:

$$\begin{aligned} \{T^0[\gamma], T^0[\delta]\} = 0, \qquad \{T^1[\gamma], T^0[\delta]\} &= \sum_i \sum_\diamond \Delta_i(\gamma, \delta) T^0[\gamma \#_i \delta]^\diamond, \\ \{T^1[\gamma], T^1[\delta]\} &= \sum_i \sum_\diamond \Delta_i(\gamma, \delta) T^1[\gamma \#_i \delta]^\diamond. \end{aligned} \tag{6}$$

Here i labels the points at which the two curves γ and δ intersect; $\#$ and $\diamond$ are used, as in section 15.2, to denote specific compositions of the two curves at the

2 As in 3+1 theory, we could have defined the momentum variable as: $T^a[\gamma](s) = \mathrm{tr}\, U_\gamma(s) \widetilde{E}^a(\gamma(s))$. However, in 2+1 dimensions, it is more convenient to integrate this vector density over the loop γ and consider instead $T^1[\gamma] := \int_\gamma ds^a \, \underset{\sim}{\eta}_{ab} T^b[\gamma](s)$.

i-th intersection; and, $\Delta_i(\gamma, \delta) := (-1)^{|o_i|}$ now simply takes values ± 1 depending on whether the dyad formed by the tangent vectors to the two curves at the i-th intersection is right or left handed. The general structure of these Poisson brackets is characteristic of the Poisson algebra of configuration and momentum observables on *any* cotangent bundle. However, as in the 3+1 theory, the important point is that now we are considering only a subset of configuration and momentum observables, namely the ones associated with closed loops. Therefore, it is somewhat remarkable that this subset is also closed under Poisson brackets. The definitions of these observables was obviously motivated by that of analogous functions –discussed in the last two chapters– first introduced by Rovelli and Smolin on the phase space of 3+1 gravity [8]. Consequently, the Poisson bracket structure of the 2+1 functions just mirrors that of their 3+1 analogs. However, there is a basic and important difference: in the 3+1 theory, the elements of the $\mathcal{T}$ algebra fail to commute with the constraints weakly whence they are *not* Dirac observables. It is this fact that causes key problems in the completion of the quantization program in 3+1 dimensions.

A number of properties of these 2+1 observables are easier to establish if one notices a certain relation between them [4]. To see this, note first that, since we are in 2+1 dimensions, the ***configuration space*** $\mathcal{C}$ itself is equipped with a natural symplectic structure: For any two vectors (δA) and $(\delta A)'$ tangent to $\mathcal{C}$ with components $(\delta A)_a^I$ and $(\delta A)'_{aI}$, set $\Omega((\delta A), (\delta A)') := \int_\Sigma d^2x\, \underset{\sim}{\eta}^{ab} (\delta A)_a^I\, (\delta A)'_{bI}$. Now, since $T^0[\gamma](A)$ is a function on $\mathcal{C}$, we can construct the Hamiltonian vector field $X[\gamma]$ it generates. We can now return to the phase space Γ. With every vector field on $\mathcal{C}$, we can associate a function on Γ which is linear in momenta (see appendices C and B). It turns out that this is precisely our momentum observable $T^1[\gamma]$ of equation (5). Thus, the symplectic structure Ω on $\mathcal{C}$ enables one to "construct" the observables $T^1[\gamma]$ starting from $T^0[\gamma]$. (It is this fact that motivated our construction of $T^1[\gamma, v]$ from $T^0[\gamma]$ in section 15.2.) Since $T^0[\gamma]$ are simply traces of holonomies, it is generally easy to analyse their properties. The above construction then enables one to extend these properties to the momentum observables $T^1[\gamma]$. Thus, the following properties:

$$\begin{aligned} T^0[0] = 2, \qquad T^0[\alpha] &= T^0[\alpha^{-1}], \qquad T^0[\alpha\#\beta] = T^0[\beta\#\alpha], \\ \text{and} \qquad T^0[\alpha]\cdot T^0[\beta] &= T^0[\alpha\#\beta] + T^0[\alpha\#\beta^{-1}], \end{aligned} \tag{7a}$$

of traces of holonomies (which follow trivially from $SO(2,1)$ trace identities), immediately imply:

$$\begin{aligned} T^1[0] = 0, \qquad T^1[\alpha] &= T^1[\alpha^{-1}], \qquad T^1[\alpha\#\beta] = T^1[\beta\#\alpha], \\ \text{and} \quad T^0[\alpha]\cdot T^1[\beta] + T^0[\beta]\cdot T^1[\alpha] &= T^1[\alpha\#\beta] + T^1[\alpha\#\beta^{-1}]. \end{aligned} \tag{7b}$$

These are the "universal" algebraic identities shared by *all* configuration and momentum observables. They will go over to the quantum theory.

A second application of the above construction of $T^1[\gamma]$ is the proof of their completeness. Again, we begin with properties of $T^0[\gamma]$. One normally thinks of traces of holonomies as containing "all the gauge invariant information" that a connection has. Indeed, had the gauge group been $SO(3)$ rather than $SO(2,1)$, we could have labelled *any* gauge equivalence class of connections A_a^I completely by the values that observables $T^0[\gamma]$ take on that equivalence class. In the present case, the situation is a little more complicated. Since this technicality will play a role in quantum theory, let us discuss it briefly. The difference from the $SO(3)$ theory arises because $SO(2,1)$ has four disconnected components, (related by the *internal* parity and time reversal mappings $\mathcal{P}$ and $\mathcal{T}$ respectively), and by a local gauge transformation we mean mappings from Σ to only the connected component. Therefore, two connections which differ from each other by the action of the operators $\mathcal{P}$ and $\mathcal{T}$ are *not* considered as gauge equivalent. The traces of holonomies, around *any* closed loop γ, of connections which are so related, on the other hand, are equal. Therefore, values of $T^0[\gamma](A)$ determine, not a unique gauge equivalence class of connections, but rather a set of equivalence classes related to each other by $\mathcal{P}$, $\mathcal{T}$ or $\mathcal{PT}$. This problem is, however, a global one; locally in the configuration space, one would expect the situation to be similar to that in the case when the gauge group is $SO(3)$. This expectation is correct. The observables $T^0[\gamma]$ form an (over)complete set in the sense that their gradients span the cotangent space of the space of gauge equivalence classes of connections almost everywhere (or, more precisely, everywhere except at the fixed points of any one of the above three discrete symmetry operators.) The construction of $T^1[\gamma]$ from $T^0[\gamma]$ now implies that these two sets of functions form an (over)complete set of gauge invariant observables on the phase space Γ. We can therefore use them as the elementary classical observables in the quantization program of chapter 10, construct a quantum $\star$-algebra based on the classical $\mathcal{T}$ algebra, and then find its $\star$-representations.

Finally, for future use let us make a note of some structure available in the classical theory. Since the constraints are first class, we can quotient the constraint surface by the Hamiltonian vector fields generated by the constraint functions and pass to the reduced phase space $\bar{\Gamma}$. Points of $\bar{\Gamma}$ represent the "true degrees of freedom". Since the constraints are either linear or independent of momenta, this is a straightforward procedure and $\bar{\Gamma}$ is again a cotangent bundle. However, while Γ is infinite dimensional, $\bar{\Gamma}$ is only finite dimensional. This comes about because, while the configuration variable A_a^I has 6 degrees of freedom per point of Σ, we also have 6 first class constraints per space point, given by Eq.4. Thus, 2+1 quantum gravity resembles quantum mechanics rather than quantum field theory.

What structure does $\bar{\Gamma}$ have? Let us begin with the reduced configuration space

$\overline{\mathcal{C}}$ over which $\overline{\Gamma}$ is the cotangent bundle. $\overline{\mathcal{C}}$ is the moduli space of flat $SO(2,1)$ connections. Thus, each point of $\overline{\mathcal{C}}$ is an equivalence class $\overline{A}$ of flat $SO(2,1)$ connections, where two are regarded as equivalent if they are related by a local $SO(2,1)$ transformation, i.e., by a smooth map from Σ to the connected component of identity of $SO(2,1)$. In general, $\overline{\mathcal{C}}$ has several *disconnected* components. To see this, note first that each element $\overline{A}$ of $\overline{\mathcal{C}}$ is completely determined by fixing a base point p on Σ and specifying the holonomies, modulo the action of $SO(2,1)$ at p, of $\overline{A}$ around the $2g$ generators of the homotopy group of Σ (assumed to have genus g.) The holonomies provide us $2g$ group elements, $(U_1, ..., U_{2g})$, and $\overline{A}$ is determined by the equivalence class $U \cdot (U_1, ..., U_{2g}) \cdot U^{-1}$ obtained by varying U in the gauge group acting *at* p. Now, each U_k is a rotation either along a time-like, null or space-like axis and the action of the gauge group at p must map that U_k to a rotation with the same type of axis. Denote by N_t, N_n and N_s the number of rotations with time-like, null, and space-like axes $(N_t + N_n + N_s = 2g)$. It is easy to check that they are independent of the initial choice of the base point p. Thus, we can associate, in an invariant way, three integers with every element $\overline{A}$ of $\overline{\mathcal{C}}$. It is clear that if two connections are labelled by distinct sets of integers, they cannot lie in the same connected component of $\overline{\mathcal{C}}$. This is why $\overline{\mathcal{C}}$ has disconnected sectors. Since $\overline{\Gamma}$ is the cotangent bundle over $\overline{\mathcal{C}}$, it is clear that the same is true of $\overline{\Gamma}$.

Let us now return to the $\mathcal{T}$ algebra. Since they are Dirac observables, the restrictions to the constraint surface of the $T^0[\gamma]$ and $T^1[\gamma]$ can be projected down unambiguously to functions $\overline{T}^0[\gamma]$ and $\overline{T}^1[\gamma]$ on the reduced phase space $\overline{\Gamma}$. As one might expect, these form an (over)complete set on $\overline{\Gamma}$ in the sense that their gradients span the cotangent space almost everywhere on $\overline{\Gamma}$. Finally, they have an interesting property which is *not* shared by the $T^0[\gamma]$ and $T^1[\gamma]$: the values of $\overline{T}^0[\gamma]$ –and hence also of $\overline{T}^1[\gamma]$– are left unchanged if the closed loop γ is replaced by a homotopic loop. Thus the $\overline{T}$-observables are labelled by the *homotopy classes* of closed loops rather than by individual closed loops. Since the notation is already rather involved, I shall not use a new symbol to denote homotopy classes; I hope the context will make it clear whether γ stands for a single loop or whether it denotes a homotopy class.

3 Quantum theory

Following the 3+1 quantization program, we shall now construct both the connection and the loop representations for 2+1 gravity.

3.1 Connection representation.

Following the procedure used in chapter 12 for 3+1 gravity, let us carry out the quantization program using the connection representation. Thus, the elementary quantum operators are $\hat{A}_a^I$ and $\hat{E}_I^a$, satisfying the canonical commutation relations. the representation space V of unconstrained states consists of complex valued functionals $\Psi(A)$ of the connection A_a^I. Since $T^0[\gamma]$ is a classical configuration observable, its quantum analog $\hat{T}^0[\gamma]$ is a multiplication operator on V. Similarly, since $T^1[\gamma]$ is linear in momentum, its quantum analog, $\hat{T}^1[\gamma]$, is a Lie derivative:

$$\hat{T}^0[\gamma]\cdot\Psi(A) := \hat{T}^0[\gamma](A)\cdot\Psi(A), \quad \text{and} \quad \hat{T}^1[\gamma]\cdot\Psi(A) := -i\hbar\mathcal{L}_{X[\gamma]}\Psi(A), \qquad (8a)$$

where $X_{[\gamma]}$ is, as before, the Hamiltonian vector field on $\mathcal{C}$ constructed from $T^0[\gamma](A)$. These expressions are special cases of the ones we found in section 10.2. It is straightforward to check that the commutators of these basic operators are $-i\hbar$ times the Poisson brackets of their classical analogs. This relation between the commutators and the Poisson brackets is quite general; it holds in any physical system for observables which are independent of or linear in momenta [9].

The next step is to impose the quantum constraints and extract the space V_{phy} of physical states. This turns out to straightforward. Since the constraints (Eq.4) are either linear or independent of momenta, there are no nontrivial factor ordering or regularization problems to solve. The quantum Gauss constraint requires that $\Psi(A)$ be gauge invariant while the second constraint requires that $\Psi(A)$ should have support *only* on flat connections. Thus, the general solution, *a la* Dirac, of the two quantum constraints is a function $\Psi(\overline{A})$ on the moduli space $\overline{\mathcal{C}}$ of flat connections on Σ. The space of these $\Psi(\overline{A})$ is the required space V_{phy}. We must now find the physical operators on V_{phy}. It is straightforward to check that all of the $\hat{T}$-operators introduced above commute with the constraints; they are all physical operators. On the physical subspace, V_{phy}, their action can be written as:

$$\hat{T}^0[\gamma]\cdot\Psi(\overline{A}) := \overline{T}^0[\gamma](\overline{A})\cdot\Psi(\overline{A}), \quad \text{and} \quad \hat{T}^1[\gamma]\cdot\Psi(\overline{A}) := -i\hbar\mathcal{L}_{\overline{X}[\gamma]}\Psi(\overline{A}), \qquad (8b)$$

where $\overline{X}[\gamma]$ is the vector field on $\overline{\mathcal{C}}$ induced by the vector field $X[\gamma]$ on $\mathcal{C}$, or, alternatively, it is the Hamiltonian vector field on $\overline{\mathcal{C}}$ generated by $\overline{T}^0[\gamma]$ via the symplectic structure $\overline{\Omega}$.

Our next task is to introduce an inner-product on V_{phy}. For this, we want to impose the "reality conditions". The classical observables $T^A[\gamma]$ are all real. Therefore, we want to find an inner product on V_{phy} which makes the quantum operators $\hat{T}^A[\gamma]$ self-adjoint. We shall first exhibit such an inner product and then discuss the issue of its uniqueness. Recall that the configuration space C is equipped with a natural symplectic structure. We can pull it back to the space of flat connections, and since it is manifestly gauge invariant, project it to the space of their gauge equivalence classes, $\bar{A}$. The result is a symplectic structure $\bar{\Omega}$ on $\bar{C}$. Denote by $d\bar{V}$ the associated Liouville volume element and introduce on V_{phy} the following inner product:

$$\langle \Psi , \Phi \rangle := \int_{\bar{C}} d\bar{V} \ \overline{\Psi(\bar{A})}\, \Phi(\bar{A}) \ . \tag{9}$$

Denote by $\mathcal{H}$ the resulting Hilbert space.[3] The $\hat{T}^0[\gamma]$, being multiplication operators, are obviously self-adjoint on $\mathcal{H}$. Since $\overline{X}[\gamma]$ are Hamiltonian vector fields, their action preserves the Liouville volume element; they are divergence-free with respect to dV. Hence it follows that $\hat{T}^1[\gamma]$ are also self-adjoint. Thus, the required reality conditions have been fulfilled and we have a $\star$-representation of the algebra of physical operators. As a curiosity, let us ask how the diffeomorphism group of the spatial 2-manifold Σ acts on these quantum states. It is clear that every (smooth) diffeomorphism on Σ gives rise to a diffeomorphism on the reduced configuration space C. Furthermore, it follows from the expression of the symplectic structure Ω on C that $\bar{\Omega}$ –and hence $d\bar{V}$– is invariant under this induced diffeomorphism. Thus, the Hilbert space $\mathcal{H}$ provides a unitary representation of the *full* diffeomorphism group of Σ.

It turns out that this $\star$-representation of the physical algebra is in fact *reducible*. To see this, recall, first, that the moduli space $\bar{C}$ has several disconnected pieces. It is clear that the subspace $\mathcal{H}_k$ of quantum states with support just on one of these pieces, say $\bar{C}_k$, is mapped to itself by the quantum $\hat{T}$-algebra. Thus, the full representation is reducible. Furthermore, the subspace $\mathcal{H}_k$ itself is in general reducible because the observables $T^A[\gamma]$ form a complete set only *almost* everywhere on $\bar{C}$. One can show that, on each irreducible piece, the inner product of Eq.9 is uniquely picked out by the reality conditions, i.e., by the requirement that each $\hat{T}^A[\gamma]$ be represented by a self-adjoint operator. Given a direct sum of irreducible $\star$-representations of an algebra, we can trivially obtain another by rescaling the inner-product on each irreducible piece by a different constant. It turns out that this is the *only* freedom available the choice of the inner product that is allowed by the reality conditions. This trivial ambiguity in the inner product arises simply because the $\hat{T}[\gamma]$ algebra is not "globally" complete. (The overall situation is similar

3 Note that we could have avoided the use of *any* volume element on C by using, as in [9], densities of weight $\frac{1}{2}$ rather than functions to represent quantum states. (see, e.g., Appendix C.) However, since $\bar{C}$ *is* endowed with a natural volume element and since its use simplifies many calculations, we have refrained from adopting this more general procedure here.

to the one we encountered in the last example in section 10.5. The superselected operators in this case are $\hat{P}$ and $\hat{T}$ defined by the classical (internal) parity and time-reversal operators via: $P \cdot \Psi(\overline{A}) := \Psi(\mathcal{P} \cdot \overline{A})$, and $T \cdot \Psi(\overline{A}) := \Psi(\mathcal{T} \cdot \overline{A})$.)

Finally, let us check that the "universal" properties of the classical T observables also carry over to the quantum theory. First, since $\overline{T}^0[\gamma]$ –and hence, also $\overline{X}[\gamma]$– depend not on the individual loop γ but rather its homotopy class, it is clear from Eqs.8 that $\hat{T}^0[\gamma]$ and $\hat{T}^1[\gamma]$ also depend only on the homotopy class of γ. Next, let us consider the algebraic relations in Eq.7. Since $\hat{T}^0$ are multiplication operators, it is clear that the identities in Eq.7*a* go over to the quantum theory simply by replacing each $T^0[\gamma]$ by the operator $\hat{T}^0[\gamma]$. What is the situation with respect to the identities in Eq.7*b* on observables $T^1[\gamma]$? Note first that Eqs.7*a* imply that the vector fields $\overline{X}[\gamma]$ satisfy the following conditions: $\overline{X}[0] = 0; \overline{X}[\alpha] = \overline{X}[\alpha^{-1}]; \overline{X}[\alpha\#\beta] = \overline{X}[\beta\#\alpha]$; and, $\overline{X}[\alpha\#\beta] + \overline{X}[\alpha\#\beta^{-1}] = T^0[\alpha]\overline{X}[\beta] + T^0[\beta]\overline{X}[\alpha]$. It now immediately follows that Eq.7b also carries over to quantum theory (where, in the last equation, we must keep $\hat{T}^0$ to the left of $\hat{T}^1$.)

3.2 Loop Representation

The main idea here is that since $\overline{T}^A[\gamma]$, with $A = 0, 1$, form a complete set of Dirac observables in the classical theory, the passage to quantum theory can be achieved by "quantization of this algebra." Therefore, although it is by no means essential, it will be simpler to eliminate the constraints classically and then quantize the resulting unconstrained system using the steps of the quantization program of section 10.2.

Let us use as the space $\mathcal{S}$ of elementary classical variables on $\overline{\Gamma}$ the complex vector space spanned by the functions $T^0[\alpha](\overline{A})$ and $T^1[\alpha](\hat{A}, hatE)$, where $[\alpha]$ is an arbitrary *homotopy class* of closed loops. Following section 10.2, let us construct the algebra $\mathcal{A}$ of quantum operators. We first construct the free algebra generated by $\mathcal{S}$ and impose on it *i*) the commutation relations: the commutator of any two $\hat{T}^A[\alpha]$ operators be $i\hbar$ times the operator associated with the Poisson bracket of the corresponding $\overline{T}^A[\alpha]$, and, *ii*) the "anti-commutation relations" which capture the algebraic properties of the classical $\overline{T}^A[\alpha]$ variables (Eq.7):

$$\begin{aligned} \hat{T}^A[\alpha] = \hat{T}^A[\alpha^{-1}]; \quad \text{and} \quad \hat{T}^A[\alpha\#\beta] &= \hat{T}^A[\beta\#\alpha]; \\ \hat{T}^0[0] = 2 \qquad \text{and} \qquad \hat{T}^1[0] &= 0; \\ \hat{T}^0[\alpha] \cdot \hat{T}^0[\beta] &= \hat{T}^0[\alpha\#\beta] + \hat{T}^0[\alpha\#\beta^{-1}] \\ \hat{T}^0[\alpha] \cdot \hat{T}^1[\beta] + \hat{T}^0[\beta]\hat{T}^1[\alpha] &= \hat{T}^1[\alpha\#\beta] + \hat{T}^1[\alpha\#\beta^{-1}]. \end{aligned} \tag{10}$$

The anti-commutation relations are necessary because, unlike in the connection representation, the set S of elementary variables is now overcomplete. The next step is to introduce the $\star$-relations on $\mathcal{A}$. Since the classical $\overline{T}^A[\alpha]$ are all real-valued, we let $(\hat{T}^A[\alpha])^\star = \hat{T}^A[\alpha]$, and extend the $\star$ operation to all of $\mathcal{A}$ using the properties of involution. As in the 3+1 theory, let us call the resulting $\star$-algebra is the *quantum $\hat{T}$-algebra.*

The next step is to find the appropriate $\star$-representation of this algebra. There are several distinct, but equivalent, methods of completing this step. Here, I will discuss a method which has not appeared in the literature but which may well extend to the 3+1 theory. Introduce a ket $|0\rangle$ and require that it be annihilated by all the $\hat{T}^1[\gamma]$: impose the condition $\hat{T}^1[\gamma]\,|0\rangle = 0$. (In the connection representation, $|0\rangle$ is thus the function $\Psi(\overline{A}) = 1$.) Now, generate the Hilbert space H by operating on $|0\rangle$ with all elements of the $\hat{T}$-algebra. To explore the structure of H, we first note that every element of $\hat{T}$ is a finite sum of finite products of $\hat{T}^0$ and $\hat{T}^1$ operators and that we ca n use the CCRs to move, one by one, the $\hat{T}^1$-operators in each product to the extreme right. In the process, we get some commutators of $\hat{T}^1$ and $\hat{T}^0$ operators. Each of these commutators is a $\hat{T}^0$. Furthermore, because of the third identity in Eq.10, each product of $\hat{T}^0$s can be reduced to a linear combination of $\hat{T}^0$s. Therefore, any element $\hat{A}\,|0\rangle$ of H can be reduced to the form $\hat{A}\,|0\rangle = \sum K_i \hat{T}^0[\alpha_i]\,|0\rangle$. This is the vector space underlying H. By its very construction, it is clear how the $\hat{T}$-algebra acts on it. Therefore, it only remains to specify the inner product.

For this, the simplest strategy would be the following. One might first express the state $A\,|\,0\rangle$ as a function on the space of homotopy classes by representing each ket $\hat{T}^0[\alpha]\,|\,0\rangle$ as the characteristic function of the class α, i.e., the function which takes the value 1 on the class α and zero elsewhere. (Thus, the ket $\hat{A}\,|\,0\rangle$ given above would be represented as the function $\Psi(\alpha) = K_i$ if $\alpha = \alpha_i$, and 0 otherwise.) One might then exploit the discreteness of the space of homotopy classes to introduce a measure on it simply by requiring that the characteristic functions form an orthonormal basis. This general strategy is indeed correct but has to be refined. The problem is that due to identities satisfied by the $\hat{T}^0$-operators, kets $\hat{T}^0[\alpha]\,|\,0\rangle$ and $\hat{T}^0[\beta]\,|\,0\rangle$ can be equal even when the homotopy classes α and β are distinct, whence it is inconsistent to assume that there is a 1-1 mapping between these elementary kets and homotopy classes. For example, β could be α^{-1}; or, α may equal $\gamma\#\delta$ and β may equal $\delta\#\gamma$. Hence, the elementary kets are in 1-1 correspondence with *equivalence classes* $\{\alpha\}$ of homotopies α, where $\alpha \sim \beta$ if and only if $\overline{T}^0[\alpha](\overline{A}) = \overline{T}^0[\beta](\overline{A})$ for all $\overline{A}$ in the reduced configuration space $\overline{\mathcal{C}}$. These equivalence classes will be referred to as *equitopies*. Let H' be the Hilbert space of functions on the (discrete) set of equitopy classes where the inner product is defined simply by requiring that the characteristic functions be orthonormal. Thus H' is generated by the elementary kets $\hat{T}^0[\alpha] \mid 0\rangle$. This is not however the physical Hilbert space H constructed above, because while each elementary ket does belong to H, the set of elementary kets is overcomplete in it. This comes about due to the third identity in (10), which has not been used in the construction of H'. Setting β equal to $\gamma\#\delta$ in this identity, we find linear relations: $\hat{T}^0[\alpha\#\gamma\#\delta] + \hat{T}^0[\alpha\#\delta^{-1}\#\gamma^{-1}] - \hat{T}^0[\alpha\#\delta\#\gamma] - \hat{T}^0[\alpha\#\gamma^{-1}\#\delta^{-1}] = 0$. Consequently, the wave function $\Psi(\{\alpha\})$ representing a general ket must satisfy the condition $\sum K_i\Psi(\{\alpha_i\}) = 0$ whenever $\sum K_i\overline{T}^0(\{\alpha_i\}) = 0$ on $\overline{\mathcal{C}}$. H is the subspace of H' where this linear relation holds. Finally, the action of $\hat{T}^A[\alpha]$ can be expressed directly in this "functional representation" as follows:

$$\begin{aligned}(\hat{T}^0[\alpha]\cdot\Psi)(\{\beta\}) &:= \sum_{\diamond}\Psi(\{[\alpha\#\beta]^{\circ}\})\\ (\hat{T}^1[\alpha]\cdot\Psi)(\{\beta\}) &:= i\hbar\sum_i\sum_{\diamond}\Delta_i(\alpha,\beta)\Psi(\{[\alpha\#_i\beta]^{\circ}\})\end{aligned} \tag{11}$$

where, to define the right side we can use any convenient loop β in the equivalence class $\{\beta\}$; the result is independent of the choice (although individual terms on the right side need not be).

The same $\star$-representation can be obtained through the Gelfand-Naimark-Segal construction directly from the $\star$-algebra. To do so, it suffices to note that, due to the conditions imposed in the construction of the $\hat{T}$-algebra, there is a unique positive linear functional f on it satisfying: $f(\hat{A}\cdot\hat{T}^1[\alpha]) = f(\hat{T}^1[\alpha]\cdot\hat{A}) = 0$ for all elements $\hat{A}$ of the $\hat{T}$-algebra and all homotopy classes α; and, ii) $f(\hat{T}^0[\alpha]) = 1$ if α is the trivial homotopy class and 0 otherwise. Using this functional as the vacuum expectation value, one recovers the representation constructed above. While this construction is not as direct as the one given above, it is applicable in more general contexts –e.g., in the loop representation of the quantum Maxwell field – and, among the available techniques, is perhaps the best suited for use in the 3+1 theory.

How does this representation compare with the connection representation? Although a conclusive, general proof does not exist, there are arguments indicating that the loop representation is irreducible. (In the example of the torus considered below, the proof *is* complete.) What is clear is that the Hilbert space H of the loop representation, as constructed here, captures only a relatively small part of the complete Hilbert space $\mathcal{H}$ of the connection representation. Whether this feature is a strength or a weakness is, however, still unclear. For, it may well turn out that only the states selected by the loop representation are physically relevant, e.g., in the sense that they admit some sort of geometrodynamical interpretation. If this turns out to be true in general, the feature would be an asset. On the other hand, it may also turn out that the loop representation misses out some physical information which is crucial to a satisfactory description of the Planck regime. In this case, one would have to generalize the above construction. The Rovelli-Smolin transform [8] provides one such avenue at least in the 2+1 theory.

3.3 Illustration

In this subsection, we illustrate the ideas introduced so far by explicitly constructing the connection and the loop representations for the case when the spatial 2-manifold, Σ, is a 2-torus.

Let α_1 and α_2 denote the generators of the homotopy group of Σ. Then, by the defining relation of the homotopy group, we have $\alpha_1\alpha_2 = \alpha_2\alpha_1$; the group is Abelian. Hence, the homotopy of *any* closed loop α is labelled just by two integers, n_1, n_2, which tell us how many times the loop winds around the two generators; $\alpha = \alpha_1^{n_1}\alpha_2^{n_2}$. Fix any base point p on Σ. Any flat connection on Σ can now be characterized just by the pair of holonomies $U[\alpha_1]$ and $U[\alpha_2]$ around the two generators, modulo the action of the gauge group at p. Since the homotopy group is Abelian, the holonomies must also commute, $U[\alpha_1]\cdot U[\alpha_2] = U[\alpha_2]\cdot U[\alpha_1]$, whence they are $SO(2,1)$ rotations around the *same* axis. As discussed in section 2, under the action of the gauge group at p, the axis itself rotates preserving only its time-like, null or space-like character. Therefore, the gauge equivalent classes $\overline{A}$ of flat connections fall into three disconnected sectors. A simple calculation reveals that the sector with time-like axis has topology $S^1 \times S^1$, the one with null axis has topology $\mathbb{R}^1$ while the one with space-like axis is given by $[0,\infty) \times [0,\infty)$. To be specific, we shall discuss the time-like sector (which, in any case, is the most interesting of the three) in detail.

In the connection representation, the time-like component $\overline{\mathcal{C}}_t$ of the reduced configuration space has topology $S^1 \times S^1$; it is again a 2-torus! Let us coordinatize it by two numbers, a_I, where $I = 1, 2$, with $a_I \in [-1, 1]$. Each element $\overline{A}$ of $\overline{\mathcal{C}}_t$ is thus labelled by two numbers, a_I. The volume element $d\overline{V}$ on $\overline{\mathcal{C}}$ now turns out to be precisely $da_1 \wedge da_2$. Thus, the Hilbert space $\mathcal{H}_t$ of quantum states are just square-integrable functions $\Psi(a_1, a_2)$ on $\overline{\mathcal{C}}_t$. Finally, it is straightforward to work out the explicit expressions of the basic $\hat{\mathcal{T}}$ operators. The operators associated with the generators α_I of the homotopy group of Σ are given by:

$$\begin{aligned}\hat{\mathcal{T}}^0[\alpha_J]\cdot\Psi(a_1,a_2) &= 2\cos(a_J\pi)\ \Psi(a_1,a_2)\\ \hat{\mathcal{T}}^1[\alpha_J]\cdot\Psi(a_1,a_2) &= 4\pi i\hbar\sin(a_J\pi)\,\epsilon_{IJ}\ \frac{\partial\Psi(a_1,a_2)}{\partial a_I}\,,\end{aligned} \tag{12}$$

where ϵ_{IJ} is the anti-symmetric symbol. Thus, $\hat{\mathcal{T}}^0[\alpha_J]$ commute among themselves and so do $\hat{T}^1[\alpha_J]$. Similarly, the two operators associated with any one generator commute with each other. The non commuting (and hence, conjugate) pairs are $\hat{T}^0[\alpha_1]$, $\hat{T}^1[\alpha_2]$, and, $\hat{T}^0[\alpha_2]$, $\hat{T}^1[\alpha_1]$.

In the loop representation, quantum states are functions of the equitopy classes. Given a closed loop γ whose homotopy class is characterized by n_1, n_2, and a gauge equivalence class of flat connections labelled by a_1, a_2, the trace of the holonomy of the connection around γ is easy to compute. We have: $\overline{T}^0[n_1, n_2](a_I) = \cos(n_1 a_1 \pi + n_2 a_2 \pi)$. Now, in the definition of equitopy classes, two loops are equivalent if the trace of the holonomy around one equals that around the other for *any* flat connection. It therefore follows that the loop (n_1, n_2) is equivalent just to the loop $(-n_1, -n_2)$. Thus, now quantum states are functions, $\Psi(n_1, n_2)$, of two integers, satisfying $\Psi(n_1, n_2) = \Psi(-n_1, -n_2)$, with finite norm : $\langle \Psi, \Psi \rangle \equiv \sum \overline{\Psi(n_1, n_2)}\, \Psi(n_1, n_2) < \infty$. (Because the homotopy group is now Abelian, the linear relations $\sum K_i \overline{T}^0[\gamma_i] = 0 \Rightarrow \sum K_i \Psi(\alpha_i) = 0$ are automatically satisfied; $H' = H$.) This is the Hilbert space H of the loop representation. From (11), the basic operators associated with the generator α_1 can be computed. We have:

$$\begin{aligned} \hat{T}^0[\alpha_1] \cdot \Psi(n_1, n_2) &= \Psi(n_1 + 1, n_2) + \Psi(n_1 - 1, n_2) \\ \hat{T}^1[\alpha_1] \cdot \Psi(n_1, n_2) &= i\hbar n_2[\Psi(n_1 + 1, n_2) - \Psi(n_1 - 1, n_2)], \end{aligned} \tag{13}$$

and similarly for operators associated with the generator α_2.

It is easy to check that the Hilbert space H has no proper subspace that is left invariant by all the $\hat{T}$-operators; in this case, the loop representation of the CCR is indeed irreducible. The connection representation, on the other hand, is reducible even on the time-like sector considered above. It contains two irreducible pieces: The subspaces of the Hilbert space $\mathcal{H}_t$ consisting of wavefunctions which are even or odd under the operation $(a_I) \rightarrow (-a_I)$. In the even sector, the internal time reversal operator T has eigenvalue $+1$ while on the odd sector, it has eigenvalue -1. (On both sectors, the internal parity operator P is identity). It is the even sector that is isomorphic with the loop representation. The odd sector of the connection representation has no counterpart at all in the loop picture.

Finally, let us display the isomorphism between the even sector of the connection representation and the loop representation. The isomorphism is a special case of the transform first introduced by Rovelli and Smolin [8] in the 3+1 theory as a formal device and now exists rigorously. Recall that the general idea was to set:

$$\Psi[\gamma] = \int_{\mathcal{C}} \text{“}d\mu(A)\text{''}\; T^0[\gamma](A)\; \Psi(A). \tag{14a}$$

In 2+1 gravity, we can restrict ourselves to the reduced configuration space $\overline{\mathcal{C}}$ and use as our measure the volume element $d\overline{V}$ introduced in subsection 3.1. With these choices, the Rovelli-Smolin transform exists rigorously. In the case when Σ is a 2-torus, the integral then reduces to:

$$\Psi[n_1, n_2] = 2\int_{-1}^{1} da_1 \int_{-1}^{1} da_2 \cos(a_1 n_1 + a_2 n_2)\pi\ \Psi(a_1, a_2). \tag{14b}$$

4 Discussion

In the last two sections, we applied the 3+1 quantization program to 2+1 gravity and found, that as far as the construction of a consistent mathematical framework is concerned, the program can indeed be completed. A number of important physical issues remain. These include the isolation of a good time parameter, the introduction of the notion of dynamics, direct interpretation of observables and states and the entire subject of quantum measurement theory. As emphasized, e.g., in [3] and [6], these are important issues and must be resolved before we can claim that a fully satisfactory quantum theory exists. A number of these problems have already been addressed by various authors [3,6,10] and further work is also in progress. Here, we focussed on the "mathematical physics" problems because our aim was to investigate whether they could be solved in the 2+1 theory and, if so, whether the methods used can be taken over to the 3+1 case.

What lessons we can draw from this 2+1 analysis for the 3+1 theory? As pointed out in the Introduction, the open problems in the 3+1 quantization program are of two types: The ones whose origin lies in the infinite dimensional, field theoretical issues and the ones which arise specifically because the theory has no background geometry. In broad terms, what the 2+1 analysis has done is to provide considerable confidence that the general program is capable of handling difficulties of the second type.

Specifically, we have learned two conceptual lessons and several technical ones. Let me begin with a few illustrative examples of the technical hints. First, the trick of "deriving" the $T^1[\gamma](A, \widetilde{E})$ from $T^0[\gamma](A)$ was first discovered [4] in the 2+1 theory and then carried over to the 3+1 case. This trick provided a simple proof of the completeness of the T variables, a property that plays an important role in quantum theory. It also allowed us to obtain algebraic conditions satisfied by the $T^1[\gamma](A, \widetilde{E})$ variables starting from those satisfied by the $T^0[\gamma](A)$, thereby providing us with a complete list of "anti-commutation" relations that must be imposed to obtain the quantum algebra. In the early work [3,8] these anti-commutation relations were overlooked. Another technical insight is the relation between the loop and the connection representations. The fact that the straightforward construction of the loop representation captures only a sector of the quantum theory in the connection representation may be quite important to quantum theory. The use of the GNS construction to obtain the inner product is also instructive since it may well extend to the 3+1 theory. Finally, we saw that if the set S of elementary classical variables is complete only *almost* everywhere, certain quantum operators are superselected and therefore quantization ambiguities occur. This subtle phenomenon is likely to occur also in the 3+1 theory.

On the conceptual front, we have learned that, for a number of reasons, connections are better suited than metrics as the basic variables for quantum gravity. First, the equations of the theory are simpler both in the classical and quantum regimes. Second, there are states in the quantum theory $-\Psi(A) = 1$ in the connection representation, or, $|\,0\rangle$ in the loop representation– in which the expectation value of the spatial metric *vanishes identically*. Furthermore, $|\,0\rangle$ is in fact the *cyclic* state from which the entire Hilbert space in the loop representation was constructed. As remarked earlier, this construction is likely to extend to the 3+1 theory. At least at first glance, an analogous construction seems difficult in quantum geometrodynamics. Once one accepts this premise, one is led to the viewpoint that the structure of quantum gravity in the Planck regime would be more transparent in terms of connections and holonomies rather than metrics and lengths. The second conceptual lesson is that closed loops are well suited to capture the diffeomorphism invariance of the theory. In the 2+1 theory, the diffeomorphism invariance led us to represent quantum states as functions on the space of homotopy classes. The discreteness of this space, in turn, simplified the problem of selecting a suitable inner product. In the 3+1 case, the homotopy classes are replaced by the (generalized) link classes. This space is again discrete. It is likely that this fact will again play an important role in the mathematical framework of the 3+1 quantum theory.

References

[1] E. Witten, Nucl. Phys.**B311**, 46 (1988).

[2] S. Deser, R. Jackiw and G. 'tHooft, Ann. Phys. (N.Y.)**152**, 220 (1984).

[3] A. Ashtekar, V. Husain, C. Rovelli, J. Samuel and L. Smolin, Class. & Quant. Grav.**6**, 185 (1989).

[4] A. Ashtekar and J. Romano, Phys. Lett.**229B**, 56 (1989).

[5] I. Bengtsson, Phys. Lett.**220B**, 51 (1989).

[6] V. Moncrief, Talk at the 12th international conference on general Relativity

and gravitation (1989); How solvable is 2+1 dimensional Einstein gravity? (Yale Pre-print, 1990).

[7] S. Martin, Nucl. Phys.**B327**, 178 (1989).

[8] C. Rovelli and L. Smolin, Phys. Rev. Lett.**61**, 1155 (1988); Nucl. Phys.**B331**, 80 (1990).

[9] A. Ashtekar, Comm. Math. Phys.**71**, 59 (1980).

[10] S. Carlip, Nucl. Phys.**324B**, 106 (1989); Observables, gauge invariance and time in 2+1 dimensional quantum gravity, pre-print (1990).

APPENDICES

Appendix A SPINORS

1 Introduction

As we saw in chapter 9, to couple Dirac fields to gravity, it is simplest to use the new variables in the spinorial form, introduced in chapter 5. Also, recall that one of the new canonical variables is a self-dual connection and the spinorial notation is especially well-suited for calculations involving self-dual fields. Therefore, in this appendix we briefly review the algebra and calculus of $SL(2,\mathbb{C})$ and $SU(2)$ spinors and the relation between the two types of spinors. Although this treatment is self-contained conceptually, detailed calculations are generally left as exercises. A more complete treatment can be found in references [1-3].

Let us begin by recalling some facts about complex vector spaces. Let W be a complex vector space and W^* be its dual, (i.e. the complex vector space of all linear mappings from W to complex numbers. As in the case of a real vector space, one can construct tensors by taking arbitrary tensor products of W and W^*. However, in this case, because of the complex nature of the vector space, we also need to consider their complex conjugate spaces $\overline{W}$ and $\overline{W^*}$. They can be defined, respectively, as the complex vector spaces of all antilinear mappings from W^* and W to complex numbers. As the name and notation suggest there exists a canonical, one-to-one antilinear mapping called *complex conjugation* (denoted by over-bar) from W onto $\overline{W}$ and from W^* onto $\overline{W^*}$. For $\alpha \in W$, $\overline{\alpha} \in \overline{W}$ is defined by:

$$\overline{\alpha}(\beta) := \overline{\beta(\alpha)}, \qquad \forall \beta \in W^*. \tag{1}$$

Complex conjugation from W^* to $\overline{W^*}$ is similarly defined. These two complex conjugate spaces $\overline{W}$ and $\overline{W^*}$ are dual to each other. Thus, in the case of a complex vector space we need to construct tensors based on all these four basic building blocks. (Note that we do not have to consider, in addition, spaces obtained by taking duals and complex conjugates of $\overline{W}$ and $\overline{W^*}$ since the resulting vector spaces are naturally isomorphic to one of the four basic vector spaces.)

Abstract index notation on real tensors [1] is extended to complex tensors as follows. To each of the building blocks, we assign upper case latin indices. As usual, elements of W carry a superscript and elements of W^* carry a subscript. Elements of their complex conjugate spaces $\overline{W}$, $\overline{W^*}$ will carry, respectively, primed superscripts and primed subscripts. Thus, for example, $\alpha^A \in W$, $\beta_A \in W^*$, $\eta^{A'} \in \overline{W}$, and $\rho_{A'} \in \overline{W^*}$. This notation is trivially extended to (complex) tensors.

Recall that in the abstract index notation for real tensors a linear mapping is represented by the contraction between a superscript and a subscript. In the case of complex tensors we should also consider antilinear mappings. However, any antilinear mapping can be decomposed into complex conjugation and a linear mapping. Thus, we need to define contraction only between two indices of the same kind.

We will often encounter the situation in which there is an internal vector space isomorphic to W *at each point* of the manifold M. More precisely, we will need to consider a fiber bundle on a manifold M whose fiber is isomorphic to the complex vector space W. In such cases we need to generalize the concept of tensor fields. The objects of interest belong to tensor products of manifold tensors with the internal tensors. We will call such objects ***generalized tensors*** [2] or ***W-valued tensors***. They are assigned both manifold tensor indices and internal indices.

2 $SL(2,\mathbb{C})$ spinors

Let W be a 2-dimensional complex vector space. Since W is 2-dimensional, the space of 2-forms, ϵ_{AB}, over W is 1-dimensional. Let us fix a non-vanishing 2-form ϵ_{AB} and define its inverse ϵ^{AB} by the relation $\epsilon^{AB}\epsilon_{AC} = \delta_C{}^B$. Let $L^A{}_B$ be a 1-1 linear mapping from W onto itself which preserves this 2-form:

$$\epsilon_{AB}L^A{}_C L^B{}_D = \epsilon_{CD}. \tag{2}$$

Then

$$\det(L^A{}_B) := \tfrac{1}{2}\,\epsilon_{AB}\epsilon^{CD}L^A{}_C L^B{}_D = 1, \tag{3}$$

so that $L^A{}_B$ is an element of the group $SL(2,\mathbb{C})$. Note that ϵ_{AB} and ϵ^{AB} provide an isomorphism between two spaces, W and W^*. Using this isomorphism we raise and lower W-indices with the following convention:

$$\xi^A = \epsilon^{AB}\xi_B, \qquad \xi^A\epsilon_{AB} = \xi_B. \tag{4}$$

Similarly, primed indices are raised and lowered using $\bar{\epsilon}_{A'B'}$ and $\bar{\epsilon}^{A'B'}$ This is quite similar to using a metric g_{ab} to raise and lower manifold tensor indices. However one should be careful about index position because of the antisymmetry of the ϵ-tensors.

Consider the space $\mathbf{V}$ of all objects of the form $\alpha^{AA'}$ satisfying:

$$\overline{\alpha}^{AA'} = -\alpha^{AA'}. \tag{5}$$

It is easy to check, by choosing a suitable basis in W that $\mathbf{V}$ has the structure of a 4-dimensional real vector space. Moreover, $\mathbf{V}$ is equipped with a natural metric: $\epsilon_{AB}\overline{\epsilon}_{A'B'}$ is a metric of signature $(-,+,+,+)$ on $\mathbf{V}$.

Fix a 4-dimensional space-time (M, g_{ab}) and consider a fiber bundle over M each of whose fibers is isomorphic to W. Since the vector space $\mathbf{V}$ is 4-dimensional and since it is equipped with a metric of signature $(-,+,+,+)$, it is natural to identify, at each point of M, the space $\mathbf{V}$ associated with the fiber over that point with the tangent space at that point. That is, it is natural to introduce an isomorphism $\sigma^a{}_{AA'}$ between $\mathbf{V}$ and the tangent space to M:

$$\sigma^a{}_{AA'}\alpha^{AA'} := \alpha^a, \tag{6a}$$

such that the metric on $\mathbf{V}$ is mapped to the metric on the tangent space of M:

$$g^{ab} = \sigma^a{}_{AA'}\,\sigma^b{}_{BB'}\,\epsilon^{AB}\,\overline{\epsilon}^{A'B'}. \tag{6b}$$

If $\sigma^a{}_{AA'}$ exists globally on M, we say that M admits an $SL(2,\mathbb{C})$ *spinor structure.* Objects of the form α^A $(\beta^{A'})$ are called unprimed (primed) $SL(2,\mathbb{C})$ *spinors* and $\sigma^a{}_{AA'}$—which glues the internal indices to the tangent space indices—is called an $SL(2,\mathbb{C})$ *soldering form.* If a given pseudo-Riemannian manifold admits a spinor structure, the soldering form $\sigma^a{}_{AA'}$ is unique up to a local $SL(2,\mathbb{C})$ transformation.

Let us make a few comments.

i) The inverse of $\sigma^a{}_{AA'}$ is $\sigma_a{}^{AA'}$, where the tensor index a is lowered using g_{ab} and the W-indices A, A' are raised using ϵ^{AB} and $\overline{\epsilon}^{A'B'}$.

ii) Because tangent vectors are real, the isomorphism satisfies $\overline{\sigma}_a{}^{AA'} = -\sigma_a{}^{AA'}$.

iii) The space $\mathbf{V}$ has a natural time orientation as well as a natural total orientation. Let us first consider the time orientation. All elements of $\mathbf{V}$ of the form $-i\,\psi^A\overline{\psi}^{A'}$ are null with respect to the metric on $\mathbf{V}$ and they all lie on the same half of the light cone, since

$$\left(-i\psi^A\overline{\psi}^{A'}\right)\left(-i\phi^B\overline{\phi}^{B'}\right)\epsilon_{AB}\,\overline{\epsilon}_{A'B'} = -\left|\psi^A\phi_A\right|^2 \tag{7}$$

is non-positive, vanishing if and only if ψ^A is proportional to ϕ^A. Similarly, all elements of the form $i\,\psi^A\overline{\psi}^{A'}$ lie on the other half of the light cone. Thus all null vectors are nicely divided into two classes. By convention we call the first class *future-directed.* This induces a time orientation on the manifold (M, g_{ab}) through the soldering form. Next, to show that a total orientation exists, we construct a globally defined nowhere vanishing 4-form from ϵ_{AB}:

$$e_{AA'BB'CC'DD'} := -i\epsilon_{AB}\epsilon_{CD}\bar{\epsilon}_{A'C'}\bar{\epsilon}_{B'D'} + c.c. \tag{8}$$

where '*c.c.*' stands for the complex conjugate. The expression is totally anti-symmetric (in pairs of indices AA') and real by inspection. Again, using the soldering form, one can construct from Eq.8 a total orientation on the manifold (M, g_{ab}). (The numerical factor in Eq.8 is so chosen that the space-time orientation form has the standard normalization.) Thus, a pseudo-Riemannian manifold (M, g_{ab}) should be at least orientable and time orientable if it is to have an $SL(2,\mathbb{C})$ spin structure on it.

iv) There is a two-to-one homomorphism from the local $SL(2,\mathbb{C})$ transformation group on spinors to the local proper Lorentz group of (M, g_{ab}). To see this, let $L^A{}_B$ be a local $SL(2,\mathbb{C})$ transformation. Then

$$L^a{}_b := \sigma^a{}_{AA'}\,\sigma_b{}^{BB'}\,L^A{}_B\bar{L}^{A'}{}_{B'} \tag{9}$$

is a Lorentz transformation with respect to the metric g_{ab}, and the mapping

$$\Lambda : \qquad L^A{}_B \mapsto L^a{}_b \tag{10}$$

is the two-to-one homomorphism between two groups. Since the kernel of the homomorphism consists only of two elements, their Lie algebras are isomorphic. To exhibit the isomorphism let us consider a parametrized curve in $SL(2,\mathbb{C})$ which passes through the identity, and the image of this curve, via Eq.9, in the Lorentz group. By taking the derivative of both sides with respect to the parameter and evaluating it at the identity element, we get:

$$\dot{L}^a{}_b = \sigma^a{}_{AM'}\sigma_b{}^{BM'}\,\dot{L}^A{}_B + \sigma^a{}_{MA'}\sigma_b{}^{MB'}\,\dot{\bar{L}}^{A'}{}_{B'} \tag{11}$$

and

$$\dot{L}^A{}_B = \tfrac{1}{2}\sigma_a{}^{AM'}\sigma^b{}_{BM'}\,\dot{L}^a{}_b. \tag{12}$$

Thus, the two Lie algebras are indeed isomorphic.

This concludes the discussion of spinor algebra. Let us now consider calculus. The first step is to extend the derivative operators that operate on space-time tensor fields to those that operate on generalized tensor fields. A derivative operator on generalized tensors, ∇, is to have the following defining properties [1-3]: *i*) While acting on tensors, it is a torsion-free derivative operator; *ii*) Its action on generalized tensors is linear and satisfies the Leibnitz rule; *iii*) It is "real", i.e., satisfies

$$\overline{\nabla_a \alpha} = \nabla_a \overline{\alpha}, \tag{13}$$

for arbitrary W-valued tensors α; and *iv*) It annihilates ϵ:

$$\nabla_a \epsilon_{AB} = 0. \tag{14}$$

How many derivative operators are there? Consider any two operators ∇ and $\tilde{\nabla}$. Due to properties *ii), iii)* listed above, their difference on any generalized tensor fields can be determined from their difference on covectors and on unprimed spinors. Let us define two tensor fields $C_{ab}{}^c$ and $C_{aA}{}^B$ by:

$$(\nabla_a - \tilde{\nabla}_a)\, k_b =: C_{ab}{}^c k_c, \quad (\nabla_a - \tilde{\nabla}_a)\lambda_A =: C_{aA}{}^B \lambda_B, \tag{15}$$

for any k_b and λ_A. These tensors satisfy:

$$C_{ab}{}^c = C_{ba}{}^c, \quad \text{and} \quad C_{a[AB]} := C_{a[A}{}^C \epsilon_{|C|B]} = 0 \quad (\text{which} \Leftrightarrow C_{aA}{}^A = 0), \tag{16}$$

due to the torsion-free property and the annihilation of ϵ, respectively. Thus, there are "as many" derivative operators as there are fields $C_{ab}{}^c$ and $C_{aA}{}^B$ satisfying Eq.16.[1]

It is well-known from Riemannian geometry that, on any space-time (M, g_{ab}), there exists a unique torsion-free derivative operator ∇ on tensors which annihilates g_{ab}. However, ∇ does not admit a unique extension to spinors. To pick out a unique extension we must impose an additional condition. The most natural condition is that the extension be compatible with the soldering form, namely:

$$\nabla_a \sigma^b{}_{AA'} = 0. \tag{17}$$

With this condition the action of ∇ on spinor fields is uniquely determined. Eq.6*b* implies that this condition is consistent with $\nabla_a g_{bc} = 0$. (We will continue to use the same symbol, ∇, for this extension.)

1 Note that Eq.12 implies: $(\nabla_a - \tilde{\nabla}_a)\lambda_{A'} = \overline{C}_{aA'}{}^{B'} \lambda_{B'}$.

Let us discuss a few properties of ∇. Define the curvature tensor ${}^4R_{abA}{}^B$ on spinors by:

$$(\nabla_a\nabla_b - \nabla_b\nabla_a)\lambda_A =: {}^4R_{abA}{}^B\lambda_B, \tag{18}$$

for all λ_A. Now, by acting with the commutator of two covariant derivatives on $\sigma_a{}^{AA'}$, we can obtain the relation between ${}^4R_{abA}{}^B$ and the Riemann tensor ${}^4R_{abc}{}^d$, the curvature tensor of ∇ on space-time tensor fields:

$$0 = (\nabla_a\nabla_b - \nabla_b\nabla_a)\sigma_c{}^{AA'} = {}^4R_{abc}{}^d\sigma_d{}^{AA'} - {}^4R_{abM}{}^A\sigma_c{}^{MA'} - {}^4\overline{R}_{abM'}{}^{A'}\sigma_c{}^{AM'}. \tag{19}$$

Contracting this equation with $\sigma^m{}_{AA'}$ we obtain:

$${}^4R_{abc}{}^d = {}^4R_{abM}{}^A\sigma_c{}^{MA'}\sigma^d{}_{AA'} + {}^4\overline{R}_{abM'}{}^{A'}\sigma_c{}^{AM'}\sigma^d{}_{AA'}. \tag{20}$$

In view of Eq.11, this result is not surprising.

Let us decompose the Riemann tensor into its self-dual and anti-self-dual parts:

$$\begin{aligned} {}^{+4}R_{abc}{}^d &:= \tfrac{1}{2}({}^4R_{abc}{}^d - i\,{}^{*4}R_{abc}{}^d) \\ {}^{-4}R_{abc}{}^d &:= \tfrac{1}{2}({}^4R_{abc}{}^d + i\,{}^{*4}R_{abc}{}^d), \end{aligned} \tag{21}$$

where ${}^{*4}R_{abc}{}^d$ is the dual of the Riemann tensor, defined by:

$${}^{*4}R_{abc}{}^d := \tfrac{1}{2}\epsilon_c{}^{dm}{}_n\,{}^4R_{abm}{}^n. \tag{22}$$

Now, using the identity

$$\tfrac{1}{2}\,\epsilon^{abcd}\,\sigma_c{}^{AM'}\sigma_{dBM'} = i\,\sigma^{[a|AM'|}\sigma^{b]}{}_{BM'}, \tag{23}$$

one can show that:

$${}^4R_{abA}{}^B\,\sigma_c{}^{AM'}\sigma^d{}_{BM'} = {}^{+4}R_{abc}{}^d. \tag{24}$$

Thus, in the right hand side of Eq.20, the first term is the self dual part of the Riemann tensor, and the second term is the anti-self dual part of the Riemann tensor.

3 $SU(2)$ spinors

While $SL(2,\mathbb{C})$ spinors are defined on a 4-dimensional pseudo-Riemannian manifold (M,g_{ab}), $SU(2)$ spinors are defined on a 3-dimensional Riemannian manifold (Σ,q_{ab}) which may or may not be regarded as a spacelike hypersurface of (M,g_{ab}). In this section we regard (Σ,q_{ab}) as an abstract Riemannian 3-manifold and define $SU(2)$ spinors on it. In the next section we will consider (Σ,q_{ab}) as a spacelike hypersurface imbedded in (M,g_{ab}), and discuss the relationship between $SL(2,\mathbb{C})$ spinors and $SU(2)$ spinors. The discussion of $SU(2)$ spinors is quite parallel to that of $SL(2,\mathbb{C})$ spinors given above. Therefore, in this section, we shall skip the details and emphasize only the points of difference between the two types of spinors.

Consider, as before, a bundle over the 3-manifold Σ, each fiber of which is isomorphic to the 2-dimensional complex vector space W. As in the case of $SL(2,\mathbb{C})$ spinors, we fix a nowhere vanishing antisymmetric object ϵ_{AB} and use it to raise and lower internal indices. To define $SU(2)$ spinors we need one more structure: a positive definite Hermitian inner product $\langle\ ,\ \rangle$ on each fiber. Define $G_{A'A}$ by

$$\overline{\psi}^{A'}G_{A'A}\phi^A := \langle\psi,\phi\rangle. \tag{25}$$

Then from the hermiticity and the positivity of the inner product it follows that

$$\overline{G}_{A'A} = G_{A'A}, \qquad \overline{\psi}^{A'}G_{A'A}\psi^A > 0 \quad \forall\psi^A \neq 0. \tag{26}$$

We will assume that ϵ_{AB} (or $G_{A'A}$) is normalized so that

$$\overline{\epsilon}^{A'B'}G_{A'A}G_{B'B} = \epsilon_{AB} \qquad (\Leftrightarrow \quad G_{A'A}G^{A'B} = \delta_A{}^B). \tag{27}$$

$G_{AA'}$ defines the '‡' operation, discussed in chapter 5, via

$$\alpha^A \mapsto (\alpha^\ddagger)_A := G_{AA'}\overline{\alpha}^{A'}. \tag{28}$$

However, as pointed out in chapter 5, it is more convenient to use ϵ_{AB} to introduce an operation '†' that maps W to itself:

$$\alpha^A \mapsto (\alpha^\dagger)^A \; := \; -\epsilon^{AB}(\alpha^\ddagger)_B \; \equiv \; -\epsilon^{AB}G_{BA'}\overline{\alpha}^{A'}, \tag{29}$$

where $\overline{\alpha}$ is the complex conjugate of α defined in Eq.1. Then it follows from Eqs.26 and 27 that

$$(\alpha^\dagger)^A\alpha_A \geq 0 \; (= 0 \Leftrightarrow \lambda_A = 0) \quad \text{and} \quad (\alpha^{\dagger\dagger})^A \equiv ((\alpha^\dagger)^\dagger)^A = -\alpha^A. \tag{30}$$

Recall that the 1-1 linear mappings from a 2-dimensional complex vector space onto itself which preserve ϵ_{AB} are the special linear mappings and the ones that preserve an Hermitian inner product are unitary. Thus, $SU(2)$ transformations are the ones that preserve both structures ϵ_{AB} and $G_{AA'}$ introduced above.

We can now extend the $\dagger$ operation to arbitrary spinors $\alpha_{A..B}{}^{C..D}$ and $\beta_{A..B}{}^{C..D}$ by requiring, as in chapters 5 and 6

$$(\alpha + c\beta)^\dagger := \alpha^\dagger + \bar{c}\beta^\dagger \quad \text{and} \quad (\alpha\beta)^\dagger := \alpha^\dagger\beta^\dagger. \tag{31}$$

With these definitions we find that Eq.27 is equivalent to the statement $(\epsilon^\dagger)_{AB} = \epsilon_{AB}$ and that given a transformation $T^A{}_B$,

$$(T^\dagger)^A{}_B = -G^A{}_{A'}G_B{}^{B'}\overline{T}^{A'}{}_{B'}. \tag{32}$$

Now consider the space $\mathbf{H}$ of all objects of the form $\alpha^A{}_B$ satisfying:

$$\alpha^A{}_A = 0, \qquad (\alpha^\dagger)^A{}_B = \alpha^A{}_B. \tag{33}$$

It is easy to verify that $\mathbf{H}$ is a 3-dimensional real vector space, equipped with a natural positive definite metric:

$$(\alpha, \beta) := -\alpha^A{}_B\,\beta^B{}_A, \tag{34}$$

for $\alpha, \beta \in \mathbf{H}$.

It turns out that there always exists a global isomorphism $\sigma^a{}_A{}^B$—called a $SU(2)$ *soldering form*—between the space $\mathbf{H}$ and the tangent space on (Σ, q_{ab}) satisfying:

$$q^{ab} = -\sigma^a{}_A{}^B\sigma^b{}_B{}^A \equiv -\operatorname{tr}(\sigma^a\sigma^b). \tag{35}$$

Eq.33 implies that $\sigma^a{}_A{}^B$ must satisfy: $\sigma^a{}_A{}^A = 0$ and $(\sigma^\dagger)^a{}_A{}^B = \sigma^a{}_A{}^B$. With this structure, ψ^A is called an $SU(2)$ *spinor* on (Σ, q_{ab}). $SU(2)$ transformations on spinors are tied to $SO(3)$ transformations on the tangent space by $\sigma_a{}^A{}_B$ through the 2-1 homomorphism:

$$U^A{}_B \quad \mapsto \quad U^a{}_b := \sigma^{aA}{}_B\,U^B{}_C\,\sigma_b{}^C{}_D(U^\dagger)^D{}_A. \tag{36}$$

The discussion of calculus for $SU(2)$ spinors is completely analogous to that of $SL(2,\mathbb{C})$ spinor calculus. In particular, there is a unique derivative operator D, satisfying properties *i–iv*, acting on tensors and spinors, which annihilates the soldering form $\sigma^a{}_{AB}$. Its spinorial and tensorial curvatures, $R_{abA}{}^B$ and $R_{abc}{}^d$ respectively, are related by a formula analogous to Eq.20:

$$R_{abc}{}^d = 2R_{abA}{}^B\,\sigma_c{}^{AM}\sigma^d{}_{BM}. \tag{37}$$

4 Relation between $SL(2,\mathbb{C})$ and $SU(2)$ spinors

In section 3 we introduced $SU(2)$ spinors on an abstract 3-dimensional Riemannian manifold (Σ, q_{ab}). Let us now suppose that (Σ, q_{ab}) is realized as a spacelike hypersurface imbedded in a 4-dimensional pseudo-Riemannian manifold (M, g_{ab}) with $SL(2,\mathbb{C})$ spinors defined on it. Then, it is useful to identify $SU(2)$ spinors on Σ with certain $SL(2,\mathbb{C})$ spinors on M, just as we identify intrinsic 3-dimensional tensors on Σ with 4-dimensional tensors on M which are tangential to Σ.

Recall that, at each point of Σ, one can identify $SO(3)$ transformations with proper Lorentz transformations which preserve the future-directed unit normal n^a. Similarly, $SU(2)$ transformations can be identified with $SL(2,\mathbb{C})$ transformations preserving $n^{AA'} := n^a \sigma_a{}^{AA'}$, where $\sigma_a{}^{AA'}$ is the $SL(2,\mathbb{C})$ soldering form. To see this, define

$$G_{A'A} := -\sqrt{2}\, i\, n_{A'A}. \tag{38}$$

This $G_{A'A}$ satisfies both properties of a Hermitian metric, Eqs.26. Hermiticity follows from $\overline{\sigma}_a{}^{AA'} = -\sigma_a{}^{AA'}$, while positivity follows from the fact that $-i\,\psi^A \overline{\psi}^{A'}$ is a future-directed null vector, whereas $n^{AA'}$ is a future-directed timelike vector. (The numerical factor $\sqrt{2}$ was inserted in the definition of $G_{A'A}$ to satisfy our normalization condition (27).) Thus, $SL(2,\mathbb{C})$ transformations preserving $n_{AA'}$ can be identified with $SU(2)$ transformations.

Let us now consider the horizontal subspace of $\mathbf{V}$ with respect to the normal $n_{AA'}$. It consists of all the elements $\alpha^{AA'}$ of $\mathbf{V}$ satisfying:

$$\alpha^{AA'} n_{AA'} = 0. \tag{39}$$

Clearly this space is 3-dimensional and the metric $\epsilon_{AB}\overline{\epsilon}_{A'B'}$ on $\mathbf{V}$ induces the following positive definite metric on the horizontal subspace:

$$\epsilon_{AB}\overline{\epsilon}_{A'B'} + n_{AA'}n_{BB'} = \epsilon_{AB}\overline{\epsilon}_{A'B'} - \tfrac{1}{2}\, G_{A'A}G_{B'B}. \tag{40}$$

Then it follows that the horizontal subspace is isometric to the space $\mathbf{H}$ we defined in Eq.33. The isometry is given by:

$$\alpha^{AA'} \quad \mapsto \quad \alpha^A{}_B := -\alpha^{AA'} G_{A'B}. \tag{41}$$

The fact that $\alpha^A{}_B$ belongs to the space $\mathbf{H}$ follows from Eq. (39) and the properties of $G_{AA'}$, Eqs.27–33. It is clearly an invertible linear mapping between the two spaces.

To show that the mapping preserves the metric, let us calculate the image of the metric on the horizontal space under the mapping:

$$G^{A'C}G^{B'D}(\epsilon_{AB}\bar{\epsilon}_{A'B'} - \tfrac{1}{2}G_{A'A}G_{B'B}) = \epsilon^{CD}\epsilon_{AB} - \tfrac{1}{2}\delta_A{}^C\delta_B{}^D, \tag{42}$$

which is precisely the metric on $\mathbf{H}$. Finally, from the $SL(2,\mathbb{C})$ soldering form $\sigma_a{}^{AA'}$ one can construct an $SU(2)$ soldering form in a trivial way:

$$\sigma_a{}^A{}_B := -q_a{}^m\sigma_m{}^{AA'}G_{A'B} \equiv i\sqrt{2}q_a{}^m\sigma_m{}^{AA'}n_{A'B}. \tag{43}$$

Thus, we can identify unprimed $SL(2,\mathbb{C})$ spinors on Σ with $SU(2)$ spinors. Finally, note that we can "invert" Eq.43 to obtain an expression for $\sigma_a^{AA'}$ in terms of $\sigma^a{}_A{}^B$and $n^{AA'}$:

$$\sigma_a^{AA'} = i\sqrt{2}\sigma_a{}^A{}_Bn^{BA'} - n_an^{AA'}. \tag{43'}$$

5 Sen connection

Given an imbedding of (Σ, q_{ab}) into (M, g_{ab}), it is natural to ask for the relation between the 3-dimensional derivative operator D on Σ (compatible with the $SU(2)$ soldering form $\sigma^a{}_A{}^B$) and the 4-dimensional derivative operator ∇ on M (compatible with $\sigma^a{}_A{}^{A'}$). Since the derivative operator D is determined by $D_a\sigma^A{}_{AB} = 0$, it annihilates the 3-metric q_{ab}. Hence, on spatial tensors, the action of D is given by:

$$D_a T_{b\ldots c}{}^{d\ldots e} = q_a{}^i q_b{}^j \ldots q_c{}^k q^d{}_m \ldots q^e{}_n \nabla_i T_{j\ldots k}{}^{m\ldots n} \quad \text{and} \quad D_b\sigma^a{}_A{}^B = 0. \tag{44}$$

Thus, the action of D on a horizontal tensor field is the same as the projection of all indices into Σ of the action of ∇ on that tensor field. Let us now ask how the action of D on $SU(2)$ spinor fields on Σ is related to the action of ∇ on $SL(2,\mathbb{C})$ spinor fields on M. To find this relation, it is convenient to first introduce a derivative operator $\mathcal{D}$ which is the ***pull-back*** of ∇ to Σ, i.e., whose action on generalized fields is defined by:

$$\mathcal{D}_a\alpha^c_{Ab} := q_a{}^l\, q_b{}^m\, q_n{}^c\, \nabla_l\alpha^n_{Am}. \tag{45}$$

It is straightforward to verify that $\mathcal{D}$ is indeed an $SU(2)$ derivative operator, i.e., that it satisfies the defining properties *(i)–(iv)* of a $SU(2)$ connection. In fact the action of $\mathcal{D}$ on tensor fields on Σ coincides with that of D. However, though $\mathcal{D}$ is compatible with the 3-metric q_{ab}, it is *not* compatible with the $SU(2)$ soldering form $\sigma^a{}_A{}^B$. We will refer to $\mathcal{D}$ as the *Sen connection* [4]. Let us explore its properties. First, it follows immediately from its definition that the difference, $\mathcal{D} - D$, is completely characterized by a generalized tensor $H_{aA}{}^B$:

$$(\mathcal{D}_a - D_a)\alpha_A =: H_{aA}{}^B \alpha_B. \tag{46}$$

A simple calculation shows that $H_{aA}{}^B$ is given just by $(-i/\sqrt{2})K_{ab}\sigma^b{}_A{}^B$, where K_{ab} is the extrinsic curvature of Σ in (M, g_{ab}), so that we have:

$$\mathcal{D}_a \alpha_A = D_a \alpha_A - \tfrac{i}{\sqrt{2}} K_{aA}{}^B \alpha_B. \tag{47}$$

(The simplest way to prove this assertion is to use Eq.42 to *define* a connection D on spinors and show that D, so defined, annihilates $\sigma^a{}_A{}^B$). Thus, while the connection D, compatible with $\sigma^a{}_A{}^B$, knows only about the intrinsic geometry of Σ, the Sen connection $\mathcal{D}$ knows about the extrinsic curvature as well. This fact plays an important role in chapters 5 and 6.

Because $\mathcal{D}$ is the restriction to Σ of ∇, we have:

$$\mathcal{D}_{[a}\mathcal{D}_{b]}\alpha_A = q_a{}^m q_b{}^n \nabla_{[m} q_{n]}{}^p \nabla_{np}\alpha_A, \tag{48}$$

so that the curvature $F_{abA}{}^B$ of $\mathcal{D}$ is related to the spinorial curvature ${}^4R_{abA}{}^B$ of ∇ via:

$$F_{abA}{}^B = q_a{}^m q_b{}^n \, {}^4R_{mnA}{}^B \tag{49}$$

This equation has interesting consequences. Using the relation (20) between the tensorial and spinorial curvature of ∇ one can show that:

$$\begin{aligned} \mathrm{tr}\,(\sigma^b F_{ab}) &\equiv \sigma^b{}_A{}^B F_{abB}{}^A = -\frac{i}{\sqrt{2}} q_a{}^b G_{bc} n^c \\ \mathrm{tr}\,(\sigma^a \sigma^b F_{ab}) &\equiv \sigma^a{}_A{}^B \sigma^b{}_B{}^C F_{abC}{}^A = G_{bc}\, n^b n^c, \end{aligned} \tag{50}$$

where G_{ab} is the Einstein tensor of g_{ab}. Thus, the constraint equations of the vacuum Einstein theory can be written simply as

$$\mathrm{tr}\,(\sigma^b F_{ab}) = 0, \qquad \mathrm{tr}\,(\sigma^a \sigma^b F_{ab}) = 0. \tag{516}$$

Furthermore, if g_{ab} satisfies the vacuum Einstein equation, the curvature $F_{abA}{}^B$ of $\mathcal{D}$ is essentially the self-dual part of the Weyl curvature. This comes about as follows. In the vacuum case, the Riemann tensor equals the Weyl tensor so that, by (24), the unprimed spinor curvature, ${}^4R_{abA}{}^B$, of ∇ has the same information as the self-dual part of the Weyl tensor. Since *any* self-dual 2-form at a point of M is completely determined by its pull-back to a spacelike, 3-dimensional subspace of the tangent space at that point, $F_{abA}{}^{B}$, in turn, has the same information as ${}^4R_{abA}{}^{B}$. In terms of 3-dimensional fields, one has:

$$\mathrm{tr}\,(F_{ab}\sigma_c\epsilon^{ab}{}_d) = -\sqrt{2}(E_{cd} - iB_{cd}), \tag{52}$$

where ϵ_{abc}, E_{ab} and B_{ab} are, respectively, the orientation 3-form on Σ, and the electric and the magnetic parts of the Weyl curvature of g_{ab} relative to Σ:

$$\begin{aligned} \epsilon_{abc} := -\sqrt{2}\,\mathrm{tr}\,(\sigma_a\sigma_b\sigma_c), \qquad E_{ab} := C_{ambn}\, n^m n^n, \\ B_{ab} := \frac{1}{2}\epsilon_{am}{}^{cd}\, C_{cdbn}\, n^m n^n. \end{aligned} \tag{53}$$

Thus, in a solution to the field equations, $\mathcal{D}$ can be thought of as the potential for the self-dual part of the 4-dimensional curvature. It is remarkable that one can feed the information about the extrinsic curvature into the Sen connection just in the way needed to code the Einstein constraints in certain (algebraically isolated) parts of its curvature and the self-dual part of the space-time curvature in the remaining parts.[2]

References

[1] R. Penrose, "Structure of space-time", in *Battelle Rencontres 1967*, edited by C. De Witt and J.A. Wheeler (Benjamin 1968).

[2] A. Ashtekar, G.T. Horowitz and A. Magnon, Gen. Rel. & Grav. **14**, 411 (1982).

[3] R. Penrose and W. Rindler, *Spinors and space-time*, vol. 1 (Cambridge University Press 1984).

[4] A. Sen, J. Math. Phys. **22**, 1781 (1981).

2 Had we projected the action of ∇ on *primed* spinors to the 3-manifold Σ, we would have obtained the second Sen connection which, in the vacuum case, is the potential for the anti-self-dual part of the Weyl tensor. Throughout these notes, we could have used that connection in place of $\mathcal{D}$. It *is* curious, however, that, in the passage to quantum theory, one has to make a choice. This asymmetry is analogous to the one that exists in twistor theory, where one works with *either* the twistor space *or* the dual twistor space. Its physical implications are not well understood.

Appendix B SYMPLECTIC FRAMEWORK

1 Introduction

The purpose of this appendix is to recall some notions from symplectic geometry, which have been used extensively throughout these notes, both as a general framework for describing the classical dynamics of general relativity (Part II), and as the basis for quantization (Part III). The symplectic framework geometrizes the Hamiltonian description of classical systems, thereby making it coordinate-independent and suggesting interesting generalizations. What we will present here is a self-contained but brief and "practically oriented" introduction to the symplectic formalism for physical theories; more extensive accounts can be found in [1–5]. We will illustrate the basic ideas by applying them to Yang-Mills theory, in sections 4 and 6. (For details on this theory itself, see [6,7].) This example is especially suited for our purposes because, as shown in Part II, there is a precise sense in which Einstein's theory can be cast into a "Yang-Mills form".

The conventions for the index notation are the same as in the main text of these lecture notes, with two additions: bold-face upper case latin letters $(\mathbf{A}, \mathbf{B}, ...)$, will denote Yang-Mills internal indices (on which the gauge group under consideration acts) and Greek letters, $(\alpha, \beta ...)$ will denote phase space indices.

2 Basic definitions

The arena for classical mechanics is a *symplectic manifold*, $(\Gamma, \Omega_{\alpha\beta})$, where Γ is an even-dimensional manifold, and $\Omega_{\alpha\beta}$ a *symplectic form*, i.e., a 2-form which is closed and non-degenerate. Given any torsion-free derivative operator ∇, one can express the closure requirement as: $\nabla_{[\alpha}\Omega_{\beta\gamma]} = 0$. The non-degeneracy condition reads: $\Omega_{\alpha\beta}v^{\alpha} = 0 \Leftrightarrow v^{\alpha} = 0$. If Γ is finite-dimensional, non-degeneracy guarantees that $\Omega_{\alpha\beta}$ has a unique inverse, $\Omega^{\alpha\beta}$, with $\Omega^{\alpha\beta}\Omega_{\beta\gamma} = \delta_{\gamma}{}^{\alpha}$, or that the mapping $\Omega : T\Gamma \rightarrow T^{*}\Gamma$ from tangent vectors to cotangent vectors, with $\Omega_{\alpha\beta}v^{\beta} = v_{\alpha}$, is an isomorphism.[1]

Each point of Γ represents a possible state of the given classical system. Dynamics can be therefore specified by introducing a vector field on Γ: integral curves of the vector field represent dynamical trajectories and the affine parameter keeps track of the passage of time. The availability of the symplectic form simplifies the task of specifying the dynamical vector field. For, as we will see, the symplectic form enables one to construct these vector fields from *functions*—the Hamiltonians—on the phase space. Thus, to specify dynamics on a symplectic manifold, it suffices to specify a function thereon.

Given a vector field v^α on Γ, we say that v^α is an *infinitesimal canonical transformation* iff it leaves the symplectic form invariant, i.e., iff

$$\mathcal{L}_v \Omega_{\alpha\beta} = 0. \tag{1}$$

The diffeomorphisms generated by these v^α are called *canonical transformations.* Since they preserve the geometrical structure of $(\Gamma, \Omega_{\alpha\beta})$, these canonical transformations are the symmetries of classical mechanics. Now, it is easy to verify that v^α satisfies Eq.1 iff there exists, locally, (and, if the first homology group of Γ is trivial, globally) a function f such that:

$$v^\alpha = X_f^\alpha := \Omega^{\alpha\beta}\nabla_\beta f. \tag{2}$$

The vector field X_f^α so constructed from f is called the *Hamiltonian vector field* of f. Thus, all Hamiltonian vector fields generate infinitesimal canonical transformations, and all one-parameter families of canonical transformations are locally generated by a function, called the *Hamiltonian* of the corresponding transformation.[2] In

1 If Γ is infinite-dimensional, one has to be careful with functional analysis. The form Ω is said to be *weakly* non-degenerate if its kernel consists only of the zero vector and *strongly* non-degenerate if the mapping it defines from the tangent space to the cotangent space is an isomorphism. In what follows, in the infinite-dimensional cases, we will assume only that Ω is weakly non-degenerate. Although weak non-degeneracy does *not* ensure that Ω admits an inverse, the main ideas to be discussed in this appendix go through in the weaker case. Roughly, equations which do not involve the inverse of the symplectic form continue to hold in the weakly non-degenerate case. Therefore, in the equations which hold in the finite-dimensional case, one first multiplies both sides by Ω with an index structure so chosen as to eliminate its inverse and *then* takes over the resulting equation to the infinite-dimensional case. However, here, we will not worry about functional analytic rigor.

2 Note that, in symplectic geometry, the term Hamiltonian has a more general meaning than in physics. *Any* function on the phase space, when used to generate a canonical transformation, is referred to as a Hamiltonian; the canonical transformation need not correspond to time evolution. In what follows, the intended sense in which the term Hamiltonian is used will be clear from the context.

particular, therefore, we have established that, in striking constrast to, say, metric manifolds, every symplectic manifold admits infinitely many independent symmetries.

Given two functions $f, g : \Gamma \to \mathbb{R}$, their *Poisson bracket* is defined by

$$\begin{aligned} \{f, g\} : &= \Omega^{\alpha\beta} \nabla_\alpha f \nabla_\beta g \\ &\equiv -\mathcal{L}_{X_f} g \equiv \mathcal{L}_{X_g} f. \end{aligned} \tag{3}$$

It is easy to verify that the Poisson bracket operation turns the vector space of functions on Γ into a Lie algebra. Using this Lie-bracket, we can now state an important property of the map $f \mapsto X_f^\alpha$ that associates to f its Hamiltonian vector field: It takes Poisson brackets of functions into commutators of vector fields:

$$X^\alpha_{\{f,g\}} = -[X_f, X_g]^\alpha. \tag{4}$$

(Note also that the map is linear and its kernel consists precisely of the constant functions on Γ.)

Let us now return to the issue of dynamics. For a large class of physically interesting systems, the dynamical vector fields are globally Hamiltonian. That is, time-evolution of physically interesting systems can be generally specified simply by fixing a function H on Γ; its Hamiltonian vector field X_H^α then provides the dynamical vector field *everywhere* on Γ. Thus, given a point in the phase space representing the initial state of the system, the dynamical trajectory is simply the integral curve of the Hamiltonian vector field X_H^α through that point. Using this fact, it is straightforward to check that the time evolution (in the Heisenberg picture) of any observable f is given by:

$$\dot{f} := \mathcal{L}_{X_H} f \equiv \{f, H\}. \tag{5}$$

Finally, we note that, since the symplectic form is closed, it can be obtained locally (and, if the second homology group of Γ is trivial, globally) from a 1-form ω_α, called the *symplectic potential*: $\Omega_{\alpha\beta} = 2\nabla_{[\alpha}\omega_{\beta]}$. For a given $\Omega_{\alpha\beta}$ the symplectic potential is thus determined up to the addition of a gradient. This potential plays an important role in geometric quantization.

3 Special cases

In simple examples, the state of a system is specified by the values of its configuration and momentum variables, $q^{\underline{i}}$ and $p_{\underline{i}}$. Note that there is a natural distinction between the two: the $q^{\underline{i}}$'s are coordinates on a configuration space C and, for each $q^{\underline{i}}$, the $p_{\underline{i}}$'s are the components of the cotangent vectors at $q^{\underline{i}}$.[3] In such cases, T has a cotangent bundle structure, $\Gamma = T^*C$, and therefore has a natural symplectic potential:

$$\omega_\alpha = p_{\underline{i}}\, \nabla_\alpha q^{\underline{i}}, \tag{6}$$

from which the natural symplectic structure can be derived:

$$\Omega_{\alpha\beta} = 2\, \nabla_{[\alpha} p_{\underline{i}}\, \nabla_{\beta]} q^{\underline{i}}. \tag{7}$$

In other words, using the dual basis $\{\nabla_\alpha q^{\underline{i}}, \nabla_\alpha p_{\underline{i}}\}$ for the cotangent space, the 2-form (Eq.7) can be written as the matrix

$$\Omega_{\underline{\alpha\beta}} = \begin{pmatrix} 0 & -I_{n\times n} \\ +I_{n\times n} & 0 \end{pmatrix}. \tag{8}$$

Notice that, in the basis $\{(\partial/\partial q^{\underline{i}})^\alpha, (\partial/\partial p_{\underline{i}})^\alpha\}$, the inverse of the symplectic form,

$$\Omega^{\alpha\beta} = 2\left(\frac{\partial}{\partial q^{\underline{i}}}\right)^{[\alpha} \left(\frac{\partial}{\partial p_{\underline{i}}}\right)^{b]} \tag{7'}$$

has components given by the negative of the matrix in Eq.8. This is due to the fact that the index contraction in our definition of the inverse $\Omega^{\alpha\beta}$ follows the usual convention of matrix multiplication.

The Poisson brackets (Eq.3) now have a familiar form

$$\{f, g\} = \frac{\partial f}{\partial q^{\underline{i}}} \times \frac{\partial g}{\partial p_{\underline{i}}} - (f \leftrightarrow g), \tag{9}$$

and Eqs.5 for the evolution imply the Hamilton equations:

$$\dot{q}^{\underline{i}} = \frac{\partial H}{\partial p_{\underline{i}}} \quad \text{and} \quad \dot{p}_{\underline{i}} = -\frac{\partial H}{\partial q^{\underline{i}}}\,. \tag{10}$$

Note, however, that in general a phase space *need not* be a cotangent bundle. An interesting example of the more general situation is provided by the choice $\Gamma = S^2$, $\Omega_{\alpha\beta} = \epsilon_{\alpha\beta}$, a volume 2-form.

3 Here, underlined indices refer to a coordinate system. Coordinates, $q^{\underline{i}}$ and $p_{\underline{i}}$ with $\underline{i} = 1, \ldots, n$ (rather than, say, $\xi^{\underline{i}}$, with $\underline{i} = 1, \ldots, 2n$), are used, to make the connection with the textbook treatment of Hamiltonian dynamics transparent.

Another special case occurs when the phase space is a vector space. Then T can be identified with its tangent space at the origin, 0, and the tangent vectors that $\Omega_{\alpha\beta}|_0$ acts on can be thought of simply as elements of Γ. Furthermore, usually, with symplectic vector spaces all one ever needs to consider is the action of the symplectic form on vectors at the origin. This is because the symplectic forms that naturally arise are usually constant tensor fields on the the vector space. Examples of such symplectic spaces are provided by linear field theories in Minkowski space and in the analysis of linear perturbations about a specific solution of a non-linear system.

Given a symplectic vector space Γ, one can look for the maximal subspaces L which are spanned by vectors $v_{\underline{i}}^\alpha$ such that $\Omega_{\alpha\beta}v_{\underline{i}}^\alpha v_{\underline{i}}^\beta = 0$. It is easy to establish that such subspaces exist and are always of half the dimension of Γ. These are the *Lagrangian subspaces* of Γ. In fact Γ can be "split" into Lagrangian subspaces Q and P which are complementary, i.e., satisfy: $Q \cap P = \{0\}$ and $Q \oplus P = \Gamma$. This splitting is, however, not unique. Nonetheless, given any splitting, one can choose bases $\{v_{\underline{i}}^\alpha\}$ and $\{w_{\underline{i}}^\alpha\}$ in Q and P, respectively, such that $\Omega_{\alpha\beta}v_{\underline{i}}^\alpha w_{\underline{j}}^\beta = -\delta_{\underline{i}\underline{j}}$. In such a basis, the symplectic form again has components given by Eq.7. One can therefore identify a set of configuration and momentum variables in Γ via: $v_{\underline{i}}^\alpha = (\partial/\partial q^{\underline{i}})^\alpha$, $w_{\underline{i}}^\alpha = (\partial/\partial p_{\underline{i}})^\alpha$.

In a symplectic vector space, interesting canonical transformations are the linear ones. These are given by linear operators $M^\alpha{}_\beta$, satisfying $M^T\Omega M = \Omega$. Infinitesimal versions of these canonical transformations can be characterized as follows. Any infinitesimal canonical transformation $U^\alpha{}_\beta$ can be thought of as the generator of a one-parameter family of canonical transformations, $M^\alpha{}_\beta(\lambda) = 1 + \lambda U^\alpha{}_\beta + O(\lambda^2)$, (i.e., $U^\alpha{}_\beta = dM^\alpha{}_\beta(\lambda)/d\lambda|_{\lambda=0}$), each of which satisfies $M(\lambda)^T\Omega M(\lambda) = \Omega$. Taking the derivative of the last equation, and evaluating at $\lambda = 0$, we get

$$U^T\Omega + \Omega U = 0, \quad \text{or} \quad \Omega_{\alpha\beta}U^\alpha{}_\mu + \Omega_{\mu\alpha}U^\alpha{}_\beta = 0. \tag{11}$$

Eq.11 is the necessary and sufficient condition for $U^\alpha{}_\beta$ to be an infinitesimal canonical transformation.

Furthermore, if $U^\alpha{}_\beta$ is an infinitesimal canonical transformation, its generating function is given by

$$\overset{\circ}{H}(v) = -\tfrac{1}{2}\,\Omega_{\alpha\beta}\,U^\alpha{}_\gamma\,v^\beta v^\gamma, \quad \forall v \in \Gamma. \qquad (12$$

4 Example

Consider Yang-Mills field with gauge group G in Minkowski (M, η_{ab}). We will be interested here in the $3+1$ formulation of the theory, i.e., in the evolution of the system on a 3-manifold Σ, which can be identified with a t = *constant* spacelike hypersurface. The configuration space of this system, $\mathcal{C} := \{\mathbf{A}_{a\mathbf{A}}{}^{\mathbf{B}}(x)\}$, is the space of Lie algebra valued connections (with gauge group G) on Σ[4]. Strictly speaking, $\mathcal{C}$, by construction, is an affine space. However, for brevity, it is convenient here to fix a gauge, i.e., to introduce a preferred origin in the space of connections, and regard the configuration space as a vector space. This allows us to identify tangent vectors at any point on $\mathcal{C}$ with tangent vectors at the origin, and these in turn with elements in $\mathcal{C}$. The fields $\mathbf{A}_{a\mathbf{A}}{}^{\mathbf{B}}(x)$ provide us with a chart on $\mathcal{C}$; the coordinate basis and its dual are given by:

$$\left\{\left(\frac{\delta}{\delta \mathbf{A}_{a\mathbf{A}}{}^{\mathbf{B}}(x)}\right)^{\underline{\alpha}}\right\} \text{ and } \left\{\left(\mathrm{d\!I}\, \mathbf{A}_{a\mathbf{A}}{}^{\mathbf{B}}(x)\right)_{\underline{\alpha}}\right\}, \tag{13}$$

where '$\mathrm{d\!I}$' is the (exterior) derivative operator on $\mathcal{C}$; and underlined Greek letters $(\underline{\alpha}, \underline{\beta}...)$ denote configuration space indices. Any tangent vector $(\mathbf{T})^{\underline{\alpha}}$ can now be expanded as

$$(\mathbf{T})^{\underline{\alpha}} \equiv \int_\Sigma d^3x\, \mathbf{T}_{a\mathbf{A}}{}^{\mathbf{B}} \left(\frac{\delta}{\delta \mathbf{A}_{a\mathbf{A}}{}^{\mathbf{B}}}\right)^{\underline{\alpha}}; \tag{14a}$$

and a cotangent vector $(\widetilde{\mathbf{C}})_{\underline{\alpha}}$ as

$$(\widetilde{\mathbf{C}})_{\underline{\alpha}} \equiv \int_\Sigma d^3x\, \widetilde{\mathbf{C}}^a{}_{\mathbf{B}}{}^{\mathbf{A}} \left(\mathrm{d\!I}\, \mathbf{A}_{a\mathbf{A}}{}^{\mathbf{B}}\right)_{\underline{\alpha}}, \tag{14b}$$

where the set of components $\mathbf{T}_{a\mathbf{A}}{}^{\mathbf{B}}(x)$ of $(\mathbf{T})^{\underline{\alpha}}$ are Lie algebra valued 1-forms on Σ and the set of components $\widetilde{\mathbf{C}}^a{}_{\mathbf{A}}{}^{\mathbf{B}}(x)$ of $(\widetilde{\mathbf{C}})_{\underline{\alpha}}$ are Lie algebra valued vector densities (of weight one) on Σ. (Since we are in Minkowski space, all such density weights can be eliminated by multiplying the fields by appropriate powers of the determinant of the metric. When the metric itself is a dynamical variable, on the other hand –as, e.g., in the coupled Einstein-Yang-Mills theory– this simple procedure is not available and one must live with density weighted momenta. We emphasize this

4 Boldface, uppercase Latin superscripts and subscripts $(\mathbf{A}, \mathbf{B})$ are indices in the representation of the Lie algebra of the gauge group G.

point by placing a ~ over fields with density weight +1.) In order that the action

$$(\widetilde{\mathbf{E}})_{\underline{\alpha}}\,(\mathbf{A})^{\underline{\alpha}} := \int_\Sigma d^3x \ \mathrm{tr}\ \widetilde{\mathbf{E}}^a\,\mathbf{A}_a \qquad (15)$$

of cotangent vectors $(\widetilde{\mathbf{E}})_{\underline{\alpha}}$ on tangent vectors $(\mathbf{A})^{\underline{\alpha}}$ be well defined, a careful choice of boundary conditions on the fields $(\widetilde{\mathbf{E}}^a{}_{\mathbf{A}}{}^{\mathbf{B}}, \mathbf{A}_{a\mathbf{A}}{}^{\mathbf{B}})$ is necessary. In addition, the boundary conditions should also ensure the finiteness of physical observables such as energy-momentum.

The phase space $\Gamma \equiv T^\star C$ can now be defined by

$$\Gamma := \{(\mathbf{A}_{a\mathbf{A}}{}^{\mathbf{B}}, \widetilde{\mathbf{E}}^a{}_{\mathbf{A}}{}^{\mathbf{B}})|\ \textit{suitable boundary conditions}\}. \qquad (16)$$

The state of the system is thus specified by giving the 3-vector potential $\mathbf{A}_{a\mathbf{A}}{}^{\mathbf{B}}(x)$, or connection 1-form, and the electric field $\widetilde{\mathbf{E}}^a{}_{\mathbf{A}}{}^{\mathbf{B}}(x)$, for all $x \in \Sigma$. $\mathbf{A}_{a\mathbf{A}}{}^{\mathbf{B}}(x)$ and $\widetilde{\mathbf{E}}^a{}_{\mathbf{A}}{}^{\mathbf{B}}(x)$ are both Lie algebra valued tensor fields on Σ. These two fields are, respectively, the configuration and momentum variables of the theory. In the natural chart on Γ, an arbitrary vector $(\mathbf{T}, \widetilde{\mathbf{C}})^\alpha$ in (the tangent space of) Γ can be expanded as

$$(\mathbf{T}, \widetilde{\mathbf{C}})^\alpha \equiv \int_\Sigma d^3x \ \mathrm{tr}\left[\mathbf{T}_a\left(\frac{\delta}{\delta \mathbf{A}_a}\right)^\alpha + \widetilde{\mathbf{C}}^a\left(\frac{\delta}{\delta \widetilde{\mathbf{E}}^a}\right)^\alpha\right]. \qquad (17)$$

As we have seen in section 3, the chart on C provides us with a natural symplectic form, Eq.7, on Γ, which for this example is

$$\Omega_{\alpha\beta} \equiv 2\int_\Sigma d^3x\, \mathrm{tr}\left(\mathrm{d\!I}\, \widetilde{\mathbf{E}}^a\right)_{[\alpha}\left(\mathrm{d\!I}\, \mathbf{A}_a\right)_{\beta]}\,. \qquad (18)$$

It follows: that the action of Ω on two tangent vectors $(\mathbf{A}', \widetilde{\mathbf{E}}')^\alpha$ and $(\mathbf{A}'', \widetilde{\mathbf{E}}'')^\alpha$ is given by

$$\Omega_{\alpha\beta}(\mathbf{A}', \widetilde{\mathbf{E}}')^\alpha(\mathbf{A}'', \widetilde{\mathbf{E}}'')^\beta \equiv \int_\Sigma d^3x \ \mathrm{tr}\ (\mathbf{A}''_a\widetilde{\mathbf{E}}'^a - \mathbf{A}'_a\widetilde{\mathbf{E}}''^a); \qquad (19)$$

from the definition, Eq.2, that the Hamiltonian vector field of a function(al) f on Γ is

$$X_f^\alpha = \int_\Sigma d^3x \ \mathrm{tr}\left[\frac{\delta f}{\delta \widetilde{\mathbf{E}}^a}\left(\frac{\delta}{\delta \mathbf{A}_a}\right)^\alpha - \frac{\delta f}{\delta \mathbf{A}_a}\left(\frac{\delta}{\delta \widetilde{\mathbf{E}}^a}\right)^\alpha\right] \qquad (20)$$

and that the Poisson bracket, Eq.9, between function(al)s f and g on Γ is

$$\{f, g\} = \int_\Sigma d^3x \ \mathrm{tr}\left[\frac{\delta f}{\delta \mathbf{A}}\frac{\delta g}{\delta \widetilde{\mathbf{E}}} - \frac{\delta f}{\delta \widetilde{\mathbf{E}}}\frac{\delta g}{\delta \mathbf{A}}\right]. \qquad (21)$$

The Hamiltonian for this system[5] , which will govern the dynamics, is given by

$$H(\mathbf{A},\widetilde{\mathbf{E}}) := \tfrac{1}{2} \int_{\Sigma} d^3x \ \mathrm{tr}\ q^{-\frac{1}{2}}(\widetilde{\mathbf{E}}^a\widetilde{\mathbf{E}}_a + \widetilde{\mathbf{B}}^a\widetilde{\mathbf{B}}_a), \tag{22}$$

where $\widetilde{\mathbf{B}}^a := \tilde{\eta}^{abc}\mathbf{F}_{bc} \equiv \tilde{\eta}^{abc}(2\partial_b\mathbf{A}_c + [\mathbf{A}_b,\mathbf{A}_c])$ is the magnetic field, and q is the determinant of the metric on Σ.

An explicit example of an expression for a Hamiltonian vector field in this theory will be given below, in section 6.

5 First class constraints

A dynamical system is said to be constrained if its physical states are restricted to lie in a submanifold $\bar{\Gamma}$ of the phase space Γ, called the *constraint surface*. One can specify $\bar{\Gamma}$ by the vanishing of a set of functions $C_{\mathbf{i}} : \Gamma \to \mathbb{R}$ called the *constraints*[6] :

$$\bar{\Gamma} := \{p \in \Gamma \mid C_{\mathbf{i}}(p) = 0, \text{ for } \mathbf{i} = 1,\ldots,m\}. \tag{23}$$

Note that $\bar{\Gamma}$ does not provide a unique choice of constraint functions $C_{\mathbf{i}}$; there is considerable "coordinate freedom" in the selection of constraint functions.

A constrained system is said to be of *first class* if for all covectors n_α normal to $\bar{\Gamma}$, $\Omega^{\alpha\beta}n_\alpha$ is tangent to $\bar{\Gamma}$. This characterization is coordinate independent. Given a set of constraint functions, one can reformulate this definition as follows: the system is of first class if the constraint functions "weakly commute", i.e.,

$$\forall \mathbf{i},\mathbf{j} \qquad \{C_{\mathbf{i}}, C_{\mathbf{j}}\} \approx 0, \tag{24}$$

where $\approx$ means "equals when restricted to the constraint surface". This implies that there exist functions $f_{\mathbf{ij}}{}^{\mathbf{k}}$ on Γ such that: $\mathbf{i},\mathbf{j},\mathbf{k} = 1,\ldots,m$, such that

$$\forall \mathbf{i},\mathbf{j} \qquad \{C_{\mathbf{i}}, C_{\mathbf{j}}\} = -f_{\mathbf{ij}}{}^{\mathbf{k}} C_{\mathbf{k}}. \tag{25}$$

These functions are called *structure functions*. If they happen to be constants, the constraint functions $C_{\mathbf{i}}$ are generators of a sub-Lie algebra of the set of functions

5 Notice that the Hamiltonian given here differs by a multiple of π from the energy of the Yang-Mills field. Had we used the energy expression for H, we would have been forced to modify the expression of Ω by a corresponding factor in order to get the correct equations of motion.

6 The boldface, lower case latin letters $(\mathbf{i},\mathbf{j}\ldots)$, used here are *numerical* indices, running over the number of constraints.

on Γ. The second definition of first class constraints is operationally more useful, but the first one is more "covariant": it shows explicitly that it is independent of the choice of constraint functions.

Let us now consider a first class constrained system $(\Gamma, \bar{\Gamma}, \Omega_{\alpha\beta}, H)$, and examine the consequences of the existence of the constraints for its dynamics. In particular, since the point representing the state of the system is required to remain on $\bar{\Gamma}$, we are interested in knowing to what extent we can consider just $\bar{\Gamma}$ as our phase space, instead of the whole of Γ.

Suppose Γ is $2n$-dimensional. By definition of first class constraints, the m vector fields $X_{\mathfrak{i}}^{\alpha} := \Omega^{\alpha\beta}\nabla_{\beta}C_{\mathfrak{i}}$, are tangential to $\bar{\Gamma}$ and, because of the non-degeneracy of $\Omega_{\alpha\beta}$, they are linearly independent. Thus, at each $p \in \bar{\Gamma}$, they span an m-dimensional subspace $\mathcal{G} \subset T_p\bar{\Gamma}$. The vector fields in $\mathcal{G}$ are called *constraint vector fields*, and $\mathcal{G}$ itself, a *gauge flat*, since, as Dirac [4] observed, motion along these directions corresponds to gauge transformations of the system.

Consider the restriction of $\Omega_{\alpha\beta}$ to $\bar{\Gamma}$ (i.e., its pullback $\bar{\Omega}_{\alpha\beta} = i^*\Omega_{\alpha\beta}$ by the inclusion map $i : \bar{\Gamma} \to \Gamma$). Then, for any n_α normal to $\bar{\Gamma}$ and any $\bar{v}^\alpha$ tangent to $\bar{\Gamma}$,

$$0 = \bar{v}^\alpha n_\alpha = \bar{v}^\alpha \Omega_{\alpha\gamma}\Omega^{\gamma\beta} n_\beta = \bar{v}^\alpha \Omega_{\alpha\gamma}\bar{n}^\gamma = -\bar{\Omega}_{\alpha\gamma}\bar{v}^\alpha \bar{n}^\gamma, \tag{26}$$

where $\bar{n}^\alpha := \Omega^{\alpha\beta} n_\beta$. However, the n^α's are in 1-1 correspondence with constraint vector fields, so all constraint vector fields are degenerate directions for $\bar{\Omega}_{\alpha\beta}$. Conversely, all degenerate directions of this tensor are of this form. Thus, $\bar{\Omega}_{\alpha\beta}$ is m-fold degenerate, and it just defines a *presymplectic structure* on $\bar{\Gamma}$. The practical significance of this fact is that $\bar{\Omega}_{\alpha\beta}$ does not have a unique inverse. One may imagine defining $\bar{\Omega}^{\alpha\beta}$ to be the inverse if it satisfies $\bar{\Omega}_{\alpha\gamma}\bar{\Omega}^{\gamma\delta}\bar{\Omega}_{\delta\beta} = \bar{\Omega}_{\alpha\beta}$. However, for any constraint vector field X^α and arbitrary $\bar{T}^\alpha$, $\bar{\Omega}^{\alpha\beta} + X^{[\alpha}\bar{T}^{\beta]}$ can then equally be considered as an inverse. In particular, using just $\bar{\Omega}_{\alpha\beta}$, we cannot associate a unique Hamiltonian vector field to a function on $\bar{\Gamma}$. We have to replace Eq.2 by

$$\bar{\Omega}_{\alpha\beta} X_H^\beta = \bar{\nabla}_\alpha H, \tag{27}$$

which determines X_H^α up to the addition of a constraint vector field. In particular, if we want to work just on $\bar{\Gamma}$, the time evolution of a system is *not* determined uniquely by the Hamiltonian. The ambiguity corresponds precisely to motions along the constraint vector fields, which therefore have the interpretation of gauge in the Hamiltonian framework.

There are however two ways to recover a well-defined evolution. The first one is to eliminate in $\bar{\Gamma}$ the variables representing the gauge degrees of freedom, and introduce the so-called *reduced phase space*, $\hat{\Gamma} := \bar{\Gamma}/\mathcal{G}$, the space of orbits of the gauge diffeomorphisms. This is possible first because he gauge flats are integrable, which follows from Frobenius' lemma and

$$[X_{\mathbf{i}}, X_{\mathbf{j}}] = -X_{\{C_{\mathbf{i}}, C_{\mathbf{j}}\}} = X_{f_{\mathbf{ij}}{}^{\mathbf{k}} C_{\mathbf{k}}} \approx f_{\mathbf{ij}}{}^{\mathbf{k}} X_{\mathbf{k}}, \tag{28}$$

and second because $\mathcal{L}_{X_{\mathbf{i}}} \Omega_{\alpha\beta} = 0 \Rightarrow \mathcal{L}_{X_{\mathbf{i}}} \bar{\Omega}_{\alpha\beta} = 0$. We now have a projection mapping $\pi : \bar{\Gamma} \to \hat{\Gamma}$, and the (non-degenerate) symplectic form $\hat{\Omega}_{\alpha\beta}$ on $\hat{\Gamma}$ is naturally defined by $\hat{\Omega}_{\alpha\beta} \hat{u}^\alpha \hat{v}^\beta := \bar{\Omega}_{\alpha\beta} \bar{u}^\alpha \bar{v}^\beta$, where $\bar{u}^\alpha$ and $\bar{v}^\alpha$ are any two vectors projected to $\hat{u}^\alpha$ and $\hat{v}^\alpha$, respectively, by the mapping π. The second way to obtain a non-degenerate symplectic form from $\bar{\Omega}_{\alpha\beta}$ is to fix a gauge, i.e., a global cross-section of $\bar{\Gamma}$ each point of which intersects the integral manifold of constraint vector fields once and only once, and restrict oneself to states which lie on this cross-section. The pull-back of $\bar{\Omega}_{\alpha\beta}$ to the gauge-fixed surface is non-degenerate. The second method is less elegant and the required global cross-section need not always exist. However, if it does exist, the method is simpler to use.

6 Example

In Yang-Mills theories, physical states are those with divergence-free electric fields, i.e., with

$$\mathbf{D}_a \widetilde{\mathbf{E}}^a{}_{\mathbf{A}}{}^{\mathbf{B}} = 0, \tag{29}$$

where, as before, the gauge covariant derivative is defined by

$$\mathbf{D}_b \widetilde{\mathbf{E}}^a{}_{\mathbf{A}}{}^{\mathbf{B}} := D_b \widetilde{\mathbf{E}}^a{}_{\mathbf{A}}{}^{\mathbf{B}} + [\mathbf{A}_b, \widetilde{\mathbf{E}}^a]_{\mathbf{A}}{}^{\mathbf{B}}. \tag{30}$$

Since the left side of Eq.29 takes values in the (Lie-algebra valued) fields on Σ, it is not a real-valued function on Γ. To use the Hamiltonian framework outlined in the previous sections directly, it is convenient to recast the left side as a functional on Γ:

$$C_{\Lambda}(\mathbf{A}, \widetilde{\mathbf{E}}) := \int_\Sigma d^3x \, \Lambda_{\mathbf{B}}{}^{\mathbf{A}}(x) \mathbf{D}_a \widetilde{\mathbf{E}}^a{}_{\mathbf{A}}{}^{\mathbf{B}}(x) = 0, \tag{31}$$

for all $\Lambda_{\mathbf{A}}{}^{\mathbf{B}} \in C_0^\infty(\Sigma)$, the space of smooth Lie algebra-valued functions of compact support on Σ. To verify that these constraints form a first class system, one must compute the Poisson brackets between the constraint functions. These are straightforward to evaluate. The result, using Eq.21, is:

$$\{C_{\Lambda}(\mathbf{A}, \widetilde{\mathbf{E}}), C_{\Phi}(\mathbf{A}, \widetilde{\mathbf{E}})\} = C_{[\Lambda, \Phi]}(\mathbf{A}, \widetilde{\mathbf{E}}), \tag{32}$$

where [,] is the (commutator) bracket in the Lie-algebra of the gauge group G. Thus, in this case, we encounter only structure *constants* rather than structure functions. The constraint algebra provides us a faithful representation of the (infinite-dimensional) Lie-algebra of *local* gauge transformations. Indeed, one can show explicitly that the Hamiltonian vector field of the constraint $C_{\mathbf{\Lambda}}$ generates gauge transformations. Substituting $f = C_{\mathbf{\Lambda}}$ into Eq.20 we get

$$X^{\alpha}_{C_{\Lambda}} = \int_{\Sigma} d^3x \ \mathrm{tr} \left[-(D_a\mathbf{\Lambda} + [\mathbf{A}_a, \mathbf{\Lambda}]) \left(\frac{\delta}{\delta \mathbf{A}_a}\right)^{\alpha} + [\mathbf{\Lambda}, \widetilde{\mathbf{E}}^a] \left(\frac{\delta}{\delta \widetilde{\mathbf{E}}^a}\right)^{\alpha} \right], \tag{33}$$

which means that the infinitesimal canonical transformation induced by this vector field is

$$\begin{aligned} \mathbf{A}_a &\mapsto \mathbf{A}_a - \epsilon(D_a\mathbf{\Lambda} + [\mathbf{A}_a, \mathbf{\Lambda}]) \\ \widetilde{\mathbf{E}}^a &\mapsto \widetilde{\mathbf{E}}^a + \epsilon[\mathbf{\Lambda}, \widetilde{\mathbf{E}}^a]. \end{aligned} \tag{34}$$

These are precisely the gauge transformations induced by the gauge function $\mathbf{\Lambda}$.

References

[1] R. Abraham and J.E. Marsden, *Foundations of mechanics*, 2nd ed. (Benjamin 1978).

[2] V.I. Arnold, *Mathematical methods of classical mechanics* (Springer-Verlag 1978).

[3] P.R. Chernoff and J.E. Marsden, *Properties of infinite-dimensional Hamiltonian systems* (Springer-Verlag 1974).

[4] V.W. Guillemin, *Symplectic techniques* (Cambridge University Press).

[5] G. Marmo, E.J. Saletan, A. Simoni and B. Vitale, *Dynamical systems: a differential geometric approach to symmetry and reduction* (Wiley 1985).

[6] C.N. Yang and R.L. Mills, Phys. Rev. **96**, 191 (1954).

[7] L.D. Faddeev and A.A. Slavnov, *Gauge fields, introduction to the quantum theory* (Benjamin 1980).

Appendix C QUANTUM MECHANICS ON MANIFOLDS

In this appendix we consider the quantum description of an unconstrained classical system with a finite number of degrees of freedom. In appendix B we saw that the symplectic framework geometrizes the Hamiltonian description of classical systems, making it coordinate-independent. We will now show that the text-book treatment of quantum mechanics can be similarly made coordinate independent. The resulting "geometric quantum mechanics" systematizes and extends the scope of the standard canonical method of quantization.

Quantum gravity in 3+1-dimensions falls outside the scope of this appendix for two obvious reasons: the system has an infinite number of degrees of freedom as well as has non-trivial first class constraints. Nonetheless, the material contained in this appendix can serve as a useful starting point for canonical quantization of gravity because of the following considerations. First, although there *are* important technical difficulties in the treatment of systems with an infinite number of degrees of freedom, *conceptually*, the procedure involved is rather similar to that outlined in this chapter. Second, although there are non-trivial constraints involved, these are ignored in the initial steps of the quantization program (of chapter 10). Thus, although the scope of this appendix is rather limited, it does in fact provide a useful introduction to the ideas and techniques that underlie quantization of a large class of physical systems, including gravity.

In this appendix, we will assume that the phase space Γ of the system under consideration is naturally a cotangent bundle over a smooth, orientable n-manifold C, the configuration space of the system. Let q denote a point in C and p_a a co-vector at q. The state of the classical system is completely specified by a point (q, p_a) Γ. Smooth, real-valued functions on Γ are called *dynamical variables*. Of particular interest to quantum theory are two types of classical dynamical variables. First, given a smooth, real-valued function f on C, its pull-back to Γ, $Q(f)(q,p_a) = f(q)$, is called a *configuration* variable. These are functions on Γ which are independent of momenta. The *momentum observables*, on the other hand, are the functions $P(V)(q,p_a)$ which are linear in momenta and can be constructed as follows: given any smooth vector field V^a on C, one sets $P(V)(q,p_a) = V^a(q)p_a$ on Γ. As we saw in appendix B, being a cotangent bundle, Γ is endowed with a natural symplectic structure. Using it, one can compute the Poisson brackets between these two classes of dynamical variables. One has:

$$\{Q(f),Q(g)\}=0;\quad \{Q(f),P(V)\}=Q(\mathcal{L}_V f);\quad \{P(V),P(W)\}=-P(\mathcal{L}_V W). \tag{1}$$

Dynamics is generated by yet another observable, the Hamiltonian H. For a large class of systems, H has the form: $H(q,p_a)=\frac{1}{2}g^{ab}p_a p_b+V^a(q)p_a+U(q)$, where g^{ab} is a metric and V^a a vector field on the configuration space C. The state of the system evolves along the orbits of the Hamiltonian vector field X_H of H.

The Schrödinger formulation of quantum mechanics is conceptually clearer if one represents quantum states by densities rather than functions on the configuration space. Therefore, we begin in section 1 with a brief detour into the definition of densities and various tensor operations defined on them. We also show that, given a non-vanishing n-form on C, one can *dedensitize* arbitrary tensor densities, i.e. construct from them tensor fields with zero density weights. In section 2, using these mathematical tools we present the geometrical formulation of quantum mechanics in terms of densities. Finally, using the dedensitizing technique, in section 3 we recast this formulation in terms of wave functions. (An algebraic formulation of quantum mechanics along these lines can be found in [1].)

1 Densities

Consider a smooth n-manifold C, and the space $\mathcal{F}$ of totally skew contravariant (i.e., type $(n,0)$) tensor fields thereon. We define (see e.g. [2,3]) a tensor density (field) T of weight m and type (p,q) on C as a map from $\mathcal{F}$ into the space of tensor fields of type (p,q) on C, such that for any function α and any $f\in\mathcal{F}$ (i.e. $f^{a_1\ldots a_n}=f^{[a_1\ldots a_n]}$),

$$\mathrm{T}\circ(\alpha f)=\alpha^m(\mathrm{T}\circ f), \tag{2}$$

where the weight m can be any real number. All operations on tensor fields can be extended to tensor densities. For example, given a tensor density T of weight m and type (p,q), its (covariant) derivative, $\nabla_a\mathrm{T}$, is a tensor density of weight m and type $(p,q+1)$ defined by

$$(\nabla_a\mathrm{T})\circ f:=\nabla_a(\mathrm{T}\circ f)-m\xi_a(\mathrm{T}\circ f), \tag{3}$$

where ξ_a is the unique co-vector field satisfying $\nabla_a f^{a_1\ldots a_n}=\xi_a f^{a_1\ldots a_n}$. Its Lie derivative (with respect to a vector field v^a), $\mathcal{L}_v\mathrm{T}$, is a tensor density of weight m

and type (p,q) defined by

$$(\mathcal{L}_v \mathrm{T}) \circ f := \mathcal{L}_v(\mathrm{T} \circ f) + m\mathrm{Div}_f v\, (\mathrm{T} \circ f), \tag{4}$$

where $\mathrm{Div}_f v$ is given by $\mathcal{L}_v f^{a_1 \dots a_n} = -(\mathrm{Div}_f v) f^{a_1 \dots a_n}$.

Given any nowhere vanishing n-form $e_{a_1 \dots a_n}$ (which will have an inverse in $\mathcal{F}$ defined by $e_{a_1 \dots a_n}(e^{-1})^{a_1 \dots a_n} = n!$), one can associate with each tensor density of weight m and type (p,q) a tensor t of the same index structure via:

$$t := \mathrm{T} \circ e^{-1}. \tag{5.a}$$

This operation will be referred to as *dedensitization* of T. Note that Eq.5.a is equivalent to the relation:

$$\mathrm{T} \circ f = \left[\tfrac{1}{n!} \epsilon_{a_1 \dots a_n} f^{a_1 \dots a_n}\right]^m t, \tag{5b}$$

for any $f \in \mathcal{F}$. Using the Leibnitz rule on Eq.5, one sees e.g., that

$$\mathcal{L}_v \mathrm{T} \circ f = \left[(\mathcal{L}_v t) + m(\mathrm{Div}_\epsilon v)\, t\right] \left[\tfrac{1}{n!} \epsilon_{a_1 \dots a_n} f^{a_1 \dots a_n}\right]^m \tag{6}$$

where $(\mathrm{Div}_\epsilon v)\epsilon := \mathcal{L}_v \epsilon$. (It is useful to note that using a derivative operator ∇_a that annihilates ϵ, $\mathrm{Div}_\epsilon v$ can be expressed as $\mathrm{Div}_\epsilon v = \nabla_a v^a$.)

2 Schrödinger formulation: States as densities

Consider the vector space $\mathcal{D}$ of all complex-valued, smooth scalar densities of weight $\frac{1}{2}$, Ψ, which have compact support on the configuration space $\mathcal{C}$, and introduce on $\mathcal{D}$ an inner product

$$\langle \Psi \mid \Psi' \rangle \quad := \int_{\mathcal{C}} \overline{\Psi} \Psi', \tag{7}$$

where the integral on the right is well-defined because the integrand is a scalar density of weight 1. The Cauchy completion of $(\mathcal{D}, \langle \mid \rangle)$ defines the Hilbert space $\mathcal{H}$. We introduce (densely defined) linear operators on $\mathcal{D}$ corresponding to classical observables following [1,2,4]. For example, given a real, smooth function f on $\mathcal{C}$, we define a configuration operator $\hat{Q}_f$ as:

$$\hat{Q}_f \Psi := f \cdot \Psi. \tag{8a}$$

Similarly, given a smooth vector field v^a on $\mathcal{C}$, we define a momentum operator $\hat{P}_v$ as:

$$\hat{P}_v \Psi := -i\hbar \mathcal{L}_v \Psi. \tag{8b}$$

These operators have the following properties:

i) They are symmetric with respect to the inner product (Eq.7).

ii) They satisfy the anti-commutation relations,

$$\begin{aligned}\hat{Q}(f)\hat{Q}(g)+\hat{Q}(g)\hat{Q}(f)&=2\hat{Q}(fg)\\ \hat{Q}(f)\hat{P}(v)+\hat{P}(v)\hat{Q}(f)&=2\hat{P}(fv),\end{aligned}\tag{9}$$

which reflect the algebraic relations between the classical configuration and momentum variables.

iii) They satisfy the canonical commutation relations

$$\begin{aligned}[\hat{Q}(f),\hat{Q}(f')]&=0\\ [\hat{Q}(f),\hat{P}(v)]&=i\hbar\hat{Q}(\mathcal{L}_v f)\\ [\hat{P}(v),\hat{P}(v')]&=-i\hbar\hat{P}(\mathcal{L}_v v').\end{aligned}\tag{10}$$

There is thus a correspondence between the Poisson algebra of the classical configuration and momentum observables (Eq.1) and the Lie algebra (under the commutator bracket) of the quantum operators. For this class of observables only, the correspondence is $[\hat{A},\hat{B}]=i\hbar\widehat{\{A,B\}}$.

To see the relation between these general configuration and momentum operators and the more familiar position and momentum operators one encounters in non-relativistic quantum mechanics, let us choose for the configuration space C, the Euclidean 3-space. Then it suffices to restrict oneself to configuration variables associated with the coordinate functions, x, y, z and the momentum variables constructed just from the Killing vectors of the Euclidean metric. The quantum configuration operators are then simply $\hat{X}, \hat{Y}$ and $\hat{Z}$ and the momentum operators reduce to $\hat{P}_x$, $\hat{P}_y$, $\hat{P}_z$, $\hat{L}_x, \hat{L}_y$,and $\hat{L}_z$. The anti-commutation relations of Eq.9 and the commutation relations of Eq.10 then reduce to the familiar relations between the position, momentum and angular momentum operators. On arbitrary manifolds, however, one does not in general have a global chart and one is naturally led to consider the much larger algebra of operators introduced above.

This completes the description of the quantum kinematics of the system. Note that in the above construction only the natural structure of C (as a smooth orientable manifold) has been used; in particular, no metric has been used in the definition of the inner product (Eq.7) or in the construction of the rest of the quantum kinematical structure.

Quantum mechanically, dynamics of the system is specified by the Hamiltonian operator $\hat{H}$. In the Schrödinger formulation, time evolution of the state of the system is given by

$$i\hbar \frac{d}{dt}\Psi(t) = \hat{H} \cdot \Psi(t). \tag{11}$$

For the class of Hamiltonians considered earlier in this chapter, $H = \frac{1}{2}g^{ab}p_a p_b + v^a p_a + U$ one can obtain a natural, symmetric quantum Hamiltonian operator, given by[1]

$$\hat{H} \cdot \Psi = \frac{-\hbar^2}{2} g^{ab}\nabla_a \nabla_b \Psi - (i\hbar)\mathcal{L}_v \Psi + U \cdot \Psi, \tag{12}$$

where ∇_a is the unique torsion free derivative operator compatible with the metric g_{ab}. For classical Hamiltonians of a more general form, there are generally factor ordering ambiguities and hence the quantum dynamics is not uniquely determined by its classical analog.

3 Schrödinger formulation: wave functions

In this section we will recast the above quantum mechanical description of the system in terms of densities to a more familiar form in which states are represented by wave functions. Fix a nondegenerate n-form $\epsilon_{a_1 \cdots a_n}$ on C, and dedensitize, as defined in Eq.5, the scalar densities Ψ introduced in section 2. Thus we have $\psi = \Psi \circ \epsilon^{-1}$. Now we can consider D to be the vector space of complex-valued, smooth, scalar functions of compact support on C, with the inner product Eq.6 expressed as :

$$\langle \psi \,|\, \psi' \rangle = \int_C \overline{\psi}\psi' \epsilon, \tag{13}$$

for all ψ and ψ' in D. (The integral is well defined since the integrand is now an n-form.) The Cauchy completion of $(D, \langle | \rangle)$ defines as before the Hilbert space $\mathcal{H}$ representing the space of states of the quantum system.

Next let us translate the expressions for the configuration and momentum operators defined in section 2. Dedensitizing Eq.8a, we obtain

$$\hat{Q}(f) \cdot \psi := f \cdot \psi. \tag{14a}$$

1 The most general, coordinate independent, expression of the Hamiltonian operator also allows for an additive term, $\frac{-\hbar^2}{2}\xi R$, where ξ is any constant and R is the Ricci scalar of the background space.

Similarly, using Eq.6, we can dedensitize Eq.8*b* to yield

$$\hat{P}(v) \cdot \psi := -i\hbar[\mathcal{L}_v\psi + \tfrac{1}{2}(\mathrm{Div}_\epsilon v) \cdot \psi], \tag{14b}$$

where as before $\mathrm{Div}_\epsilon v = \nabla_a v^a$ is the divergence of v^a with respect to any derivative operator ∇_a compatible with the chosen n-form e. By construction, we see that the operators $\hat{Q}$ and $\hat{P}$ are symmetric with respect to the inner product (Eq.13), that they generate an associative algebra, and that they satisfy the anti-commutation and commutation relations of Eqs.8 and 9.

Note that so far the choice of the volume form e is quite arbitrary; nothing is gained by choosing one over another. However, the expression of the Hamiltonian operator on wave functions becomes somewhat cumbersome unless we make a specific choice. Let us therefore choose ϵ to be the volume form defined by the metric g^{ab} that appears in the expression of the classical Hamiltonian. Then, the derivative operator ∇ that appears in Eq.12 is compatible not only with the metric but also with the volume form. Eq.12 yields the following familiar Hamiltonian operator on wave functions:

$$\hat{H} \cdot \psi = [\frac{-\hbar^2}{2} g^{ab}\nabla_a\nabla_b - (i\hbar)(v^a\nabla_a + \tfrac{1}{2}\nabla_a v^a) + f] \cdot \psi, \tag{15}$$

and the dynamical evolution is given by:

$$i\hbar\frac{d}{dt}\psi(t) = \hat{H} \cdot \psi(t). \tag{11'}$$

References

[1] A.Ashtekar, Comm. Math. Phys. **71**, 59 (1980).

[2] R.Geroch, *Geometrical Quantum Mechanics Lecture Notes*, Univ. of Chicago, (1974), (unpublished).

N.Woodhouse, *Geometric Quantization*, Clarendon Press, Oxford, see ch. 5.

[3] J.D.Romano and R.S.Tate, Class. & Quant. Grav. **6**, page (1989).

[4] A.Ashtekar and R.Geroch, Rep. Prog. Phys. **37**, 1211 (1974), see appendix 2.

Appendix D QUANTIZATION OF CONSTRAINED SYSTEMS

In appendix C we saw how one can quantize systems without constraints. The procedure we followed can be restated as follows. One begins with a subset S of the set of smooth real-valued functions on the phase space T. Elements of S may be called the *elementary variables*; they are to be promoted to quantum operators directly. The subset has to be large enough to generate the full algebra of functions on T, and at the same time, small enough to be closed under Poisson brackets. Quantization then consists of finding a representation of the elements of S by operators on a Hilbert space, such that the Poisson bracket between any two elementary variables is represented by ($i\hbar$ times) the commutator of the corresponding operators, and real elementary variables are represented by symmetric operators.

The above procedure is not immediately applicable to *constrained* systems. Recall from appendix B that, in presence of constraints, not all points of the classical phase space are accessible to the physical system; only those configurations and momenta are allowed which satisfy certain conditions, the *constraints* of the theory. The presence of constraints makes the task of selecting suitable elementary variables difficult. One must therefore modify the quantization strategy. In this appendix we consider systems with a finite number of degrees of freedom which are subject to a set of first class constraints and outline two strategies that one can employ to obtain their quantum description.[1] The first is called the *reduced phase space method*, and the second, the *Dirac Quantization procedure*. We will see that, in general, the two methods provide *distinct* quantum theories. While one can offer guidelines as to which choice is likely to be appropriate in a given context, as far as I know, there does not exist an universal principle that dictates the choice.

Our discussion of these two approaches will be very brief; we will attempt only to convey the basic ideas underlying each approach and to point out the posssible differences in the final quantum descriptions that they lead to. Details can be found, e.g., in appendix 3 of [1] and in [2].

1 The two approaches treat the second class constraints in the same way; these are eliminated at the classical level prior to quantization. It is for this reason that we focus on first class constraints.

1 Reduced space quantization

Let C, an n-manifold, be the classical configuration space of a physical system. Its phase space (Γ, Ω) is then the cotangent bundle over C. Let the system have m first class constraints. The basic idea in the first method is to eliminate *all* the constraints classically, and then quantize the resulting constraint-free Hamiltonian system. Since the constraints are of first class, the Hamiltonian vector fields generated by the constraints are tangential to the constraint surface and represent "gauge" directions. Therefore, as in section 5 of appendix B, one first constructs the reduced phase space $\hat{\Gamma}$ by factoring out the constraint surface by the orbits of these gauge vector fields. $\hat{\Gamma}$ inherits a natural symplectic structure $\hat{\Omega}$ from the full phase space. Each point of the reduced phase space represents the "true degrees of freedom" of our classical system. One can therefore attempt to quantize this unconstrained system $(\hat{\Gamma}, \hat{\Omega})$ following the procedure of appendix C.

Recall, however, that in appendix C we assumed that the underlying phase space is a cotangent bundle. Unfortunately, in general, there is no guarantee that the reduced phase space will have this structure. If it fails to be a cotangent bundle, one must modify the procedure of appendix C. In the very first step, that of construction of the elementary variables, a more sophisticated technique is now needed. Such techniques are provided by, e.g., the geometric quantization framework. (See, e.g., [3]). There is, however, no general procedure applicable to all such systems; typically, the selection of elementary variables has to be made differently in each case.

A simple illustration of this situation is provided by the following example. Let the system consist of two harmonic oscillators of equal mass and spring constant subject to the constaint that their total energy, $H(q_i, p_i) := \frac{1}{2}(p_1^2 + q_1^2 + p_2^2 + q_2^2)$ is constrained to be a fixed constant, say, k^2. In this case, the phase space is topologically R^4, and the constraint surface is the 3-sphere, S^3, of radius k. The Hamiltonian vector field of the constraint is easy to construct. It provides the Hopf fibration of S^3. The reduced phase space, $\hat{\Gamma}$, the quotient manifold, has therefore the topology of a 2-sphere, S^2. Since $\hat{\Gamma}$ is compact, it cannot be a cotangent bundle over any configuration space. To quantize the system, therefore, we cannot follow appendix C and use as elementary variables the configuration and momentum functions. A new stategy is needed. In this specific example, we can use properties of the 2-sphere to construct the set S. An obvious choice is to use as elementary variables, the first 4 spherical harmonics, $Y_{lm}(\theta, \phi)$ with $l = 0, 1$ and $m = 0, \pm 1$. Standard results on spherical harmonics suffice to guarantee that this set S is sufficiently large; any function on S^2 can be expressed as a (limit of a) sum of products of elements of S. Furthermore, using the symplectic structure, $\Omega = \sin\theta \, d\theta \wedge d\phi$, it is straightforward to check that this set is closed under Poisson brackets. The Poisson algebra is essentially that of the rotation group $SO(3)$. In quantum theory then one looks for representations of this group.

Thus, the general situation is rather complicated. There *is* however a large class of systems in which a major simplification results: the case when the constraints are at most linear in momenta. Now, $\hat{\Gamma}$ has the additional structure of a cotangent bundle over a *reduced* configuration space $\hat{C}$. For example, for constraints independent of momenta (i.e. consisting of purely configuration variables), $\hat{C}$ is of course just the constraint surface in C. When the constraints are linear in momenta, $\hat{C}$ is the space of orbits (in C) of the vector fields defining the constraints. In both these cases, in the construction of $\hat{C}$ the "gauge degrees" are classically eliminated. Since there are originally n configuration degrees of freedom and m constraints, $\hat{C}$ is $n - m$ dimensional; we have isolated the $(n - m)$ true degrees of freedom of the system. One is now left with a classical system free of constraints and can quantize it exactly as in appendix C.

2 Dirac quantization

We now consider the classical system with which we began in section 2 and carry out its Dirac quantization. As in the case with unconstrained systems, we begin with the construction of a suitable Poisson algebra generated by elementary variables on the full phase space. The next step is to find a representation of this algebra on some complex vector space V. A choice for the carrier space of the representation is the space of complex-valued smooth functions, or, as in appendix C, the space of complex-valued smooth scalar densities of weight $\frac{1}{2}$ on the n-dimensional configuration space C. We now single out the *physical* states, i.e. states that lie in the kernel of *all* the constraint operators. A necessary condition that the kernel is "sufficently large" – i.e., that it encompasses the "correct number" of degrees of freedom – is that the commutator of any two constraints be *weakly zero*, i.e., be an operator in which a constraint appears on the extreme right. Now, since the classical constraints are asssumed to be of first class, we know that they are closed under the Poisson bracket. Unfortunately, this fact, in itself, does not suffice to ensure that the quantum constaint operators are weakly closed. In particular, care is needed in the choice of factor ordering. If a factor ordering ensuring weak closure does exist, then, generically, the kernel of the constraints has the "correct" number of degrees of freedom. In this case, there are just m operator constraint conditions –commutators do not generate new constraints– whence the physical states have, as expected, $n - m$ degrees of freedom.

Note that not all of the original elementary variables carry over to operators on the space of *physical states*, V_{phy}. Physical operators are the ones that leave this space invariant, i.e. commute with the constraints weakly. To extract physics from these physical states and operators, however, one needs to equip V_{phy} with an Hermitian inner product. Unfortunately, the theory originally given by Dirac [2] does not address this issue. In this sense, it is incomplete.

A possible extension of Dirac's procedure is the following. One may first find an inner product on the full carrier space V (endowing it with a Hilbert space structure) such that the real variables are represented by symmetric operators, and then solve the constraints. Since V_{phy} is a subspace of V, one may thus hope to automatically get an inner product on V_{phy}. Unfortunately, the procedure does not work in general. The problem is that physical states may not have finite norm with respect to the inner product which makes the unconstrained, real variables symmetric. A simple example is provided by the free particle in $\mathbb{R}^3$, subject to the momentum constraint $p_z = 0$. If the representation space V is chosen to be the space of smooth functions on the unconstained configuration space $\mathbb{R}^3$, the inner product that makes the unconstrained, real-valued elementary variables symmetric is just the standard L^2 one. Now, the constraint is represented by the operator $-i\hbar(\partial/\partial z)$, whence the only state in its kernel –i.e. in V_{phy}– with finite norm is the trivial function, 0!

An alternative strategy (advocated in chapter 10) is motivated by the fact that to calculate any physical expectation values one needs an inner product only on the physical states. Therefore, one first solves the constraints in the chosen representation, and then looks for an inner product on the space V_{phy} of solutions. The inner product is to be chosen by requiring that all real *physical* variables be represented by symmetric operators on V_{phy}. Note that in this framework, it is not even meaningful to ask for expectation values of unphysical operators and the problem we came across with the constraint $p_z = 0$ is avoided. However, to obtain the "correct" inner-product on V_{phy}, one must now isolate a "sufficient number" of real, physical variables. In general, this is a difficult task to carry out.

3 Discussion

Let us now consider a few examples to illustrate the differences in the two approaches outlined in the last two sections. These examples have been discussed in detail in the literature. We will therefore only summarize the results of those calculations. (Some further details can be found in chapter 10.)

It is known ([4]) that if one allows constraints to be quadratic in momenta, the two methods can give dramatically different results: even the Hilbert spaces of physical states in the final descriptions can be inequivalent! Typically, the Hilbert space constructed via reduced phase space quantization is a proper subspace of the one obtained by Dirac quantization. Intuitively, one can interpret the inequivalence by saying that the Dirac method allows a rather extreme form of quantum tunneling in which the entire support of a physical state can lie in the classically forbidden regions. (see [4] and section 10.4) Whether such effects actually occur in Nature is unclear and the issue of which of the two procedures is the physically correct one can ultimately be settled only by experiments.

If the constraints are at worst linear in momenta, the differences in the two quantum theories, if any, are more subtle. In this case, there is a natural isomorphism between two Hilbert spaces of physical states. Thus, one can say that the two procedures are equivalent *kinematically*. There may still be a difference: *dynamically* the two procedures can give different physical results. This can happen even in the simple case when the Hamiltonian of the unconstrained system is a function of the type $H(q,p) = \frac{1}{2m}p^2 + V(q)$ which commutes weakly with the constraints [5,6]. The differences can be quite important: even the spectra of the Hamiltonian can be different. Contrary to what one might expect at first, it is the Dirac quantization procedure that agrees with the Faddeev-Popov path integral method with its appropriate determinant factors [6].

What are the ramifications of this difference to quantum gravity? For linearized gravity, the two procedures yield the same quantum theory. It seems unlikely, however, that the equivalence will survive once non-linearities are brought in. Indeed, the two methods appear to differ already in the truncated models such as the ones used in quantum cosmology. In the Bianchi cosmologies, for example, one often solves the vector constaints of general relativity classically (i.e. via the reduced phase space method) and then imposes the scalar constraint *a la* Dirac. The resulting Wheeler-DeWitt equation is a condition on the physically permissible quantum states. Since the Wheeler-DeWitt operator is analogous to the Hamiltonian given above, it is likely that one would obtain an inequivalent quantum theory if one uses the Dirac procedure for *all* constraints.

The general program for quantizing general relativity in the new variables introduced in chapter 10 is closer in spirit to the Dirac procedure.

References

[1] A.Ashtekar and R.Geroch, Rep. Prog. Phys. **37**, 1211 (1974).

[2] P.A.M.Dirac, *Lectures on Quantum Mechanics*, Yeshiva University (1964).

[3] N. Woodhouse, *Geometric Quantization*, (Oxford University Press, Oxford, 1980).

[4] A.Ashtekar and G.T.Horowitz, Phys. Rev. **D26**, 3342 (1982).

[5] J.D.Romano and R.S.Tate, Class. & Quant. Grav. **6**, 1487 (1989).

[6] K.Schleich, Class. & Quant. Grav. **7**, 1529 (1990).

BIBLIOGRAPHY

BIBLIOGRAPHY

Books and dissertations

Abhay Ashtekar (with invited contributions), *New Perspectives in Canonical Gravity*. Lecture Notes. Bibliopolis, Napoli, Italy, February 1988. (Errata published as Syracuse University preprint by Joseph D. Romano and Ranjeet S. Tate, June 1989.)

Joseph D. Romano. *Geometrodynamics vs. Connection Dynamics*. Ph.D. thesis, Syracuse University, 1991.

Papers

1980-1986

[1] Paul Sommers. Space spinors. J. Math. Phys. 21:2567–2571, 1980.

[2] Amitabha Sen. On the existence of neutrino "zero-modes" in vacuum spacetimes. J. Math. Phys. 22:1781–1786, 1981.

[3] Abhay Ashtekar and G.T. Horowitz. On the canonical approach to quantum gravity. Phys. Rev. D26:3342–3353, 1982.

[4] Amitabha Sen. Gravity as a spin system. Phys. Lett. B119:89–91, 1982.

[5] Abhay Ashtekar. On the Hamiltonian of general relativity. Physica A124:51–60, 1984.

[6] E. T. Newman. Report of the workshop on classical and quantum alternate theories of gravity. In B. Bertotti, F. de Felice, and A. Pascolini, editors, *The*

This bibliography was first prepared by Peter Hübner. It is now being updated by Gabriella Gonzalez. While the lecture notes contain material only upto January '90, we believe that the bibliography is current upto June '90. Please send additional references to "gonzalez@suhep" (bitnet). The papers and preprints are arranged alphabetically by author within each year.

Proceedings of the 10th International Conference on General Relativity and Gravitation, Amsterdam, 1984.

[7] Abhay Ashtekar. New variables for classical and quantum gravity. Phys. Rev. Lett. 57:2244–2247, 1986.

[8] Abhay Ashtekar. Self-duality and spinorial techniques in the canonical approach to quantum gravity. In C. J. Isham and R. Penrose, editors, *Quantum Concepts in Space and Time*, pages 303–317. Oxford University Press, 1986.

[9] Robert M. Wald. Non-existence of dynamical perturbations of Schwarzschild with vanishing self-dual part. Class. & Quant. Grav. 3:55–63, 1986.

1987

[10] Abhay Ashtekar. New Hamiltonian formulation of general relativity. Phys. Rev. D36:1587–1602, 1987.

[11] Abhay Ashtekar. Einstein constraints in the Yang-Mills form. In G. Longhi and L Lusanna, editors, *Constraint Theory and Relativistic Dynamics*, Singapore, 1987. World Scientific.

[12] Abhay Ashtekar, Pawel Mazur, and Charles G. Torre. BRST structure of general relativity in terms of new variables. Phys. Rev. D36:2955–2962, 1987.

[13] John L. Friedman and Ian Jack. Formal commutators of the gravitational constraints are not well-defined: A translation of Ashtekar's ordering to the Schrödinger representation. Phys. Rev. D37:3495–3504, 1987.

[14] Ted Jacobson and Lee Smolin. The left-handed spin connection as a variable for canonical gravity. Phys. Lett. B196:39–42, 1987.

[15] Joseph Samuel. A Lagrangian basis for Ashtekar's reformulation of canonical gravity. Pramāna-J Phys. 28:L429-L432, 1987.

[16] N. C. Tsamis and R. P. Woodard. The factor ordering problem must be regulated. Phys. Rev. D36:3641–3650, 1987.

1988

[17] Abhay Ashtekar. A 3 + 1 formulation of Einstein self-duality. In J. Isenberg, editor, *Mathematics and General Relativity*, Providence, 1988. American Mathematical Society.

[18] Abhay Ashtekar. Microstructure of space-time in quantum gravity. In K. C. Wali, editor, *Proceedings of the Eight Workshop in Grand Unification*, Singapore, 1988. World Scientific.

[19] Abhay Ashtekar. New perspectives in canonical quantum gravity. In B. R. Iyer, A. Kembhavi, J. V. Narlikar, and C. V. Vishveshwara, editors, *Highlights in Gravitation and Cosmology*. Cambridge University Press, 1988.

[20] Abhay Ashtekar, Ted Jacobson, and Lee Smolin. A new characterization of half-flat solutions to Einstein's equation. Comm. Math. Phys. 115:631–648, 1988.

[21] Ingemar Bengtsson. Note on Ashtekar's variables in the spherically symmetric case. Class. & Quant. Grav. 5:L139–L142, 1988.

[22] J. N. Goldberg. A Hamiltonian approach to the strong gravity limit. Gen. Rel. & Grav. 20:881–891, 1988.

[23] J. N. Goldberg. Triad approach to the Hamiltonian of general relativity. Phys. Rev. D37:2116–2120, 1988.

[24] Viqar Husain. The $G_{\mathrm{Newton}} \to \infty$ limit of quantum gravity. Class. & Quant. Grav. 5:575–582, 1988.

[25] Ted Jacobson. Fermions in canonical gravity. Class. & Quant. Grav. 5:L143–L148, 1988.

[26] Ted Jacobson. New variables for canonical supergravity. Class. & Quant. Grav. 5:923–935, 1988.

[27] Ted Jacobson. Superspace in the self-dual representation of quantum gravity. In J. Isenberg, editor, *Mathematics and General Relativity*, Providence, 1988. American Mathematical Society.

[28] Ted Jacobson and Lee Smolin. Covariant action for Ashtekar's form of canonical gravity. Class. & Quant. Grav. 5:583–594, 1988.

[29] Ted Jacobson and Lee Smolin. Nonperturbative quantum geometries. Nucl. Phys. B299:295–345, 1988.

[30] Hideo Kodama. Specialization of Ashtekar's formalism to Bianchi cosmology. Prog. Theor. Phys. 80:1024–1040, 1988.

[31] Carlo Rovelli and Lee Smolin. Knot theory and quantum gravity. Phys. Rev. Lett. 61:1155–1158, 1988.

[32] Joseph Samuel. Gravitational instantons from the Ashtekar variables. Class. & Quant. Grav. 5:L123–L125, 1988.

[33] Lee Smolin. Quantum gravity in the self-dual representation. In J. Isenberg, editor, *Mathematics and General Relativity*, Providence, 1988. American Mathematical Society.

[34] C. G. Torre. The propagation amplitude in spinorial gravity. Class. & Quant. Grav. 5:L63–L68, 1988.

[35] Edward Witten. (2+1) dimensional gravity as an exactly soluble system. Nucl. Phys. B311:46–78, 1988.

1989

[36] Abhay Ashtekar. Non-pertubative quantum gravity: A status report. In M. Cerdonio, R. Cianci, M. Francaviglia, and M. Toller, editors, *General Relativity and Gravitation*, Singapore, 1989. World Scientific.

[37] Abhay Ashtekar. Recent developments in Hamiltonian gravity. In B. Simon, I. M. Davies, and A. Truman, editors, *The Proceedings of the IXth International Conference on Mathematical Physics*, Bristol, 1989. Adam-Higler.

[38] Abhay Ashtekar. Recent developments in quantum gravity. In E. J. Fenyves, editor, *Proceedings of the Texas Symposium on Relativistic Astrophysics*. New York Academy of Science, 1989.

[39] Abhay Ashtekar, A. P. Balachandran, and S. G. Jo. The CP-problem in quantum gravity. Int. Journ. Theor. Phys. A4:1493–1514, 1989.

[40] Abhay Ashtekar, Viqar Husain, Carlo Rovelli, Joseph Samuel, and Lee Smolin. $2+1$ quantum gravity as a toy model for the $3+1$ theory. Class. & Quant. Grav. 6:L185–L193, 1989.

[41] Abhay Ashtekar and Joseph D. Romano. Chern-Simons and Palatini actions and $(2+1)$-gravity. Phys. Lett. B229:56–60, 1989.

[42] Abhay Ashtekar, Joseph D. Romano, and Ranjeet S. Tate. New variables for gravity: Inclusion of matter. Phys. Rev. D40:2572–2587, 1989.

[43] Abhay Ashtekar and Joseph D. Romano. Key $(3+1)$-equations in terms of new variables (for numerical relativity). Syracuse University Report (1989).

[44] Ingemar Bengtsson. Yang-Mills theory and general relativity in three and four dimensions. Phys. Lett. B220:51–53, 1989.

[45] Ingemar Bengtsson. Some remarks on space-time decomposition, and degenerate metrics, in general relativity. Int. J. Mod. Phys. A4:5527–5538, 1989.

[46] Riccardo Capovilla, John Dell, and Ted Jacobson. General relativity without a metric. Phys. Rev. Lett. 63:2325–2328, 1989.

[47] Steven Carlip. Exact quantum scattering in 2+1 dimensional gravity. Nucl. Phys. B324:106–122, 1989.

[48] B. P. Dolan. On the generating function for Ashtekar's canonical transformation. Phys. Lett. B233:89-92 , 1989.

[49] Tevian Dray, Ravi Kulkarni, and Joseph Samuel. Duality and conformal structure. J. Math. Phys. 30:1306–1309, 1989.

[50] N. N. Gorobey and A. S. Lukyanenko. The closure of the constraint algebra of complex self-dual gravity. Class. & Quant. Grav. 6:L233–L235, 1989.

[51] M. Henneaux, J. E. Nelson, and C. Schomblond. Derivation of Ashtekar variables from tetrad gravity. Phys. Rev. D39:434–437, 1989.

[52] Viqar Husain. Intersecting loop solutions of the Hamiltonian constraint of quantum general relativity. Nucl. Phys. B313:711–724, 1989.

[53] Viqar Husain and Lee Smolin. Exactly solvable quantum cosmologies from two Killing field reductions of general relativity. Nucl. Phys. B327:205–238, 1989.

[54] V. Khatsymovsky. Tetrad and self-dual formulation of Regge calculus. Class. & Quant. Grav. 6:L249–L255, 1989.

[55] Sucheta Koshti and Naresh Dadhich. Degenerate spherical symmetric cosmological solutions using Ashtekar's variables. Class. & Quant. Grav. 6:L223–L226, 1989.

[56] Stephen P. Martin. Observables in 2+1 dimensional gravity. Nucl. Phys. 327:78–204, 1989.

[57] L. J. Mason and E. T. Newman. A connection between Einstein and Yang-Mills equations. Comm. Math. Phys. 121:659–668, 1989.

[58] J. E. Nelson and T. Regge. Group manifold derivation of canonical theories. Int. J. Mod. Phys. A4:2021, 1989.

[59] Paul Renteln and Lee Smolin. A lattice approach to spinorial quantum gravity. Class. & Quant. Grav. 6:275–294, 1989.

[60] Amitabha Sen and Sharon Butler. The quantum loop. *The Sciences*, pages 32–36, November/December 1989.

[61] Lee Smolin. Invariants of links and critical points of the Chern-Simon path integrals. Mod. Phys. Lett. A4:1091–1112, 1989.

[62] Sanjay M. Wagh and Ravi V. Saraykar. Conformally flat initial data for general relativity in Ashtekar's variables. Phys. Rev. D39:670–672, 1989.

[63] Edward Witten. Gauge theories and integrable lattice models. Nucl. Phys. B322:629–697, 1989.

[64] Edward Witten. Topology-changing amplitudes in (2+1) dimensional gravity. Nucl. Phys. B323:113–122, 1989.

1990

[65] C. Aragone and A. Khouder . Vierbein gravity in the light-front gauge. Class. & Quant. Grav. 7:1291–1298, 1990.

[66] Abhay Ashtekar. Old problems in the light of new variables. In *Conceptual Problems in Quantum Gravity*, Boston, 1990. Birkhäuser.

[67] Abhay Ashtekar. Self duality, quantum gravity, Wilson loops and all that. In N. Ashby, D. F. Bartlett, and W. Wyss, editors, *Proceedings of the 12th International Conference on General Relativity and Gravitation*. Cambridge University Press, 1990.

[68] Abhay Ashtekar and Jorge Pullin. Bianchi cosmologies: A new description. Proc. Phys. Soc. Israel 9:65-76, 1990.

[69] Ingemar Bengtsson. A new phase for general relativity? Class. & Quant. Grav. 7:27–39, 1990.

[70] Ingemar Bengtsson. P, T, and the cosmological constant. Int. J. Mod. Phys. A5:3449-3459, 1990.

[71] Ingemar Bengtsson. Self-Dual Yang-Mills fields and Ashtekar variables. Class. & Quant. Grav. 7:L223-L228, 1990.

[72] M. P. Blencowe. The Hamiltonian constraint in quantum gravity. Nucl. Phys. B341:213, 1990.

[73] L. Bombelli and R. J. Torrence. Perfect fluids and Ashtekar variables, with applications to Kantowski-Sachs models. Class. & Quant. Grav. 7:1747, 1990.

[74] Riccardo Capovilla, John Dell, and Ted Jacobson. Gravitational instantons as SU(2) gauge fields. Class. & Quant. Grav. 7:L1–L3, 1990.

[75] Steven Carlip. Observables, gauge invariance and time in 2+1 dimensional gravity. Phys. Rev. D42:2647-2654, 1990.

[76] S. Carlip and S. P. de Alwis. Wormholes in (2+1)-gravity. Nucl. Phys. B337:681-694, 1990.

[77] G.Esposito. Mathematical structures of space-time. Cambridge preprint DAMTP-R-gols, to appear in Fortschritte der Physik.

[78] R. Floreanini and R. Percacci. Canonical algebra of GL(4)-invariant gravity. Class. & Quant. Grav. 7:975–984, 1990.

[79] N. N. Gorobey and A. S. Lukyanenko. The Ashtekar complex canonical transformation for supergravity. Class. & Quant. Grav. 7:67–71, 1990.

[80] C. Holm. Connections in Bergmann manifolds. Int. Journ. Theor. Phys. A29:23-36, 1990.

[81] Viqar Husain and Jorge Pullin. Quantum theory of space-times with one Killing field. Modern Phys. Lett. A5:733-741, 1990.

[82] H. Kodama. Holomorphic wavefunction of the universe. Phys. Rev. D42:2548-2565, 1990.

[83] Sucheta Koshti and Naresh Dadhich. On the self-duality of the Weyl tensor using Ashtekar's variables. Class. & Quant. Grav. 7:L5–L7, 1990.

[84] Noah Linden. New designs on space-time foams. *Physics World* 3:30-31, 1990.

[85] N.Manojlovic. Alternative loop variables for canonical gravity. Class. & Quant. Grav. 7:1633-1645, 1990.

[86] E. W. Mielke. Generating functional for new variables in general relativity and Poincare gauge theory. Phys. Lett. A149:345-350, 1990.

[87] E. W. Mielke. Positive gravitational energy proof from complex variables? Phys. Rev. D, in print 1990.

[88] D. Rayner. A formalism for quantising general relativity using non-local variables. Class. & Quant. Grav. 7:111–134, 1990.

[89] D. Rayner. Hermitian operators on quantum general relativity loop space. Class. & Quant. Grav. 7:651–661, 1990.

[90] Paul Renteln. Some results of SU(2) spinorial lattice gravity. Class. & Quant. Grav. 7:493–502, 1990.

[91] Carlo Rovelli and Lee Smolin. Loop representation of quantum general relativity. Nucl. Phys. B331:80-152, 1990.

[92] D.C. Robinson and C. Soteriu. Ashtekar's new variables and the vacuum constraint equations. Class. & Quant. Grav., to appear, 1990.

[93] M. Seriu and H. Kodama. New canonical formulation of the Einstein theory. Prog. Theor. Phys. 83:7-12, 1990.

[94] Lee Smolin. Loop representation for quantum gravity in 2 + 1 dimensions. In *Proceedings of the 12th John Hopkins Workshop: Topology and Quantum Field Theory (Florence, Italy)*, 1990.

[95] Lee Smolin. Nonperturbative quantum gravity via the loop representation. In A. Ashtekar and J. Stachel, editors, *Proceedings of the Osgood Hill Conference on Conceptual Problems in Quantum Gravity*, Boston, 1990. Birkhäuser, to appear.

[96] C. G. Torre. Perturbations of gravitational instantons. Phys. Rev. D41:3620-3621, 1990.

[97] C. G. Torre. A topological field theory of gravitational instantons. Phys. Lett. B, in print 1990.

[98] C. G. Torre. On the linearization stability of the conformally (anti)self dual Einstein equations, J. Math. Phys., in print 1990.

[99] H. Waelbroeck. 2+1 lattice gravity. Class. & Quant. Grav. 7:751–769, 1990.

[100] R. P. Wallner. New variables in gravity theories. Phys. Rev. D42:441-448, 1990.

[101] R. S. Ward. The SU(∞) chiral model and self-dual vaccuum spaces. Class. & Quant. Grav. 7:L217-L222, 1990.

Preprints

[102] Abhay Ashtekar. Lessons from 2+1 dimensional quantum gravity. Syracuse preprint, to appear in "Strings 90" edited by R. Arnowitt et al, (World Scientific).

[103] Abhay Ashtekar. Canonical Quantum Gravity. To appear in the Proceedings of the 1990 Banff Summer School on Gravitation, ed. by R. Mann. World Scientific, Singapore, 1990.

[104] Ingemar Bengtsson. Complex actions, real slices, and Ashtekar's variables. S-412 96 Preprint 1989, Inst. of Theor. Physics, Göteburg, Sweden.

[105] Ingemar Bengtsson. The cosmological constants. Preprint 1990, Inst. of Theor. Physics, Göteburg, Sweden.

[106] Ingemar Bengtsson. Another "cosmological" constant. Preprint 1990, Inst. of Theor. Physics, Göteburg, Sweden.

[107] B. Brügmann The method of loops applied to lattice gauge theory. Phys. Rev. D, in print 1990.

[108] B. Brügmann and J. Pullin. Intersecting N loop solutions of the Hamiltonian constraint of quantum gravity. Syracuse University pre-print 1990.

[109] Greorgy Burnett, Joseph D. Romano, and Ranjeet S. Tate. Polynomial coupling of matter to gravity using Ashtekar variables. Syracuse University pre-print 1990.

[110] R. Capovilla, J. Dell, T. Jacobson and L. Mason. Self dual forms and gravity. UMDGR 90-251/ TJHSST 90-1. Preprint 1990.

[111] R. Capovilla. Generally covariant gauge theories. UMDGR 90-253. Preprint 1990.

[112] R. Capovilla, J. Dell and T. Jacobson. A pure spin-connection formulation of gravity. UMDGR 90-252/TJHSST 90-2. Preprint 1990.

[113] Steven Carlip. 2+1 dimensional quantum gravity and the Braid group. Talk given at the Workshop on Physics, Braids and Links, Banff Summer School in Theoretical Physics, August 1989.

[114] Steven Carlip. Measuring the metric in 2+1 dimensional quantum gravity. IASSNS/HEP/90-46. Preprint 1990.

[115] N. Dadhich, S. Koshti and A. Kshirsagar. On constraints of pure connection formulation of General Relativity for non-zero cosmological constant. Preprint 1990.

[116] B. P. Dolan. The extension of chiral gravity to SL(2,C). DIAS STP 89-21 Preprint (1989). To appear in the Proceedings of the 1990 Banff Summer School on gravitation, ed. by R. Mann. World Scientific, Singapore, 1990.

[117] R. Floreanini and R. Percacci. Palatini formalism and new canonical variables for GL(4)-invariant gravity. SISSA 26-EP Preprint, 1990.

[118] R. Floreanini and R. Percacci. GL(3) invariant gravity without metric. Class. & Quant. Grav., to appear 1990.

[119] R. Floreanini and R. Percacci. Topological pregeometry. Mod. Phys. Lett., to appear 1990.

[120] Takeshi Fukuyama and Kiyoshi Kaminura. Complex action and quantum gravity. Preprint, September 1989.

[121] J. Gegenberg, G. Kunstatter, H.P. Levio. Topological Matter coupled to gravity in 2+1 dimensions. Univ. of Winnipeg preprint, 1990.

[122] Kazuo Ghoroku. New variable formalism of higher derivative gravity.

[123] A. Giannopoulos and V. Daftardar. The direct evaluation of the Ashtekar variables for any metric using the algebraic computing system STENSOR. Imperial TP/89-90/11; submitted to Class. & Quant. Grav..

[124] G. Gonzalez and J. Pullin. BRST quantization of 2+1 gravity. Phys. Rev. D, to appear 1990.

[125] B. Grossmann. General relativity and a non-topological phase of topological Yang-Mills theory. Inst. for Advanced Studies, Princeton. Preprint, 1990.

[126] Andrzej Herdegen. Canonical gravity from a variation principle in a copy of tangent bundle. Inst. of Physics, Jagellonian Univ, Krakow, Poland. Preprint, 1989.

[127] V. Husain and K. Kuchař. General covariance, the new variables, and dynamics without dynamics. Univ. of Utah preprint, 1990, (submitted to Phys. Rev.D). Talk presented at the Quantum Gravity Workshop at the Discussion Conference on Recent Advances in General Relativity, Pittsburgh, May 1990.

[128] S. Koshti and N. Dadhich. Gravitational instantons with matter sources using Ashtekar variables. Inter-Univ. Centre for Astron. and Astrophysics, Pune, India. Preprint, 1990.

[129] R. Loll. A new quantum representation for canonical gravity and SU(2) Yang-Mills theory. BONN-HE-90-02 Bonn University preprint, 1990. ISSN-0172-8733

[130] L. J. Mason and Jörg Frauendiener. The Sparling 3-form, Ashtekar variables and quasi-local mass, 1989 preprint.

[131] N. O'Murchadha and M. Vandyck. Gravitational degrees of freedom in Ashtekar's formulation of general relativity. Univ. of Cork preprint, 1990.

[132] C. Nayak. Einstein-Maxwell theory in 2+1 dimensions, 1990.

[133] C. Nayak. The loop space representation of 2+1 quantum gravity : physical observables and the issue of time, 1990.

[134] Peter Peldán. Gravity coupled to matter without the metric. Göteborg 90-20. Preprint, 1990.

[135] Paul Renteln. Some notes on spinorial quantum gravity. Preprint, 1990.

[136] Carlo Rovelli. Holonomies and loop representation in quantum gravity. To appear in the Newman Festschrift, ed. by A. Janis and J. Porter. (Birkhäuser, Boston, in press), 1990.

[137] Carlo Rovelli and Lee Smolin. Loop representation for lattice gauge theory. Pittsburgh and Syracuse preprint, 1990.

[138] Joseph Samuel. Self-duality in Classical Gravity. To appear in the Newman Festschrift, ed. by A. Janis and J. Porter. (Birkhäuser, Boston, in press), 1990.

[139] R. P. Wallner. A new form of Einstein's equations. Univ of Cologne, West Germany, preprint 1990.

[140] K. Yamagishi and G.F. Chapline. Induced 4-d self-dual quantum gravity: $\hat{W}_\infty$ algebraic approach. Class. & Quant. Grav., in print, 1990.

www.ingramcontent.com/pod-product-compliance
Lightning Source LLC
Chambersburg PA
CBHW061621220326
41598CB00026BA/3841
9789810205744